Everyday Mathematics®

The University of Chicago School Mathematics Project

Student Reference Book

Everyday Mathematics®

The University of Chicago School Mathematics Project

Student Reference Book

McGraw Hill Education

Chicago, IL • Columbus, OH • New York, NY

UCSMP Elementary Materials Component

Max Bell, Director, UCSMP Elementary Materials Component; Director, *Everyday Mathematics* First Edition
James McBride, Director, *Everyday Mathematics* Second Edition
Andy Isaacs, Director, *Everyday Mathematics* Third Edition
Amy Dillard, Associate Director, *Everyday Mathematics* Third Edition
Rachel Malpass McCall, Associate Director, *Everyday Mathematics* Common Core State Standards Edition

Authors

Max Bell, Jean Bell, John Bretzlauf, Amy Dillard, James Flanders, Robert Hartfield, Andy Isaacs, Deborah Arron Leslie, James McBride, Kathleen Pitvorec, Peter Saecker

Assistants

Lance Campbell (Research), Adam Fischer (Editorial), John Saller (Research)

Technical Art

Diana Barrie

 The *Student Reference Book* is based upon work supported by the National Science Foundation under Grant No. ESI-9252984. Any opinions, findings, conclusions, or recommendations expressed in this material are those of the authors and do not necessarily reflect the views of the National Science Foundation.

everyday**math**.com

 **Education**

STEM McGraw-Hill is committed to providing instructional materials in Science, Technology, Engineering, and Mathematics (STEM) that give all students a solid foundation, one that prepares them for college and careers in the 21st century.

Send all inquiries to:
McGraw-Hill Education
STEM Learning Solutions Center
P.O. Box 812960
Chicago, IL 60681

ISBN: 978-0-07-657652-4
MHID: 0-07-657652-3

Printed in the United States of America.

1 2 3 4 5 6 7 8 9 QDB 17 16 15 14 13 12 11

The *McGraw-Hill* Companies

Contents

Whole Numbers — 1

Decimals and Percents — 25

Mathematics ... Every Day
The U.S. Census — 61

Contents

Geometry and Constructions   157

Mathematics ... Every Day
**Mathematics in Weaving
and Textiles**   **201**

Measurement   207

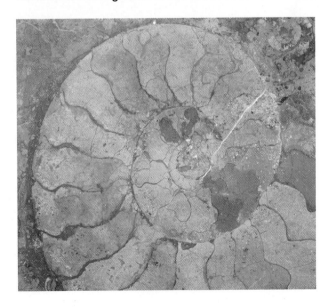

Contents

Games   301

Mathematics ... Every Day
Computer-Generated Art   **339**

Art and Design Activities   345

About the *Student Reference Book*

A reference book is organized to help people find information quickly and easily. Dictionaries, encyclopedias, atlases, cookbooks, and even telephone books are examples of reference books. Unlike novels and biographies, which are usually read in sequence from beginning to end, reference books are read in small segments to find specific information at the time it is needed.

You can use this *Student Reference Book* to look up and review information on topics in mathematics. It consists of the following sections:

♦ A **table of contents** that lists the topics covered and shows how the book is organized.

♦ Essays on **mathematical topics,** such as whole numbers, decimals, percents, fractions, data analysis, geometry, measurement, algebra, and problem solving.

♦ A collection of **photo essays** called **Mathematics... Every Day,** which show in words and pictures some of the ways that mathematics is used.

♦ Descriptions of how to use a **calculator** to perform various mathematical operations and functions.

♦ Directions on how to play **mathematical games** that will help you practice math skills.

♦ A set of **tables and charts** that summarize information, such as a place-value chart, prefixes for names of large and small numbers, tables of equivalent measures and of equivalent fractions, decimals, and percents, and a table of formulas.

♦ A **glossary** of mathematical terms consisting of brief definitions of important words.

♦ An **answer key** for every Check Your Understanding problem in the book.

♦ An **index** to help you locate information quickly.

This reference book also contains descriptions of **Art and Design Activities** you can do.

How to Use the *Student Reference Book*

Suppose you are asked to solve a problem and you know that you have solved problems like it before. But at the moment, you are having difficulty remembering how to do it. This is a perfect time to use the *Student Reference Book*.

You can look in the **table of contents** or the **index** to find the page that gives a brief explanation of the topic. The explanation will often show a step-by-step sample solution.

There is a set of problems at the end of most essays, titled **Check Your Understanding.** It is a good idea to solve these problems and then turn to the **answer key** at the back of the book. Check your answers to make sure that you understand the information presented in the section you have been reading.

Always read mathematical text with paper and pencil in hand. Take notes, and draw pictures and diagrams to help you understand what you are reading. Work the examples. If you get a wrong answer in the Check Your Understanding problems, try to find your mistake by working back from the correct answer given in the answer key.

It is not always easy to read text about mathematics, but the more you use the *Student Reference Book,* the better you will become at understanding this kind of material. You should find that your skills as an independent problem-solver are improving. We are confident that these skills will serve you well as you undertake more advanced mathematics courses.

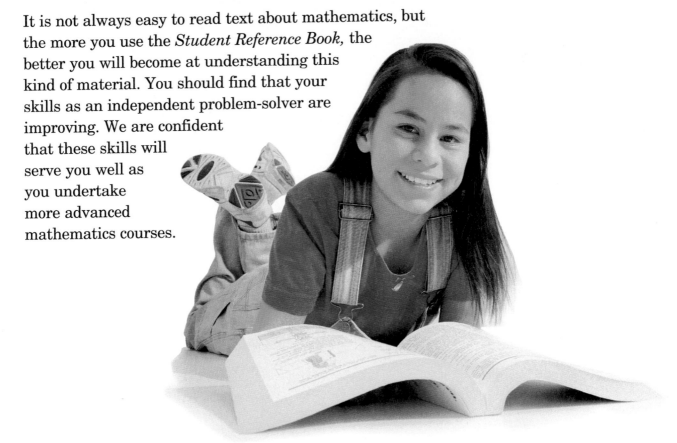

Whole Numbers

Uses of Numbers

It is hard to live even one day without using or thinking about numbers. Numbers are used on clocks, calendars, car license plates, rulers, scales, and so on. The major ways that numbers are used are listed below.

♦ Numbers are used for **counting.**

Examples Students sold 139 tickets to the school play.

The first U.S. Census counted 3,929,326 people.

The population of Downey is 110,441.

♦ Numbers are used for **measuring.**

Examples Jill swam the length of the pool in 29.8 seconds.

The package is 26 inches long and weighs $3\frac{9}{16}$ pounds.

♦ Numbers are used to show where something is in a **reference system.**

Examples

Situation	Reference System
Normal room temperature is 21°C.	Celsius temperature scale
Mick was born on July 23, 1996.	Calendar
The time is 10:13 A.M.	Clock time
Detroit is located at 42°N and 83°W.	Earth's latitude and longitude system

♦ Numbers are used to **compare** measures or counts.

Examples The cat weighs $\frac{1}{2}$ as much as the dog.

There were 3 times as many girls as boys at the game.

♦ Numbers are used for **identification** and as **codes.**

Examples phone number: (709) 555-1212   ZIP code: 60637
driver's license number: M286-423-2061
bar code (to identify product and manufacturer):
9 780076 000616
car license plate: 520 992

DOWNEY
CITY LIMIT
POP 110,441   ELEV 120

9 780076 000616

INDIANA-39
520 992

Kinds of Numbers

The **counting numbers** are the numbers used to count things. The set of counting numbers is 1, 2, 3, 4, and so on.

The **whole numbers** are any of the numbers 0, 1, 2, 3, 4, and so on. The whole numbers include all of the counting numbers and zero (0).

Counting numbers are useful for counting. But they do not always work for measures because most measures fall between whole numbers. **Fractions** and **decimals** were invented to keep track of such in-between measures.

Fractions are often used in recipes for cooking and for measures in carpentry and other building trades. Decimals are used for almost all measures in science and industry. Money amounts are usually written as decimals.

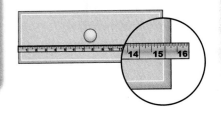

Examples The package weighs 4 pounds 15.3 ounces.

The recipe calls for $2\frac{2}{3}$ cups of molasses.

The drawer measures 1 foot $3\frac{1}{4}$ inches wide.

Negative numbers are used to describe some locations when there is a zero point.

Examples A temperature of 40 degrees below zero is written as −40°F or as −40°C. (Note that 40 degrees below zero is the only temperature at which Celsius and Fahrenheit thermometers show the same reading.)

A depth of 179 feet below sea level is written as −179 feet.

Negative numbers are also used to indicate changes in quantities.

Examples A weight loss of $6\frac{1}{2}$ pounds is recorded as $-6\frac{1}{2}$ pounds.

A decrease in income of $1,500 is recorded as −$1,500.

Place Value for Whole Numbers

Any number, no matter how large or small, can be written using one or more of the **digits** 0, 1, 2, 3, 4, 5, 6, 7, 8, and 9.

A **place-value chart** is used to show how much each digit in a number is worth. The **place** for a digit is its position in the number. The **value** of a digit is how much it is worth according to its place in the number.

Study the place-value chart below. Look at the numbers that name the places. As you move from right to left along the chart, each number is **10 times as large as the number to its right.**

10,000s ten thousands	1,000s thousands	100s hundreds	10s tens	1s ones
9	4	8	0	5

Example   The number 94,805 is shown in the place-value chart above.

The value of the 9 is 90,000 (9 * 10,000).
The value of the 4 is 4,000 (4 * 1,000).
The value of the 8 is 800 (8 * 100).
The value of the 0 is 0 (0 * 10).
The value of the 5 is 5 (5 * 1).

94,805 is read as "ninety-four thousand, eight hundred five."

In larger numbers, groups of 3 digits are separated by commas. Commas help identify the thousands, millions, billions, trillions, quadrillions, quintillions, and so on.

Example   The number 246,357,026,909,389 is shown in the place-value chart.

trillions				billions				millions				thousands				ones		
100	10	1	,	100	10	1	,	100	10	1	,	100	10	1	,	100	10	1
2	4	6	,	3	5	7	,	0	2	6	,	9	0	9	,	3	8	9

This number is read as 246 **trillion**, 357 **billion**, 26 **million**, 909 **thousand**, 389.

Check Your Understanding

Read each number to yourself. What is the value of the 6 in each number?

1. 26,482   **2.** 45,678,910   **3.** 207,464   **4.** 8,765,432

Check your answers on page 414.

Powers of 10

Numbers like 10, 100, and 1,000 are called **powers of 10.**
They are numbers that can be written as products of 10s.

100 can be written as 10 * 10 or 10^2.
1,000 can be written as 10 * 10 * 10 or 10^3.

The raised number is called an **exponent.** The exponent tells
how many 10s are multiplied. 10^2 is read "10 to the second
power" or "10 squared." 10^6 is read "10 to the sixth power."

A number written with an exponent, like 10^3, is in **exponential
notation.** A number written in the usual place-value way, like
1,000, is in **standard notation.**

The chart below shows powers of 10 from ten through one billion.

Powers of 10		
Standard Notation	Product of 10s	Exponential Notation
10	10	10^1
100	10*10	10^2
1,000 (1 thousand)	10*10*10	10^3
10,000	10*10*10*10	10^4
100,000	10*10*10*10*10	10^5
1,000,000 (1 million)	10*10*10*10*10*10	10^6
10,000,000	10*10*10*10*10*10*10	10^7
100,000,000	10*10*10*10*10*10*10*10	10^8
1,000,000,000 (1 billion)	10*10*10*10*10*10*10*10*10	10^9

> **Note**
>
> 10^2 is read
> "10 to the second power"
> or "10 squared."
>
> 10^3 is read
> "10 to the third power"
> or "10 cubed."
>
> 10^4 is read
> "10 to the fourth power."

> **Note**
>
> Any number that can
> be written as a product
> of $\frac{1}{10}$s is also called a
> power of 10.
>
> For example, $\frac{1}{10}$,
>
> $\frac{1}{100} = \frac{1}{10} * \frac{1}{10}$, and
>
> $\frac{1}{1,000} = \frac{1}{10} * \frac{1}{10} * \frac{1}{10}$
>
> are all powers
> of 10.

Example 1,000 * 1,000 = ?

Use the table above to write 1,000 as 10*10*10.
1,000 * 1,000 = (10 * 10 * 10) * (10 * 10 * 10)
= 10^6
= 1 million

So, 1,000 * 1,000 = 1 million.

Example 1,000 millions = ?

Write 1,000 * 1,000,000 as (10*10*10) * (10*10*10*10*10*10).
This is a product of nine 10s, or 10^9.

1,000 millions = 1 billion

Exponential Notation

A **square array** is an arrangement of objects into rows and columns that form a square. All rows and columns must be filled, and the number of rows must equal the number of columns. A counting number that can be represented by a square array is called a **square number.** Any square number can be written as the product of a counting number with itself.

two square arrays

> **Example** 16 is a square number. It can be represented by an array consisting of 4 rows and 4 columns. $16 = 4 * 4$

square array for 16

Here is a shorthand way to write the square number 16: $16 = 4 * 4 = 4^2$. 4^2 is read as "4 times 4," "4 squared," or "4 to the second power." The raised 2 is called an **exponent.** It tells that 4 is used as a factor two times (two 4s are multiplied). The 4 is called the **base.** Numbers written with an exponent are said to be in **exponential notation.**

Exponents are also used to show that a factor is used more than twice.

> **Examples**
>
> $2^3 = 2 * 2 * 2$
> The number 2 is used as a factor 3 times.
> 2^3 is read "2 cubed" or "2 to the third power."
>
> $9^5 = 9 * 9 * 9 * 9 * 9$
> The number 9 is used as a factor 5 times.
> 9^5 is read "9 to the fifth power."
>
> Any number raised to the first power is equal to itself. For example, $5^1 = 5$.

Some calculators have special keys for changing numbers written in exponential notation to standard notation.

> **Example** Use a calculator. Find the value of 2^6.
>
> On Calculator A, key in 2 (∧) 6 (Enter). Answer: 64
> On Calculator B, key in 2 (xʸ) 6 (=). Answer: 64
> $2^6 = 64$ You can verify this by keying in 2 (×) 2 (×) 2 (×) 2 (×) 2 (×) 2 (=).

Check Your Understanding

Write each number in standard notation. Do not use a calculator to solve Problems 1–4.

1. 6^2 **2.** 4^3 **3.** 10^5 **4.** 9^1 **5.** 225^2 **6.** 11^5

Check your answers on page 414.

Scientific Notation

Scientific Notation for Big Numbers

The population of the world is about 6 billion people. The number 6 billion can be written as 6,000,000,000 or as $6 * 10^9$.

The number 6,000,000,000 is written in **standard notation.** The number $6 * 10^9$ is written in **scientific notation.** $6 * 10^9$ is read as "6 times 10 to the ninth power."

Scientific notation is a way to represent big and small numbers with only a few symbols. A number in scientific notation is written as the product of two factors. The first factor is at least 1 but less than 10. The second factor is a power of 10.

Earth weighs about $1.3 * 10^{25}$ pounds.

Example Write $4 * 10^8$ in standard notation.

First, look at the power of 10. It is 10 to the eighth power, so it is the product of 10 used as a factor 8 times:

$10^8 = 10 * 10 * 10 * 10 * 10 * 10 * 10 * 10$
$ = 100,000,000$
$ = 100$ million

So, $4 * 10^8 = 4 * 100,000,000$
$ = 400,000,000$
$ = 400$ million

Note

For all numbers n (except 0), $n^0 = 1$

$5^0 = 1$ $13.2^0 = 1$

$1^0 = 1$ $10^0 = 1$

$\pi^0 = 1$ $(\frac{1}{5})^0 = 1$

Often the first factor of a number in scientific notation has digits to the right of the decimal point.

Example The nearest star beyond the sun is about $2.5 * 10^{13}$ miles away.

In standard notation,
$2.5 * 10^{13} = 2.5 * 10,000,000,000,000 = 25,000,000,000,000$.

$2.5 * 10^{13}$ is best read "two and five-tenths times ten to the thirteenth power."

It can be read more briefly as "two point five times ten to the thirteenth."

Scientific Notation for Small Numbers

A positive number less than 1 can be written in scientific notation using 10 raised to a **negative exponent power.**

A number raised to a negative exponent power is equal to 1 over the number raised to the positive exponent power. Using variables, this means that $b^{-n} = \frac{1}{b^n}$. For example, $10^{-5} = \frac{1}{10^5}$.

Examples of scientific notation for small numbers:

$4 * 10^{-3} = 4 * \frac{1}{10^3}$
$\phantom{4 * 10^{-3}} = \frac{4}{1,000} = 0.004$

$6 * 10^{-1} = 6 * \frac{1}{10^1}$
$\phantom{6 * 10^{-1}} = \frac{6}{10} = 0.6$

$3 * 10^{-5} = 3 * \frac{1}{10^5}$
$\phantom{3 * 10^{-5}} = \frac{3}{100,000} = 0.00003$

Converting between Scientific Notation and Standard Notation

Examples Convert to standard notation.

$8.7 * 10^6$

- Note the exponent in the power of 10.
- If the exponent is positive, as in $8.7 * 10^6$, move the decimal point in the other factor that many places to the right.

(Insert the decimal point if necessary, and attach 0s as you move it.)

8 . 7 0 0 0 0 0 .

(6 places)

$8.7 * 10^6 = 8,700,000$

$5.6 * 10^{-4}$

- Note the exponent in the power of 10.
- If the exponent is negative, as in $5.6 * 10^{-4}$, move the decimal point in the other factor that many places to the left.

(Insert the decimal point if necessary, and attach 0s as you move it.)

0 . 0 0 0 5 . 6

(4 places)

$5.6 * 10^{-4} = 0.00056$

Examples Convert from standard notation to scientific notation.

1. Locate the decimal point. Write or imagine the decimal point if it isn't there.

   8,700,000.   0.00056

2. Move the decimal point so that you get a number with only one digit (not 0) to the left of the decimal point (in the ones place). Count the number of places you moved the decimal point.

   8 . 7 0 0 0 0 0 .   0 . 0 0 0 5 . 6

   (6 places)   (4 places)

3. The number of places you moved the decimal point tells which exponent to use. If the original number was between 0 and 1, the exponent is negative.

   10^6 10^{-4}

4. Use the number you got in Step 2 and the power of 10 you got in Step 3 to write the number in scientific notation. Omit any 0s you don't need.

   $8,700,000 = 8.7 * 10^6$ $0.00056 = 5.6 * 10^{-4}$

Check Your Understanding

Write in scientific notation.

1. 200,000   **2.** 10 million   **3.** 430,000,000   **4.** 0.00006   **5.** 0.035

Write in standard notation.

6. $4 * 10^7$ **7.** $2.8 * 10^4$ **8.** $6.62 * 10^8$ **9.** $3 * 10^{-2}$ **10.** $1.23 * 10^{-3}$

Check your answers on page 414.

Comparing Numbers and Amounts

When two numbers or amounts are compared, there are two possible results: They are equal, or they are not equal because one is larger than the other.

Different symbols are used to show that numbers and amounts are equal or not equal.

♦ Use an **equal sign** (=) to show that the numbers or amounts *are equal.*

♦ Use a **not-equal sign** (≠) to show that they are *not equal.*

♦ Use a **greater-than symbol** (>) or a **less-than symbol** (<) to show that they are *not equal* and to show which is larger.

Symbol	=	≠	>	<
Meaning	"equals" or "is the same as"	"is not equal to"	"is greater than"	"is less than"
Examples	$\frac{1}{2} = 0.5$ $3^3 = 27$ $2 * 5 = 9 + 1$	$2 \neq 3$ $3^2 \neq 6$ $1 \text{ m} \neq 100 \text{ mm}$	$1.42 > 1.4$ 16 ft 9 in. > 15 ft 11 in. $10^3 > 100$	$3 < 5$ $100 - 2 < 99 + 2$ $\frac{1}{10^3} < 1$

When you compare amounts that include units, use the *same unit* for both amounts.

Example   Compare 30 yards and 60 feet.

The units are different—yards and feet. Change yards to feet, then compare.

1 yd = 3 ft, so 30 yd = 30 * 3 ft, or 90 ft. Now compare feet. 90 ft > 60 ft

Therefore, 30 yd > 60 ft.

Or, change feet to yards, and then compare. 1 ft = $\frac{1}{3}$ yd, so 60 ft = 60 * $\frac{1}{3}$ yd, or 20 yd.

Now compare yards. 30 yd > 20 yd

Therefore, 30 yd > 60 ft.

Check Your Understanding

True or false?

1. $8^2 < 16$ **2.** 37 in. > 3 ft   **3.** $6 * 5 \neq 90 / 3$ **4.** $20 - 1 > 20 - 1$

Check your answers on page 414.

Factors of a Counting Number

A **rectangular array** is an arrangement of objects into rows and columns that form a rectangle. All rows and columns must be filled. Each row has the same number of objects. And each column has the same number of objects. A rectangular array can be represented by a multiplication **number model.**

Push buttons on a phone form a 4-by-3 rectangular array.

Example   This rectangular array has 14 red dots.

It has 2 rows with 7 dots in each row.
$2 * 7 = 14$ is a number model for this array.
2 and 7 are **factors** of 14. 14 is the **product** of 2 and 7.
2 and 7 are a **factor pair** for 14.

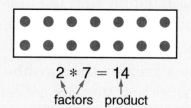

$2 * 7 = 14$
factors   product

Counting numbers can have more than one factor pair.
1 and 14 are another factor pair for 14 because $1 * 14 = 14$.

To test whether a counting number a is a **factor of a counting number** b, divide b by a. If the result is a counting number and the remainder is 0, then a is a factor of b.

Note

Whenever you are asked to find the factors of a counting number:

(1) each factor *must* be a counting number, and

(2) the other number in its factor pair *must* also be a counting number.

Examples   4 is a factor of 12 because 12 / 4 gives 3 with a remainder of 0.

6 is *not* a factor of 14 because 14 / 6 gives 2 with a remainder of 2.

One way to find all the factors of a counting number is to find all the factor pairs for that number.

Example   Find all the factors of the number 64.

Number Models	Factor Pairs
64 = 1 * 64	1, 64
64 = 2 * 32	2, 32
64 = 4 * 16	4, 16
64 = 8 * 8	8, 8

The factors of 64 are 1, 2, 4, 8, 16, 32, and 64.

Did You Know?

Square numbers always have an odd number of factors. All other counting numbers have an even number of factors.

Check Your Understanding

List all the factors of each number.

1. 14   2. 16   3. 45   4. 72   5. 19   6. 100

Check your answers on page 414.

Divisibility

When one counting number is divided by another counting number and the quotient is a counting number with a remainder of 0, then the first number is **divisible by** the second number. If the quotient is a whole number with a non-zero remainder, then the first number is *not divisible by* the second number.

Examples   128 / 4 → 32 R0. The remainder is 0, so 128 is divisible by 4.

92 / 5 → 18 R2. The remainder is not 0, so 92 is not divisible by 5.

It is possible to test for divisibility without actually dividing.

Here are a few such **divisibility tests:**

♦ All counting numbers are **divisible by 1.**
♦ Counting numbers with a 0, 2, 4, 6, or 8 in the ones place are **divisible by 2.** They are the **even numbers.**
♦ Counting numbers with 0 in the ones place are **divisible by 10.**
♦ Counting numbers with 0 or 5 in the ones place are **divisible by 5.**
♦ If the sum of the digits in a counting number is divisible by 3, then the number is **divisible by 3.**
♦ If the sum of the digits in a counting number is divisible by 9, then the number is **divisible by 9.**
♦ If a counting number is divisible by both 2 and 3, it is **divisible by 6.**

Did You Know❓

In books, magazines, and newspapers, the pages on the left side are almost always even-numbered.

Example   Tell some numbers 324 is divisible by. 324 is divisible by:

♦ 2 because 4 in the ones place is an even number.
♦ 3 because the sum of its digits is 9, which is divisible by 3.
♦ 9 because the sum of its digits is divisible by 9.
♦ 6 because it is divisible both by 2 and by 3.

324 is not divisible by 10 or by 5 because it does not have a 0 in the ones place.

Check Your Understanding

Which numbers are divisible by 2? By 3? By 5? By 6? By 9? By 10?

1. 105   **2.** 4,470   **3.** 526   **4.** 621   **5.** 13,680

Check your answers on page 414.

Prime and Composite Numbers

A **prime number** is a counting number greater than 1 that has exactly two factors: 1 and the number itself. A prime number is divisible only by 1 and itself.

A **composite number** is a counting number that has more than two factors. A composite number is divisible by at least three different counting numbers.

> **Did You Know?**
>
> The Babylonian number system was based on the number 60. 60 is a very useful composite number because it has so many factors: 1, 2, 3, 4, 5, 6, 10, 12, 15, 20, 30, and 60.

Examples 13 is a prime number because its only factors are 1 and 13. 13 has exactly two factors.

18 is a composite number because it has more than two factors. Its factors are 1, 2, 3, 6, 9, and 18.

Every counting number greater than 1 can be renamed as a product of prime numbers. This is called the **prime factorization** of that number.

> **Note**
>
> The only factor of 1 is 1 itself. So the number 1 is neither prime nor composite.

Example Find the prime factorization of 80.

The number 80 can be renamed as the product $2 * 2 * 2 * 2 * 5$. The prime factorization of 80 can be written as $2^4 * 5$.

> **Note**
>
> The prime factorization of a prime number is that number. For example, the prime factorization of 11 is 11.

One way to find the prime factorization of a number is to make a **factor tree.** First, write the number. Then, below it, write any two factors whose product is that number. Repeat the process for these two factors. Continue until all the factors are prime numbers.

Example Find the prime factorization of 36.

No matter which two factors are used to start the tree, the tree will always end with the same prime factors.

$36 = 2 * 2 * 3 * 3$

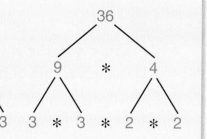

The prime factorization of 36 is $2 * 2 * 3 * 3$, or $36 = 2^2 * 3^2$.

Check Your Understanding

Make a factor tree to find the prime factorization of each number.

1. 12   **2.** 32   **3.** 42   **4.** 24   **5.** 50   **6.** 100

Check your answers on page 414.

Addition Algorithms

Partial-Sums Method

The **partial-sums method** is used to find sums mentally or with paper and pencil.

To use the partial-sums method, add from left to right, one column at a time. Then add the partial sums.

Example   679 + 345 = ?

		100s	10s	1s
		6	7	9
	+	3	4	5
Add the 100s.	600 + 300 →	9	0	0
Add the 10s.	70 + 40 →	1	1	0
Add the 1s.	9 + 5 →		1	4
Add the partial sums.	900 + 110 + 14 →	**10**	**2**	**4**

679 + 345 = 1,024

Note

Larger numbers with 4 or more digits are added in the same way.

Column-Addition Method

The **column-addition method** may be used to find sums with paper and pencil, but it is not a good method for finding sums mentally.

To add numbers using the column-addition method:

♦ Draw lines to separate the 1s, 10s, 100s, and any other places.

♦ Add the numbers in each column. Write each sum in its column.

♦ If the sum of any column is a 2-digit number, adjust that column sum. Trade part of the sum into the column to its left.

Example   467 + 764 = ?

	100s	10s	1s
	4	6	7
+	7	6	4

Add the numbers in each column.
Adjust the 1s and 10s:

	100s	10s	1s
	11	12	11

11 ones = 1 ten and 1 one
Trade the 1 ten into the tens column.

	100s	10s	1s
	11	13	1

Adjust the 10s and 100s:
13 tens = 1 hundred and 3 tens
Trade the 1 hundred into the hundreds column.

	100s	10s	1s
	12	**3**	**1**

467 + 764 = 1,231

Whole Numbers

A Short Method

This is how most adults in the United States were taught to add. Add from right to left. Add one column at a time, without displaying the partial sums.

Example   359 + 298 = ?

Step 1:
Add the ones.

```
  1
  3 5 9
+ 2 9 8
-------
      7
```

9 ones + 8 ones = 17 ones = 1 ten + 7 ones

Step 2:
Add the tens.

```
1 1
  3 5 9
+ 2 9 8
-------
    5 7
```

1 ten + 5 tens + 9 tens = 15 tens = 1 hundred + 5 tens

Step 3:
Add the hundreds.

```
1 1
  3 5 9
+ 2 9 8
-------
  6 5 7
```

1 hundred + 3 hundreds + 2 hundreds = 6 hundreds

359 + 298 = 657

The Opposite-Change Rule

Here is the **opposite-change rule:** If you subtract a number from one addend and add the same number to the other addend, the sum is the same.

Use this rule to make a problem easier by changing either of the addends to a number that has 0 in the ones place. Make the *opposite change* to the other addend. Then add.

Note

Addends are numbers that are added. In 8 + 4 = 12, the numbers 8 and 4 are addends.

Examples   69 + 36 = ?

One way: Add and subtract 1.

```
  69      (add 1)        70
+ 36      (subtract 1)  + 35
                        -----
                         105
```

69 + 36 = 105

Another way: Subtract and add 4.

```
  69      (subtract 4)    65
+ 36      (add 4)       + 40
                        -----
                         105
```

Check Your Understanding

Add.

1. 235 + 54
2. 64 + 49
3.   646
   + 317
4. 578 + 292 + 857
5.   2,864
   + 4,063

Check your answers on page 414.

SRB
14   fourteen

Subtraction Algorithms

Trade-First Subtraction Method

The **trade-first method** is similar to the method for subtracting that most adults in the United States were taught.

◆ If each digit in the top number is greater than or equal to the digit below it, subtract separately in each column.

◆ If any digit in the top number is less than the digit below it, adjust the top number before doing any subtracting. Adjust the top number by "trading."

Did You Know?

The method of subtraction called "trading" or "borrowing" dates back at least to the 1400s. The word "borrow" was not used until around 1600.

Example   574 − 386 = ?

100s	10s	1s
5	7	4
− 3	8	6

Look at the 1s place. You cannot remove 6 ones from 4 ones.

100s	10s	1s
	6	14
5	7̸	4̸
− 3	8	6

So trade 1 ten for 10 ones. Now look at the 10s place. You cannot remove 8 tens from 6 tens.

100s	10s	1s
	16	
4	6̸	14
5̸	7̸	4̸
− 3	8	6
1	8	8

So trade 1 hundred for 10 tens.

Now subtract in each column.

574 − 386 = 188

Larger numbers with 4 or more digits are subtracted in the same way.

Check Your Understanding

Subtract.

1. 84 − 38   2. 653 − 362   3. 535 − 293   4. 818
                                                  − 746

5. 7,622
   − 2,077

Check your answers on page 414.

Counting-Up Method

You can subtract two numbers by counting up from the smaller number to the larger number. First, count up to the nearest multiple of 10. Next, count up by 10s and 100s. Then count up to the larger number.

Example $525 - 58 = ?$

Write the smaller number, 58.

$$
\begin{array}{r}
5\ 8 \\
+\ \ \textcircled{2} \quad \text{Count up to the nearest 10.} \\
\hline
6\ 0
\end{array}
$$

As you count from 58 up to 525, circle each number that you count up.

$$
\begin{array}{r}
+\textcircled{4\ 0} \quad \text{Count up to the nearest 100.} \\
\hline
1\ 0\ 0
\end{array}
$$

Add the numbers you circled:
$2 + 40 + 400 + 25 = 467$

$$
\begin{array}{r}
+\textcircled{4\ 0\ 0} \quad \text{Count up to the largest possible hundred.} \\
\hline
5\ 0\ 0
\end{array}
$$

You counted up by 467.

$$
\begin{array}{r}
+\ \ \textcircled{2\ 5} \quad \text{Count up to the larger number.} \\
\hline
5\ 2\ 5
\end{array}
$$

So, $525 - 58 = 467$.

Left-to-Right Subtraction Method

Starting at the left, subtract column by column.

Examples $932 - 356 = ?$ $673 - 286 = ?$

	$932 - 356 = ?$	$673 - 286 = ?$
Subtract the 100s.	$\begin{array}{r} 9\ 3\ 2 \\ -\ 3\ 0\ 0 \\ \hline 6\ 3\ 2 \end{array}$	$\begin{array}{r} 6\ 7\ 3 \\ -\ 2\ 0\ 0 \\ \hline 4\ 7\ 3 \end{array}$
Subtract the 10s.	$\begin{array}{r} -\ \ \ 5\ 0 \\ \hline 5\ 8\ 2 \end{array}$	$\begin{array}{r} -\ \ \ 8\ 0 \\ \hline 3\ 9\ 3 \end{array}$
Subtract the 1s.	$\begin{array}{r} -\ \ \ \ \ \ 6 \\ \hline 5\ 7\ 6 \end{array}$	$\begin{array}{r} -\ \ \ \ \ \ 6 \\ \hline 3\ 8\ 7 \end{array}$

$932 - 356 = 576$ $673 - 286 = 387$

Check Your Understanding

Subtract.

1. $366 - 84$ 2. $537 - 455$ 3. $844 - 66$ 4. $605 - 281$

Check your answers on page 414.

Partial-Differences Method

1. Subtract from left to right, one column at a time.

2. In some cases, the larger number is on the bottom and the smaller number is on top. When this happens and you subtract, the difference will be a negative number.

Example $7,465 - 2,639 = ?$

$$\begin{array}{r} 7,465 \\ - \ 2,639 \\ \hline \end{array}$$

Subtract the 1,000s.	$7,000 - 2,000 \rightarrow$	$5\,0\,0\,0$
Subtract the 100s.	$400 - 600 \rightarrow$	$- \quad 2\,0\,0$
Subtract the 10s.	$60 - 30 \rightarrow$	$3\,0$
Subtract the 1s.	$5 - 9 \rightarrow$	$- \quad \quad 4$
Find the total.	$5,000 - 200 + 30 - 4 \rightarrow$	$\mathbf{4,8\,2\,6}$

$7,465 - 2,639 = 4,826$

Same-Change Rules

Here are the **same-change rules** for subtraction problems:

♦ If you add the same number to both numbers in the problem before subtracting, the answer is the same.

♦ If you subtract the same number from both numbers in the problem, the answer is the same.

Use these rules to change the second number in the problem to a number that has 0 in the ones place. Make the *same change* to the first number. Then subtract.

Examples $83 - 27 = ?$

One way: Add 3.

$$\begin{array}{rl} 8\,3 & \text{(add 3)} \\ - \ 2\,7 & \text{(add 3)} \\ \hline \end{array} \qquad \begin{array}{r} 8\,6 \\ - \ 3\,0 \\ \hline 5\,6 \end{array}$$

$83 - 27 = 56$

Another way: Subtract 7.

$$\begin{array}{rl} 8\,3 & \text{(subtract 7)} \\ - \ 2\,7 & \text{(subtract 7)} \\ \hline \end{array} \qquad \begin{array}{r} 7\,6 \\ - \ 2\,0 \\ \hline 5\,6 \end{array}$$

Check Your Understanding

Subtract.

1. $647 \quad 54$ 2. $751 - 347$ 3. $449 - 275$ 4. $5,216 - 1,418$

Check your answers on page 414.

Extended Multiplication Facts

Numbers such as 10, 100, and 1,000 are called **powers of 10.**

It is easy to multiply a whole number, n, by a power of 10. To the right of the number n, write as many zeros as there are zeros in the power of 10.

Examples

$10 * 91 = 910$	$10 * 40 = 400$	$100 * 380 = 38,000$
$100 * 91 = 9,100$	$100 * 40 = 4,000$	$10,000 * 42 = 420,000$
$1,000 * 91 = 91,000$	$1,000 * 40 = 40,000$	$1,000,000 * 5 = 5,000,000$

If you have memorized the basic multiplication facts, you can solve problems such as $8 * 80$ and $6,000 * 3$ mentally.

Examples

$8 * 80 = ?$

Think: 8 [8s] = 64

Then 8 [80s] is 10 times as much.

$8 * 80 = 10 * 64 = 640$

$6,000 * 3 = ?$

Think: 6 [3s] = 18

Then 6,000 [3s] is 1,000 times as much.

$6,000 * 3 = 1,000 * 18 = 18,000$

You can use a similar method to solve problems such as $50 * 50$ and $400 * 90$ mentally.

Examples

$50 * 50 = ?$

Think: 5 [50s] = 250

Then 50 [50s] is 10 times as much.

$50 * 50 = 10 * 250 = 2,500$

$400 * 90 = ?$

Think: 4 [90s] = 360

Then 400 [90s] is 100 times as much.

$400 * 90 = 100 * 360 = 36,000$

Check Your Understanding

Solve these problems mentally.

1. $6 * 100$ **2.** $1,000 * 89$ **3.** $7 * 900$ **4.** $8,000 * 8$ **5.** $600 * 800$ **6.** $400 * 90$

Check your answers on page 414.

Multiplication Algorithms

The symbols × and * are both used to indicate multiplication. In this book, the symbol * is used more often.

Partial-Products Method

In the **partial-products method,** you must keep track of the place value of each digit. It may help to write 1s, 10s, and 100s above the columns. Each partial product is either a basic multiplication fact or an extended multiplication fact.

Example 4 * 236 = ?

	100s	10s	1s
	2	3	6
*			4

Think of 236 as 200 + 30 + 6.

Multiply each part of 236 by 4.

4 * 200 → 8 0 0 } extended multiplication facts
4 * 30 → 1 2 0
4 * 6 → 2 4   basic multiplication fact

Add the three partial products. 9 4 4

4 * 236 = 944

Example 43 * 26 = ?

	100s	10s	1s
		2	6
*		4	3

Think of 26 as 20 + 6.

Think of 43 as 40 + 3.

Multiply each part of 26 by each part of 43.

40 * 20 → 8 0 0 } extended multiplication facts
40 * 6 → 2 4 0
3 * 20 → 6 0
3 * 6 → 1 8   basic multiplication fact

Add these four partial products. 1,1 1 8

43 * 26 = 1,118

Check Your Understanding

Multiply. Write each partial product. Then add the partial products.

1. 179 * 4   **2.** 37 * 64   **3.** 60 * 59   **4.** 87 * 45   **5.** 273 * 70

Check your answers on page 415.

Lattice Method

The **lattice method** for multiplying has been used for hundreds of years. It is very easy to use if you know the basic multiplication facts.

Example 4 * 915 = ?

The box with cells and diagonals is called a **lattice**.
Write 915 above the lattice.
Write 4 on the right side of the lattice.

Multiply 4 * 5. Then multiply 4 * 1. Then multiply 4 * 9.
Write the answers as shown.

Add the numbers along each diagonal, starting at the right.

Read the answer. 4 * 915 = 3,660

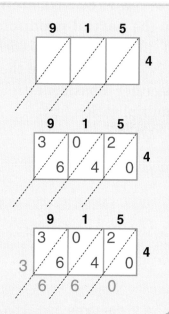

Example 86 * 37 = ?

Write 37 above the lattice.
Write 86 on the right side of the lattice.

Multiply 8 * 7. Then multiply 8 * 3.
Multiply 6 * 7. Then multiply 6 * 3.
Write the answers as shown.

Add the numbers along each diagonal, starting at the right.

When the numbers along a diagonal add up to 10 or more:

• record the ones digit in the sum.
• add the tens digit to the sum along the next diagonal above.

Read the answer. 86 * 37 = 3,182

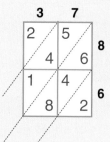

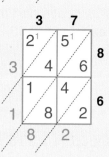

Check Your Understanding

Draw a lattice for each problem. Then multiply.

1. 6 * 59   **2.** 77 * 86   **3.** 76 * 98   **4.** 7 * 648   **5.** 879 * 4

Check your answers on page 415.

Extended Division Facts

Numbers such as 10, 100, and 1,000 are called **powers of 10.**

In the examples below, use the following method to divide a whole number that ends in zeros by a power of 10:

♦ Cross out zeros in the number, starting in the ones place.

♦ Cross out as many zeros as there are zeros in the power of 10.

Examples

90,000 / **10** = 9000~~0~~	63,000 / **10** = 6300~~0~~	860,000 / **10,000** = 86~~0000~~
90,000 / **100** = 900~~00~~	63,000 / **100** = 630~~00~~	7,000,000 / **100,000** = 70~~00000~~
90,000 / **1,000** = 90~~000~~	63,000 / **1,000** = 63~~000~~	

If you know the basic division facts, you can solve problems such as 540 / 9 and 18,000 / 3 mentally.

Examples

540 / 9 = ?	18,000 / 3 = ?
Think: 54 / 9 = 6	*Think:* 18 / 3 = 6
Then 540 / 9 is 10 times as much.	Then 18,000 / 3 is 1,000 times as much.
540 / 9 = 10 * 6 = 60	18,000 / 3 = 1,000 * 6 = 6,000

You can use a similar method to solve problems such as 18,000 / 30 and 32,000 / 400 mentally.

Examples

18,000 / 30 = ?	32,000 / 400 = ?
Think: 18,000 / 3 = 6,000	*Think:* 32,000 / 4 = 8,000
Then 18,000 / 30 is $\frac{1}{10}$ as much.	Then 32,000 / 400 is $\frac{1}{100}$ as much.
18,000 / 30 = $\frac{1}{10}$ * 6,000 = 600	32,000 / 400 = $\frac{1}{100}$ * 8,000 = 80

Check Your Understanding

Solve these problems mentally.

1. 84,000 / 1,000   2. 56,000 / 8   3. 4,500 / 90   4. 45,000 / 900

Check your answers on page 415.

Division Algorithms

Different symbols may be used to indicate division. For example, "94 divided by 6" may be written as $94 \div 6$, $6\overline{)94}$, $94 / 6$, or $\frac{94}{6}$.

♦ The number that is being divided is called the **dividend.**

♦ The number that divides the dividend is called the **divisor.**

♦ The answer to a division problem is called the **quotient.**

♦ Some numbers cannot be divided evenly. When this happens, the answer includes a quotient and a **remainder.**

> **Four ways to show "123 divided by 4"**
>
> $123 \div 4$ $\qquad$ $123 / 4$
>
> $4\overline{)123}$ $\qquad$ $\frac{123}{4}$
>
> 123 is the dividend.
> 4 is the divisor.

Partial-Quotients Method

In the **partial-quotients method,** it takes several steps to find the quotient. At each step, you find a partial answer (called a **partial quotient**). These partial answers are then added to find the quotient.

Study the example below. To find the number of 6s in 1,010, first find partial quotients and then add them. Record the partial quotients in a column to the right of the original problem.

Example $\qquad$ $1,010 / 6 = ?$

Write partial quotients in this column.

```
6)1,010        ↓    Think: How many [6s] are in 1,010? At least 100.
 − 600   100        The first partial quotient is 100. 100 * 6 = 600
   410              Subtract 600 from 1,010. At least 50 [6s] are left in 410.

 − 300    50        The second partial quotient is 50. 50 * 6 = 300
   110              Subtract. At least 10 [6s] are left in 110.

 −  60    10        The third partial quotient is 10. 10 * 6 = 60
    50              Subtract. At least 8 [6s] are left in 50.

 −  48     8        The fourth partial quotient is 8. 8 * 6 = 48
     2   168        Subtract. Add the partial quotients.
     ↑     ↑
```

Remainder   Quotient

The answer is 168 R2. Record the answer as $6\overline{)1,010}^{\,168\ R2}$ or write $1,010 / 6 \rightarrow 168$ R2.

The partial-quotients method works with a 2-digit or a 1-digit divisor. It may help to write some easy facts for the divisor.

Example   Divide 800 by 22.

Some facts for 22
(to help find partial quotients):

$1 * 22 = 22$
$2 * 22 = 44$
$5 * 22 = 110$
$10 * 22 = 220$

```
22)800
  -440   20    (20 [22s] in 800)
   360
  -220   10    (10 [22s] in 360)
   140
  -110    5    (5 [22s] in 140)
    30
  - 22    1    (1 [22s] in 30)
     8   36
```

Record the answer as $22\overline{)800}^{\,36\ R8}$ or write $800 / 22 \rightarrow 36$ R8.

There are different ways to find partial quotients. Study the example below. The answer is the same for each way.

Example $391 / 4 = ?$

One way:
```
4)391
-200   50
 191
-120   30
  71
- 40   10
  31
- 20    5
  11
-  8    2
   3   97
```

A second way:
```
4)391
-200   50
 191
-160   40
  31
- 20    5
  11
-  8    2
   3   97
```

A third way:
```
4)391
-360   90
  31
- 20    5
  11
-  8    2
   3   97
```

The answer, 97 R3, is the same for each way.

Check Your Understanding

Divide.

1. $6\overline{)92}$ **2.** $645 / 5$ **3.** $637 \div 7$ **4.** $4\overline{)834}$

Check your answers on page 415.

Column-Division Method

The best way to understand column division is to think of a division problem as a money-sharing problem. In the example below, think of sharing $863 equally among 5 people.

Example $5\overline{)863} = ?$

1. Draw lines to separate the digits in the dividend (the number being divided).
   Work left to right. Begin in the left column.

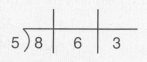

2. Think of the 8 in the hundreds column as 8 $100 bills to be shared by 5 people.
   Each person gets 1 $100 bill. There are 3 $100 bills remaining.

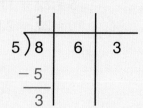

3. Trade the 3 $100 bills for 30 $10 bills.
   Think of the 6 in the tens column as 6 $10 bills. That makes 30 + 6 = 36 $10 bills.

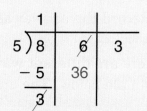

4. If 5 people share 36 $10 bills, each person gets 7 $10 bills. There is 1 $10 bill remaining.

5. Trade the 1 $10 bill for 10 $1 bills.
   Think of the 3 in the ones column as 3 $1 bills. That makes 10 + 3 = 13 $1 bills.

6. If 5 people share 13 $1 bills, each person gets 2 $1 bills. There are 3 $1 bills remaining.

Record the answer as 172 R3.
Each person receives $172 and $3 are left over.

U.S. Traditional Addition: Whole Numbers

You can add numbers with **U.S. traditional addition.**

Did You Know?

Writing numbers above the addends in U.S. traditional addition is sometimes called "carrying." Numbers that are carried, such as the 1 in the 10s column in Step 1 below, are sometimes called "carry marks" or "carries."

Example 8,286 + 7,217 = ?

Step 1: Start with the 1s: 6 + 7 = 13.
13 ones = 1 ten + 3 ones
Write 3 in the 1s place below the line.
Write 1 above the numbers in the 10s place.

$$\begin{array}{r} {}^{1} \\ 8\,2\,8\,6 \\ +\;7\,2\,1\,7 \\ \hline 3 \end{array}$$

Step 2: Add the 10s: 1 + 8 + 1 = 10.
10 tens = 1 hundred + 0 tens
Write 0 in the 10s place below the line.
Write 1 above the numbers in the 100s place.

$$\begin{array}{r} {}^{1\,1} \\ 8\,2\,8\,6 \\ +\;7\,2\,1\,7 \\ \hline 0\,3 \end{array}$$

Step 3: Continue adding through the 1,000s place.

$$\begin{array}{r} {}^{1\,1} \\ 8\,2\,8\,6 \\ +\;7\,2\,1\,7 \\ \hline 1\,5\,5\,0\,3 \end{array}$$

8,286 + 7,217 = 15,503

Check Your Understanding

Add.

1. 2,960 + 4,328 = ?

2. $\begin{array}{r} 8{,}649 \\ +\;\;792 \\ \hline \end{array}$

3. $\begin{array}{r} 18{,}766 \\ +\;9{,}560 \\ \hline \end{array}$

4. $\begin{array}{r} 14{,}900{,}637 \\ +\;1{,}967{,}088 \\ \hline \end{array}$

Check your answers on page 424.

U.S. Traditional Subtraction: Whole Numbers

You can subtract numbers with **U.S. traditional subtraction.**

Example $705 - 458 = ?$

Step 1: Start with the 1s.
Since $8 > 5$, you need to regroup.
There are no tens in 705, so trade 1 hundred
for 10 tens and then trade 1 ten for 10 ones:
$705 = 6$ hundreds $+ 9$ tens $+ 15$ ones.
Subtract the 1s: $15 - 8 = 7$.

$$\begin{array}{r} 6 \;\; \overset{9}{\cancel{1\kern-0.3em0}} \;\; 15 \\ \cancel{7} \;\; \cancel{0} \;\; \cancel{5} \\ -\; 4 \;\; 5 \;\; 8 \\ \hline 7 \end{array}$$

Step 2: Go to the 10s.
Subtract the 10s: $9 - 5 = 4$.

$$\begin{array}{r} 6 \;\; \overset{9}{\cancel{1\kern-0.3em0}} \;\; 15 \\ \cancel{7} \;\; \cancel{0} \;\; \cancel{5} \\ -\; 4 \;\; 5 \;\; 8 \\ \hline 4 \;\; 7 \end{array}$$

Step 3: Go to the 100s.
Subtract the 100s: $6 - 4 = 2$.

$$\begin{array}{r} 6 \;\; \overset{9}{\cancel{1\kern-0.3em0}} \;\; 15 \\ \cancel{7} \;\; \cancel{0} \;\; \cancel{5} \\ -\; 4 \;\; 5 \;\; 8 \\ \hline 2 \;\; 4 \;\; 7 \end{array}$$

$705 - 458 = 247$

Check Your Understanding

Subtract.

1. $435 - 379 = ?$

2. $\begin{array}{r} 650 \\ -\; 465 \\ \hline \end{array}$

3. $\begin{array}{r} 8{,}604 \\ -\; 5{,}907 \\ \hline \end{array}$

4. $\begin{array}{r} 890{,}074 \\ -\; 79{,}082 \\ \hline \end{array}$

Check your answers on page 424.

U.S. Traditional Multiplication: Whole Numbers

You can use **U.S. traditional multiplication** to multiply.

Example $5 * 347 = ?$

Step 1: Multiply the ones.
    $5 * 7$ ones $= 35$ ones $= 3$ tens $+ 5$ ones
    Write 5 in the 1s place below the line.
    Write 3 above the 4 in the 10s place.

$$
\begin{array}{ccc}
 & 3 & \\
3 & 4 & 7 \\
* & & 5 \\
\hline
 & & 5 \\
\end{array}
$$

Step 2: Multiply the tens.
    $5 * 4$ tens $= 20$ tens
    Remember the 3 tens from Step 1.
    20 tens $+ 3$ tens $= 23$ tens in all
    23 tens $= 2$ hundreds $+ 3$ tens
    Write 3 in the 10s place below the line.
    Write 2 above the 3 in the 100s place.

$$
\begin{array}{ccc}
2 & 3 & \\
3 & 4 & 7 \\
* & & 5 \\
\hline
 & 3 & 5 \\
\end{array}
$$

Step 3: Multiply the hundreds.
    $5 * 3$ hundreds $= 15$ hundreds
    Remember the 2 hundreds from Step 2.
    15 hundreds $+ 2$ hundreds $= 17$ hundreds in all
    17 hundreds $= 1$ thousand $+ 7$ hundreds
    Write 7 in the 100s place below the line.
    Write 1 in the 1,000s place below the line.

$$
\begin{array}{cccc}
 & 2 & 3 & \\
 & 3 & 4 & 7 \\
* & & & 5 \\
\hline
1 & 7 & 3 & 5 \\
\end{array}
$$

$5 * 347 = 1,735$

Check Your Understanding

Multiply.

1. $3 * 539 = ?$

2. $\begin{array}{r} 684 \\ * \quad 3 \\ \hline \end{array}$

3. $\begin{array}{r} 3,884 \\ * \quad\quad 4 \\ \hline \end{array}$

4. $89,645 * 8 = ?$

Check your answers on page 424.

You can use U.S. traditional multiplication to multiply by two-digit numbers.

Example   62 * 735 = ?

Step 1: Multiply 735 by the 2 in 62, as if the problem were 2 * 735.

```
        1
    7   3   5
*       6   2
─────────────
    1   4   7   0
```
← The partial product
2 * 735 = 1,470

Step 2: Multiply 735 by the 6 in 62, as if the problem were 6 * 735.

The 6 in 62 stands for 6 tens, so write this partial product one place to the left.

Write a 0 in the 1s place to show you are multiplying by tens.

Write the new carries above the old carries.

```
        2   3
            1
    7   3   5
*       6   2
─────────────
    1   4   7   0
4   4   1   0   0
```
← 60 * 735 = 44,100

Step 3: Add the two partial products to get the final answer.

```
        2   3
            1
    7   3   5
*       6   2
─────────────
    1   4   7   0
+ 4 4   1   0   0
─────────────────
  4 5   5   7   0
```
← 62 * 735 = 45,570

62 * 735 = 45,570

Check Your Understanding

Multiply.

1. 48 * 438 = ?

2.   965
      * 43

3.   1,854
      * 87

4. 3,805 * 35 = ?

Check your answers on page 424.

U.S. Traditional Long Division: Single-Digit Divisors

U.S. traditional long division is another method you can use to divide.

Example Share $957 equally among 5 people.

Step 1: Share the $100 s.

$$
\begin{array}{r}
1 \\
5)\overline{957} \\
-5 \\
\hline
4
\end{array}
$$

← Each person gets 1 $100 .

← 1 $100 each for 5 people

← 4 $100 s are left.

Step 2: Trade 4 $100 s for 40 $10 s.

That makes 45 $10 s in all.

$$
\begin{array}{r}
1 \\
5)\overline{957} \\
-5 \downarrow \\
\hline
45
\end{array}
$$

← 45 $10 s are to be shared.

Step 3: Share the $10 s.

$$
\begin{array}{r}
19 \\
5)\overline{957} \\
-5 \\
\hline
45 \\
-45 \\
\hline
0
\end{array}
$$

← Each person gets 9 $10 s.

← 9 $10 s each for 5 people

← 0 $10 s are left.

Step 4: Share the $1 s.

$$
\begin{array}{r}
191 \\
5)\overline{957} \\
-5 \downarrow \\
\hline
45 \\
-45 \downarrow \\
\hline
07 \\
-5 \\
\hline
2
\end{array}
$$

← Each person gets 1 $1 .

← 7 $1 s are to be shared.

← 1 $1 each for 5 people

← 2 $1 s are left.

$957 / 5 → $191 R$2
Each person gets $191; $2 are left over.

Check Your Understanding

Divide.

1. 840 / 7 **2.** 6)984 **3.** 4)539 **4.** 5,280 / 6

Check your answers on page 424.

U.S. traditional long division is not limited to dividing money.

Note

The "leading" 0 in the quotient is shown in the problem to help you understand the long division method. It should not be included in the answer.

Example   3,628 / 5

Think about the problem as dividing 3,628 into 5 equal shares.

Step 1: Start with the thousands.

$$\begin{array}{r} 0 \\ 5\overline{)3628} \end{array}$$ ← There are not enough thousands to share 5 ways.

Step 2: So trade 3 thousands for 30 hundreds. Share the hundreds.

$$\begin{array}{r} 07 \\ 5\overline{)3628} \\ -35 \\ \hline 1 \end{array}$$
← Each share gets 7 hundreds.
← 36 hundreds
← 7 hundreds * 5 shares
← 1 hundred is left.

Step 3: Trade 1 hundred for 10 tens. Share the tens.

$$\begin{array}{r} 072 \\ 5\overline{)3628} \\ -35 \\ \hline 12 \\ -10 \\ \hline 2 \end{array}$$
← Each share gets 2 tens.
← 10 tens + 2 tens
← 2 tens * 5 shares
← 2 tens are left.

Step 4: Trade 2 tens for 20 ones. Share the ones.

$$\begin{array}{r} 0725 \\ 5\overline{)3628} \\ -35 \\ \hline 12 \\ -10 \\ \hline 28 \\ -25 \\ \hline 3 \end{array}$$
← Each share gets 5 ones.

← 20 ones + 8 ones
← 5 ones * 5 shares
← 3 ones are left.

3,628 / 5 → 725 R3

Check Your Understanding

Divide.

1. 5,376 / 6   **2.** 6)8,586   **3.** 4)6,923   **4.** 8,029 / 3

Check your answers on page 424.

U.S. Traditional Long Division: Multidigit Divisors

You can use **U.S. traditional long division** to divide by larger numbers.

Example   Share $681 equally among 21 people.

Make a table of easy multiples of the divisor.
This can help you decide how many to share at each step.

1 * 21	21
2 * 21	42
3 * 21	63
4 * 21	84
5 * 21	105
6 * 21	126
8 * 21	168
10 * 21	210

Double 21.
Add 2 * 21 and 1 * 21.
Double 2 * 21.
Halve 10 * 21.
Double 3 * 21.
Double 4 * 21.
Move decimal point one place to the right.

Step 1: There are not enough [$100]s to share 21 ways, so trade 6 [$100]s for 60 [$10]s.

Share the 68 [$10]s.

```
      3        ← Each person gets 3 [$10]s.
21)681         ← There are 68 [$10]s to share.
  -63          ← 3 [$10]s * 21
    5          ← 5 [$10]s are left.
```

Step 2: Trade the 5 [$10]s for 50 [$1]s.

Share the 51 [$1]s.

```
     32        ← Each person gets 2 [$1]s.
21)681
  -63↓
    51         ← 50 [$1]s + 1 [$1]
   -42         ← 2 [$1]s * 21
     9         ← 9 [$1]s are left.
```

$681 / 21 → $32 R$9

Example   7720 / 25

Make a table of easy multiples of the divisor.

1 * 25	25
2 * 25	50
3 * 25	75
4 * 25	100
5 * 25	125
6 * 25	150
8 * 25	200
10 * 25	250

Double 25.

Add 2 * 25 and 1 * 25.

Double 2 * 25.

Halve 10 * 25.

Double 3 * 25.

Double 4 * 25.

Move decimal point one place to the right.

Step 1: There are not enough thousands to share 25 ways, so trade the thousands for hundreds. Share the hundreds.

```
      3     ← Each share gets 3 hundreds.
25)7720     ← 77 hundreds
  −75        ← 3 hundreds * 25 shares
    2        ← 2 hundreds are left.
```

Step 2: Trade the hundreds for tens. Share the tens.

```
     30     ← There are not enough tens to share.
25)7720
 −75
   22       ← 20 tens + 2 tens
```

Step 3: Trade the tens for ones. Share the ones.

```
    308     ← Each share gets 8 ones.
25)7720
 −75
   220      ← 22 tens + 0 ones
 −200       ← 8 ones * 25 shares
   20       ← 20 ones are left.
```

7720 / 25 → 308 R20

Did You Know?

Beginning in the late 1920s and early 1930s, the U.S. Treasury issued a small number of large bills, including $500, $1,000, $5,000, $10,000, and $100,000 bills. By the mid-1940s, the Treasury stopped making these bills, and in 1969 President Nixon removed them from circulation because they were rarely used and were attractive to counterfeiters.

Check Your Understanding

Divide.

1. 650 / 25

2. 7,720 / 25

3. 13)5,819

4. 48)5,286

Check your answers on page 424.

Decimals and Percents

Decimals

Mathematics in everyday life uses more than **whole numbers.** Other numbers, called **decimals** and **fractions,** name numbers that are between whole numbers. Decimals and fractions are used to name a part of a whole thing or a part of a collection. We use decimals and fractions to make more precise measurements than can be made using only whole numbers.

You probably see many uses of decimals every day.

♦ Fractional parts of a dollar are almost always written as decimals.

♦ Weather reports give rainfall amounts in decimals.

♦ Digital scales in supermarkets show the weight of fruits, vegetables, and meat with decimals.

♦ Many sports events are timed to a tenth or hundredth of a second, and the times are reported as decimals.

♦ Sports statistics often use decimals. For example, batting averages and average points scored per game are usually reported as decimals.

Decimals are another way to write fractions. Many fractions have denominators of 10, 100, 1,000, and so on. It is easy to write the decimal names for fractions like these.

Some fractions between 1 and 2:

$\frac{3}{2}, \frac{7}{4}, \frac{11}{8}, \frac{19}{16}$

Some decimals between 1 and 2:

1.5, 1.75, 1.875, 1.9999

Did You Know?

Jim Hines was the first person to complete the 100-meter dash in less than 10 seconds. His time at the 1968 Olympics was 9.95 seconds and his record stood for 15 years.

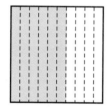

$$\frac{6}{10} = 0.6$$

This square is divided into 10 equal parts. Each part is $\frac{1}{10}$ of the square. The decimal name for $\frac{1}{10}$ is 0.1.

$\frac{6}{10}$ of the square is shaded. The decimal name for $\frac{6}{10}$ is 0.6.

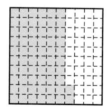

$$\frac{62}{100} = 0.62$$

This square is divided into 100 equal parts. Each part is $\frac{1}{100}$ of the square. The decimal name for $\frac{1}{100}$ is 0.01.

$\frac{62}{100}$ of the square is shaded. The decimal name for $\frac{62}{100}$ is 0.62.

Like mixed numbers, decimals are used to name numbers greater than one.

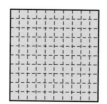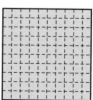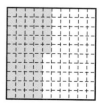

$$2\frac{45}{100} = 2.45$$

In a decimal, the dot is called the **decimal point.** It separates the whole-number part from the decimal part. A decimal with one digit after the decimal point names *tenths.* A decimal with two digits after the decimal point names *hundredths.* A decimal with three digits after the decimal point names *thousandths.*

decimal point

$$12 . 105$$

whole-number part   decimal part

Examples

tenths	hundredths	thousandths
$0.5 = \frac{5}{10}$	$0.43 = \frac{43}{100}$	$0.291 = \frac{291}{1,000}$
$0.7 = \frac{7}{10}$	$0.75 = \frac{75}{100}$	$0.003 = \frac{3}{1,000}$
$0.9 = \frac{9}{10}$	$0.08 = \frac{8}{100}$	$0.072 = \frac{72}{1,000}$

Did You Know ?

Decimals were invented by the Dutch scientist Simon Stevin in 1585. But there is no single, worldwide form for writing decimals. For 3.25 (American notation), the British write 3·25, and the Germans and French write 3,25.

Reading Decimals

One way to read a decimal is to say it as you would a fraction or mixed number. For example, $0.001 = \frac{1}{1,000}$ and can be read as "one thousandth." $7.9 = 7\frac{9}{10}$, so 7.9 can be read as "seven and nine tenths."

You can also read decimals by first saying the whole-number part, then saying "point," and finally saying the digits in the decimal part. For example, 6.8 can be read as "six point eight"; 0.15 can be read as "zero point one five." This way of reading decimals is often useful when there are many digits in the decimal.

Examples   0.26 is read as "26 hundredths" or "zero point two six."

34.5 is read as "34 and 5 tenths" or "34 point 5."

0.004 is read as "4 thousandths" or "0 point zero zero four."

Check Your Understanding

Read each decimal to yourself in two ways.

1. 0.4   **2.** 1.65   **3.** 0.872   **4.** 16.04   **5.** 0.003   **6.** 59.061

Check your answers on page 415.

Place Value for Decimals

A place-value chart works the same way for decimals as it does for whole numbers.

Example

1,000s	100s	10s	1s		0.1s	0.01s	0.001s
thousands	hundreds	tens	ones	.	tenths	hundredths	thousandths
		4	7	.	8	0	5

In the number 47.805,

8 is in the **tenths** place; its value is 8 tenths, or $\frac{8}{10}$, or 0.8.

0 is in the **hundredths** place; its value is 0.

5 is in the **thousandths** place; its value is 5 thousandths, or $\frac{5}{1,000}$, or 0.005.

Study the place-value chart below. Look at the numbers that name the places. As you move from *right to left* along the chart, each number is **10 times as large as the number to its right.**

Example

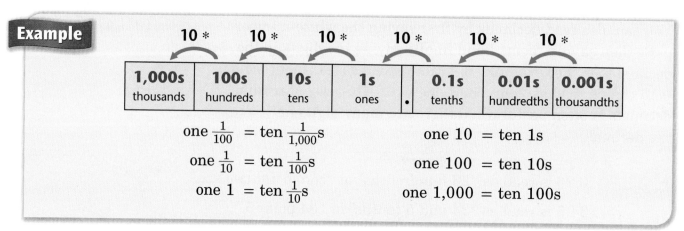

one $\frac{1}{100}$ = ten $\frac{1}{1,000}$s   one 10 = ten 1s

one $\frac{1}{10}$ = ten $\frac{1}{100}$s   one 100 = ten 10s

one 1 = ten $\frac{1}{10}$s   one 1,000 = ten 100s

As you move from *left to right* along the place-value chart below, each number is $\frac{1}{10}$ **as large as the number to its left.**

Example

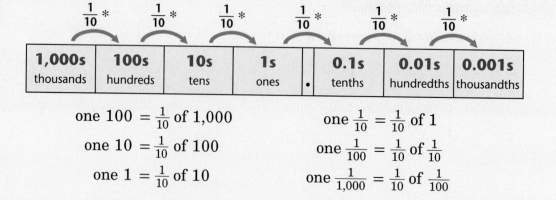

one 100 = $\frac{1}{10}$ of 1,000   one $\frac{1}{10}$ = $\frac{1}{10}$ of 1

one 10 = $\frac{1}{10}$ of 100   one $\frac{1}{100}$ = $\frac{1}{10}$ of $\frac{1}{10}$

one 1 = $\frac{1}{10}$ of 10   one $\frac{1}{1,000}$ = $\frac{1}{10}$ of $\frac{1}{100}$

Powers of 10

Study this base-ten place-value chart.

1,000s	100s	10s	1s	.	0.1s	0.01s	0.001s
thousands	hundreds	tens	ones	.	tenths	hundredths	thousandths
3	1	3	9	.	0	7	6
three thousand, one hundred thirty-nine				and	seventy-six thousandths		

Look at the numbers across the top of the chart that name the places. A whole number that can be written using only 10s as factors is called a **power of 10.** A power of 10 can be written in exponential notation.

Powers of 10 (greater than 1)		
Standard Notation	**Product of 10s**	**Exponential Notation**
10	10	10^1
100	10 * 10	10^2
1,000	10 * 10 * 10	10^3
10,000	10 * 10 * 10 * 10	10^4
100,000	10 * 10 * 10 * 10 * 10	10^5

> **Note**
>
> A number written in the usual place-value way, like 100, is in **standard notation.** A number written with an exponent, like 10^2, is in **exponential notation.**

Decimals that can be written using only 0.1s as factors are also called powers of 10. They can be written in exponential notation with negative exponents.

Powers of 10 (less than 1)		
Standard Notation	**Product of 0.1s**	**Exponential Notation**
0.1	0.1	10^{-1}
0.01	0.1 * 0.1	10^{-2}
0.001	0.1 * 0.1 * 0.1	10^{-3}
0.0001	0.1 * 0.1 * 0.1 * 0.1	10^{-4}
0.00001	0.1 * 0.1 * 0.1 * 0.1 * 0.1	10^{-5}

> **Note**
>
> A number raised to a negative exponent power is equal to the fraction 1 over the number raised to the positive exponent power. For example,
> $$10^{-2} = \frac{1}{10^2}$$
> $$= \frac{1}{10 * 10}$$
> $$= \frac{1}{100}.$$

The number 1 is also called a power of 10 because $1 = 10^0$. The pattern in the table below shows why mathematicians define 10^0 to be equal to 1.

100,000s	10,000s	1,000s	100s	10s	1s	.	0.1s	0.01s	0.001s	0.0001s	0.00001s
10^5	10^4	10^3	10^2	10^1	10^0	.	10^{-1}	10^{-2}	10^{-3}	10^{-4}	10^{-5}

All the numbers across the top of a place-value chart are powers of 10.

Note the pattern in the exponents: Each exponent is 1 less than the exponent in the place to its left.

Comparing Decimals

You may use place value to compare decimals in the same way you compare whole numbers. Start by comparing digits in the leftmost place. Continue to the right until the digits in a place do not match.

Note

Decimals and fractions are often shown on a number line. For any pair of numbers on the number line, the number to the left is less than the number to the right.

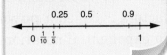

| **Examples** | Compare 1.35 and 1.288. | Compare 0.5 and 0.105. |

1.35
1.288
  ↑ different
3 tenths > 2 tenths.

So, 1.35 > 1.288.

0.5
0.105
  ↑ different
5 tenths > 1 tenth.

So, 0.5 > 0.105.

You can attach one or more 0s at the end of the decimal part of a number without changing the value of the number.

Examples   1.4 = 1.40   0.6 = 0.600   3.09 = 3.090   14.200 = 14.20 = 14.2

Another way to compare decimals is to model them with base-10 blocks. If you don't have the blocks, draw shorthand pictures.

Base-10 Blocks and Their Shorthand Pictures

▫ cube ▫ | long | flat ☐   big cube ☐

Example   Compare 2.3 and 2.16.

2.3

2.16

2 flats and 3 longs are more than 2 flats, 1 long, and 6 cubes.

So, 2.3 is more than 2.16.
2.3 > 2.16

For this example:

A flat ☐ is worth 1

A long | is worth 0.1.

A cube ▫ is worth 0.01.

Check Your Understanding

Compare the numbers in each pair.

1. 0.78, 0.079   2. 1.099, 1.1   3. 0.99, 0.10   4. $\frac{5}{4}$, 1.3

Check your answers on page 415.

30 thirty

Addition and Subtraction of Decimals

There are many ways to add and subtract decimals. One way is to use base-10 blocks. When working with decimals, we usually say that the flat is worth 1.

To add with base-10 blocks, count out blocks for each number, and put all the blocks together. Make any trades for larger blocks that you can. Then count the blocks for the sum.

To subtract with base-10 blocks, count out blocks for the larger number. Take away blocks for the smaller number, making trades as needed. Then count the remaining blocks.

Using base-10 blocks is a good idea, especially at first. However, drawing shorthand pictures is usually easier and quicker.

For the examples on this page:
A flat ☐ is worth 1
A long | is worth 0.1.
A cube ▫ is worth 0.01.

Example   1.61 + 4.7 = ?

First, draw pictures for each number.

This means that 1.61 + 4.7 = 6.31. This makes sense because 1.61 is near $1\frac{1}{2}$ and 4.7 is near $4\frac{1}{2}$. So, the answer should be near 6, which it is.

1.61 + 4.7 = 6.31

Next, draw a ring around 10 longs and trade them for 1 flat.

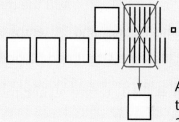

After the trade, there are 6 flats, 3 longs, and 1 cube.

Example   4.07 − 2.7 = ?

The picture for 4.07 does not show any longs.

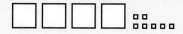

You want to take away 2.7 (2 flats and 7 longs). To do this, trade 1 flat for 10 longs.

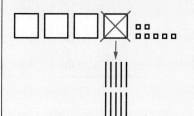

Now remove 2 flats and 7 longs (2.7).

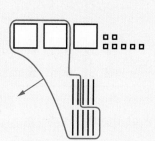

1 flat, 3 longs, and 7 cubes are left. These blocks show 1.37.

4.07 − 2.7 = 1.37

Most paper-and-pencil methods for adding and subtracting whole numbers also work for decimals. The main difference is that you have to line up the places correctly, either by attaching 0s to the end of the numbers or by lining up the ones place.

Examples $32.5 + 19.6 = ?$

Partial-Sums Method:

	10s	1s	0.1s
	3	2 .	5
+	1	9 .	6

Add the tens.	$30 + 10 \rightarrow$	4	0 . 0
Add the ones.	$2 + 9 \rightarrow$	1	1 . 0
Add the tenths.	$0.5 + 0.6 \rightarrow$		1 . 1
Add the partial sums.	$40.0 + 11.0 + 1.1 \rightarrow$	5	2 . 1

Column-Addition Method:

	10s	1s	0.1s
	3	2	. 5
+	1	9	. 6
Add the numbers in each column. →	4	11	. 11
Rename 11 ones and 11 tenths as 12 ones and 1 tenth. →	4	12	. 1
Rename 4 tens and 12 ones as 5 tens and 2 ones. →	5	2	. 1

$32.5 + 19.6 = 52.1$, using either method.

Example $7.4 - 2.65 = ?$

Trade-First Method:
Write the problem in columns. Be sure to line up the places correctly. Since 2.65 has two decimal places, write 7.4 as 7.40.

1s	0.1s	0.01s
7 .	4	0
− 2 .	6	5

Look at the 0.01s place.

You cannot remove 5 hundredths from 0 hundredths.

1s	0.1s	0.01s
	3	10
7 .	4̸	0̸
− 2 .	6	5

So trade 1 tenth for 10 hundredths.

Now look at the 0.1s place. You cannot remove 6 tenths from 3 tenths.

1s	0.1s	0.01s
	13	
6	3̸	10
7̸ .	4̸	0̸
− 2 .	6	5
4 .	7	5

So trade 1 one for 10 tenths. Now subtract in each column.

$7.4 - 2.65 = 4.75$

Example   7.4 − 2.65 = ?

Left-to-Right Subtraction Method:

Since 2.65 has two decimal places, write 7.4 as 7.40.

$$
\begin{array}{r}
\mathbf{7.40} \\
\text{Subtract the ones.} \quad -\ \mathbf{2.00} \\
\hline
5.40 \\
\text{Subtract the tenths.} \quad -\ 0.\mathbf{60} \\
\hline
4.80 \\
\text{Subtract the hundredths.} \quad -\ 0.0\mathbf{5} \\
\hline
\mathbf{4.75}
\end{array}
$$

7.4 − 2.65 = 4.75

Example   7.4 − 2.65 = ?

Counting-Up Method:

Since 2.65 has two decimal places, write 7.4 as 7.40.
There are many ways to count up from 2.65 to 7.40. Here is one.

$$
\begin{array}{r}
2.65 \\
\boxed{+\ 0.35} \\
\hline
3.00 \\
\boxed{+\ 4.00} \\
\hline
7.00 \\
\boxed{+\ 0.40} \\
\hline
7.40
\end{array}
$$

Add the numbers you circled and counted up by:

$$
\begin{array}{r}
0.35 \\
4.00 \\
+\ 0.40 \\
\hline
4.75
\end{array}
$$

You counted up by 4.75

7.4 − 2.65 = 4.75

Calculator:

If you use a calculator, it's important to check your answer by estimating because it's easy to press a wrong key accidentally.

Did You Know?

The *abacus* is an ancient and powerful calculating tool that was probably invented in the Middle East. It can be used to add, subtract, multiply, and divide both whole numbers and decimals.

A skilled user with an efficient abacus (like the Japanese soroban shown above) can calculate almost as quickly as someone using a calculator.

Check Your Understanding

Add or subtract.

1. 2.69 + 7.35   2. 21.5 − 8.8   3. 7.4 + 3.082   4. 10 − 1.79

Check your answers on page 415.

Units and Precision in Decimal Addition and Subtraction

Counts and measures always have units. For addition or subtraction, all the numbers must have the same unit. If they do not, before solving the problem you will have to convert at least one of the numbers so that all units are the same.

Examples Find the perimeter of the triangle.

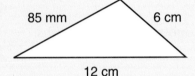

Method 1
Convert the centimeter measures to millimeters, then add.

6 cm = 60 mm   12 cm = 120 mm

Perimeter = 60 mm + 120 mm + 85 mm = 265 mm

Method 2
Convert the millimeter measure to centimeters, then add.

85 mm = 8.5 cm

Perimeter = 6 cm + 12 cm + 8.5 cm = 26.5 cm

In most practical situations, all measures have the same degree of precision. If some measures are more precise than others, convert them so that they have the same precision as the *least precise* measure.

Example The winning times in the men's 100-meter dash in the 1936 and 1988 Olympic Games are shown at the right.

Year	Winner	Time
1936	Jesse Owens, U.S.A.	10.3 seconds
1988	Carl Lewis, U.S.A.	9.92 seconds

How much faster did Carl Lewis run than Jesse Owens?
Carl Lewis was timed to the nearest hundredth of a second.
Jesse Owens was timed to the nearest tenth of a second.
Round the more precise measure, 9.92 seconds, to match the less precise measure, 10.3 seconds. 9.92 seconds rounded to the nearest tenth of a second is 9.9 seconds.

Since 10.3 − 9.9 = 0.4, Carl Lewis ran the 100-meter dash about 0.4 second faster than Jesse Owens.

Check Your Understanding

1. Which measurement is less precise: 4.7 meters or 3.62 meters?
2. Solve. Use the degree of precision of the less precise measure.
   a. 4.7 m − 3.62 m
   b. 5.765 sec − 5.73 sec

Check your answers on page 415.

Multiplying by Powers of 10

These are some powers of 10.

$10 * 10 * 10 * 10$	$10 * 10 * 10$	$10 * 10$	10	1	$\cdot$	$\frac{1}{10}$	$\frac{1}{10} * \frac{1}{10}$	$\frac{1}{10} * \frac{1}{10} * \frac{1}{10}$	$\frac{1}{10} * \frac{1}{10} * \frac{1}{10} * \frac{1}{10}$
10,000	1,000	100	10	1	$\cdot$	0.1	0.01	0.001	0.0001

Multiplying decimals by a power of 10 that is greater than 1 is easy. One way is to use partial-products multiplication.

Example Solve $1,000 * 45.6$ by partial-products multiplication.

Step 1: Solve the problem as if there were no decimal point.

Step 2: To place the decimal point where it belongs, estimate the answer to $1,000 * 45.6$.

$1,000 * 45 = 45,000$, so $1,000 * 45.6$ must be near 45,000.

So, the answer to $1,000 * 45.6$ is 45,600.

```
              1000
         *     456
400 * 1000 →  400000
 50 * 1000 →   50000
  6 * 1000 →    6000
              ------
              456000
```

Here is another method. This works for any power of 10.

Examples

$1,000 * 45.6 = ?$ | $0.001 * 45.6 = ?$

Step 1: Locate the decimal point in the power of 10.

$1,000 = 1000.$ | 0.001

Step 2: Move the decimal point LEFT or RIGHT until it is right of the number 1.

$1.000.$ | $0.001.$

Step 3: Count the number of places you moved the decimal point.

3 places to the left | 3 places to the right

†Step 4: Move the decimal point in the other factor the same number of places but in the OPPOSITE direction. Insert 0s as needed.

$45.600.$ | $0.045.6$

$1,000 * 45.6 = 45,600$ | $0.001 * 45.6 = 0.0456$

† To decide whether to move the decimal point to the right or left, think: "Should the answer be *greater than* or *less than* the decimal I started with?" If the answer should be greater, move the decimal point to the right. If the answer should be less, move the decimal point to the left.

Some Powers of 10

10^4	10^3	10^2	10^1	10^0	.	10^{-1}	10^{-2}	10^{-3}	10^{-4}
10,000	1,000	100	10	1	.	0.1	0.01	0.001	0.0001

Here is another method for multiplying by a power of 10.

Examples

$1,000 * 45.6 = ?$ | $0.001 * 45.6 = ?$

Step 1: Think of the power of 10 in exponential notation.

$1,000 = 10^3$ | $0.001 = 10^{-3}$

Step 2: Note the number in the exponent.

$10^{\overset{\uparrow}{3}}$ | $10^{\overset{\uparrow}{-3}}$

†Step 3: If the exponent is POSITIVE, move the decimal point in the other factor that number of places to the RIGHT. Insert 0s as needed.

4 5.6 0 0.

If the exponent is NEGATIVE, move the decimal point in the other factor that number of places to the LEFT. Insert 0s as needed.

0.0 4 5.6

$1,000 * 45.6 = 45,600$ | $0.001 * 45.6 = 0.0456$

† To decide whether to move the decimal point to the right or left, think: "Should the answer be *greater than* or *less than* the decimal I started with?" If the answer should be greater, move the decimal point to the right. If the answer should be less, move the decimal point to the left.

Check Your Understanding

Multiply.

1. $100 * 4.56$
2. $0.01 * 4.56$
3. $0.26 * 10,000$
4. $4.07 * 0.1$

5. $0.44 * 0.001$
6. $1,000 * \$7.50$
7. $1.01 * 10$
8. $0.01 * 45.3$

Check your answers on page 415.

Multiplication of Decimals

You can use the same procedures for multiplying decimals as you use for whole numbers. The main difference is that with decimals you have to decide where to place the decimal point in the product.

Here is one way to multiply with decimals.

Step 1: Make a magnitude estimate of the product.

Step 2: Multiply the factors as if they were whole numbers, ignoring any decimal points. Use a multiplication method you would use for whole numbers. The answer will be a whole number.

Step 3: Use your estimate of the product from Step 1 to place the decimal point in the answer.

Note

A *magnitude estimate* is a very rough estimate that answers questions like: *Is the solution in the ones? Tens? Hundreds? Thousands?*

Example $16.3 * 4.7 = ?$

Step 1: Estimate the product.
Round 16.3 to 16 and 4.7 to 5.
Since $16 * 5 = 80$, the product will be in the tens.
(*In the tens* means between 10 and 100.)

Step 2: Multiply. Ignore the decimal points.

$$
\begin{array}{rcr}
 & & 163 \\
 & & * \ 47 \\
\hline
40 * 100 & \rightarrow & 4000 \\
40 * 60 & \rightarrow & 2400 \\
40 * 3 & \rightarrow & 120 \\
7 * 100 & \rightarrow & 700 \\
7 * 60 & \rightarrow & 420 \\
7 * 3 & \rightarrow & + \ 21 \\
\hline
 & & 7661 \\
\end{array}
$$

Step 3: Use the estimate to place the decimal point in the product.

Since the magnitude estimate is in the tens, the product must be in the tens. Place the decimal point between the 6s in 7661.

So, $16.3 * 4.7 = 76.61$.

Did You Know?

In 1674, Gottfried Leibniz invented one of the first mechanical calculating machines that could multiply numbers. None of these machines have survived. The photograph below shows a reproduction that was built using Leibniz's plans.

Example $3.27 * 0.8 = ?$

Step 1: Make a magnitude estimate.
Round 3.27 to 3 and 0.8 to 1.
Since $3 * 1 = 3$, the product will be in the ones.
(*In the ones* means between 1 and 10.)

Step 2: Multiply as you would with whole numbers.
Ignore the decimal points. $327 * 8 = 2616$

Step 3: Place the decimal point correctly in the answer.

Since the magnitude estimate is in the ones, the
product must be in the ones. Place the decimal
point between the 2 and the 6 in 2616.

So, $3.27 * 0.8 = 2.616$.

Note

A magnitude estimate
may be on the "borderline"
and you need to be
more careful.

For example, a magnitude
estimate for 3.4 * 3.4 is
$3 * 3 = 9$. This estimate
is "in the 1s." But 9 is
close to 10, so that the
exact answer may be
"in the 10s." You should
place the decimal point
so that the answer is
close to 10.

Since $34 * 34 = 1156$,
the exact answer must be
$3.4 * 3.4 = 11.56$.

There is another way to find where to place the decimal point in
the product. This method is especially useful when the factors
are less than 1 and have many decimal places.

Example $0.05 * 0.0062 = ?$

Step 1: Count the decimal places to the right
of the decimal point in each factor.

    2 decimal places in 0.05
    4 decimal places in 0.0062

Step 2: Add the number of decimal places.
    $2 + 4 = 6$

Step 3: Multiply the factors as if they
were whole numbers.
    $5 * 62 = 310$

Step 4: Start at the right of the product. Move the
decimal point LEFT by the number of
decimal places found in Step 2.

    $0.000310.$

Note that when these two numbers are multiplied as if they were
whole numbers, there are only 3 digits in the product ($5 * 62 = 310$).
It is necessary to insert 3 zeros in front of 310 in order to move the
decimal point 6 times.

$0.05 * 0.0062 = 0.000310$

Check Your Understanding

Multiply.

1. $2.8 * 4.6$ **2.** $3.05 * 6.5$ **3.** $0.52 * 3.03$ **4.** $0.2 * 0.022$

Check your answers on page 415.

Lattice Multiplication with Decimals

Lattice multiplication can be used to multiply decimals.

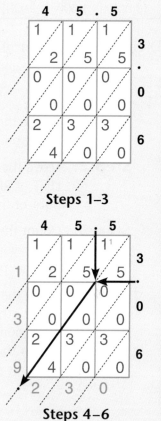

Example Find 45.5 * 3.06 using lattice multiplication.

Step 1: Estimate the answer. 45.5 * 3.06 ≈ 45 * 3 = 135
(The symbol ≈ means *is about equal to.*)

Step 2: Draw the lattice and write the factors, including the decimal points, at the top and right side. In the factor above the grid, the decimal point should be above a column line. In the factor on the right side of the grid, the decimal point should be to the right of a row line.

Step 3: Find the products inside the lattice.

Step 4: Add along the diagonals, moving from right to left.

Step 5: Locate the decimal point in the answer as follows. Slide the decimal point in the factor above the grid down along the column line. Slide the decimal point in the factor on the right side of the grid across the row line. When the decimal points meet, slide the decimal point down along the diagonal line. Write a decimal point at the end of the diagonal line.

Step 6: Compare the result with the estimate from Step 1.

The product, 139.230, is very close to the estimate of 135.

Steps 1–3

Steps 4–6

Example Find 84.5 * 11.6 using lattice multiplication.

A good magnitude estimate is 84.5 * 11.6 ≈ 85 * 10 = 850.

The answer to 84.5 * 11.6 should be near 850 (in the hundreds).

The product, 980.20, and the magnitude estimate, 850 are both in the hundreds.

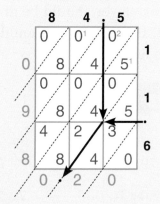

Did You Know?

The lattice method of multiplication was used by Persian scholars as long ago as the year 1010. It was often called the "grating" method.

Check Your Understanding

Draw a lattice for each problem and multiply.

1. 16.5 * 4.5   **2.** 4.03 * 17   **3.** 8.3 * 34.1

Check your answers on page 416.

Dividing by Powers of 10

Some Powers of 10

10^4	10^3	10^2	10^1	10^0	.	10^{-1}	10^{-2}	10^{-3}	10^{-4}
10,000	1,000	100	10	1	.	0.1	0.01	0.001	0.0001

Dividing by a power of 10 is easy. Here is one method.

Examples

$45.6 / 1,000 = ?$

$45.6 / 0.001 = ?$

Step 1: Locate the decimal point in the power of 10.

$1,000 = 1000.$
↑

0.001
↑

Step 2: Move the decimal point LEFT or RIGHT until it is right of the number 1.

$1.000.$

$0.001.$

Step 3: Count the number of places you moved the decimal point.

3 places to the left

3 places to the right

†**Step 4:** Move the decimal point in the other factor the same number of places and in the SAME direction. Insert 0s as needed.

$0.045.6$

$45.600.$

$45.6 / 1,000 = 0.0456$

$45.6 / 0.001 = 45,600$

† To decide whether to move the decimal point to the right or left, think: "Should the answer be *greater than* or *less than* the decimal I started with?" If the answer should be greater, move the decimal point to the right. If the answer should be less, move the decimal point to the left.

Check Your Understanding

Divide.

1. $79.8 / 10$
2. $79.8 / 0.1$
3. $0.78 / 100$
4. $\$360 / 1,000$
5. $6.93 / 0.001$
6. $80 / 10,000$
7. $26.5 / 0.01$
8. $0.03 / 0.001$

Check your answers on page 416.

Here is another method for dividing by a power of 10.

Examples

$45.6 / 1{,}000 = ?$ | $45.6 / 0.001 = ?$

Step 1: Think of the power of 10 in exponential notation.

$1{,}000 = 10^3$ | $0.001 = 10^{-3}$

Step 2: Note the number in the exponent.

10^3 ↑ | 10^{-3} ↑

†Step 3: If the exponent is POSITIVE, move the decimal point in the other factor that number of places to the LEFT. Insert 0s as needed.

0 . 0 4 5 . 6

If the exponent is NEGATIVE, move the decimal point in the other factor that number of places to the RIGHT. Insert 0s as needed.

4 5 . 6 0 0 .

$45.6 / 1{,}000 = 0.0456$ | $45.6 / 0.001 = 45{,}600$

† To decide whether to move the decimal point to the right or left, think: "Should the answer be *greater than* or *less than* the decimal I started with?" If the answer should be greater, move the decimal point to the right. If the answer should be less, move the decimal point to the left.

Check Your Understanding

Divide.

1. 34.5 / 10
2. 34.5 / 0.1
3. 0.13 / 100
4. $760 / 1,000
5. 8.25 / 0.001
6. 50 / 10,000
7. 62.5 / 0.01
8. 0.01 / 0.001
9. 0.001 / 0.01

Check your answers on page 416.

Did You Know?

The slide rule was invented around 1620–1630 by William Oughtred. It may be used to multiply and divide numbers. For most problems, it will give 3-digit answers that are accurate. But the user must make a magnitude estimate to place the decimal point correctly in the answer.

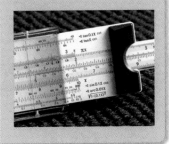

Division of Decimals

The method that is used to divide whole numbers may also be used to divide decimals. However, you must estimate the quotient in order to place the decimal point in the answer.

dividend / divisor = quotient

Step 1: Estimate the quotient.

Step 2: Ignore any decimal points. Divide as if the divisor and dividend were whole numbers.

Step 3: Use your estimate from Step 1 to place the decimal point in the quotient correctly.

Example   97.24 / 23 = ?

Step 1: Estimate the quotient.

97.24 is about 100, and 23 is about 25.
100 / 25 = 4, so 97.24 / 23 should be about 4.

Step 2: Divide, ignoring the decimal point.

If there is a remainder, rewrite the remainder as a fraction. Then round the answer to the nearest whole number.

```
23) 9724
  - 9200    400
    524
  -  460     20
     64
  -   46      2
     18      422
```

$9724 / 23 = 422\frac{18}{23} \approx 423$

Step 3: Use the estimate from Step 1 to place the decimal point.

Since the estimate is that the quotient should be about 4, the decimal point should be placed between the 4 and the 2 in 423.

So, 97.24 / 23 = 4.23.

Note

People sometimes believe that "division makes numbers smaller" and "multiplication makes numbers larger." This is true when you multiply or divide by numbers greater than 1. But it is not true when you multiply or divide by numbers between 0 and 1.

• When a positive number n is multiplied by a number that is less than 1, the product is smaller than n. For example, $8 * 0.5 = 4$.

• When a positive number n is divided by a number between 0 and 1, the quotient will be greater than n. For example, $8 / 0.5 = 16$.

Check Your Understanding

Divide.

1. 148.8 / 6   **2.** 23.21 ÷ 11   **3.** 43.4 / 7   **4.** 666.6 ÷ 15

Check your answers on page 416.

You can use the same method when dividing by a decimal.

Example   8.25 / 0.3 = ?

Step 1: Estimate the quotient.

8.25 is about 8 and 0.3 is about $\frac{1}{3}$.
So, 8.25 / 0.3 is about the same as 8 / $\frac{1}{3}$.

Think: How many many $\frac{1}{3}$s are in 8? If each of
8 pieces is divided into thirds, there will be 24 pieces.
So 8 / $\frac{1}{3}$ = 24. And 8.25 / 0.3 should be about 24.

Step 2: Divide, ignoring the decimal points.

Step 3: Use the estimate from Step 1 to place the decimal point.

Since the estimate was about 24, place the decimal
point between the 7 and the 5 in 275.

$$
\begin{array}{r|r}
3\overline{)\,825} & \\
-\ 600 & 200 \\
\hline
225 & \\
-\ 210 & 70 \\
\hline
15 & \\
-\ 15 & 5 \\
\hline
0 & 275 \\
\end{array}
$$

825 / 3 = 275

So, 8.25 / 0.3 = 27.5.

To rename a fraction as a decimal, you can divide the
numerator by the denominator.

Example   Rename $\frac{3}{4}$ as a decimal.

Step 1: Estimate the quotient. It must be less
than 1 but greater than $\frac{1}{2}$.

Step 2: Rewrite 3 as 3.00, and divide 3.00 / 4.

Step 3: Divide, ignoring the decimal point.

Step 4: Use the estimate from Step 1
to place the decimal point
in the quotient. The quotient
must be 0.75.

$$
\begin{array}{r|r}
4\overline{)\,300} & \\
-\ 200 & 50 \\
\hline
100 & \\
-\ 100 & 25 \\
\hline
0 & 75 \\
\end{array}
$$

300 / 4 = 75

So, $\frac{3}{4}$ = 0.75.

Note

Fractions can be used to
show division problems.
For example, $\frac{3}{4}$ is another
way to write 3 ÷ 4.
So one way to rename
$\frac{3}{4}$ as a decimal is
to divide 3 by 4.

Check Your Understanding

Divide.

1. 8.4 / 2.1   **2.** 6.72 / 2.1   **3.** 47.3 / 0.1   **4.** 3 / 8

Check your answers on page 416.

Division with Many Decimal Places

Sometimes division problems involve decimals with many decimal places. Estimating the quotient is difficult. In such cases, find an equivalent division problem that is easier to solve.

Step 1: Think about the division problem as a fraction.

Step 2: Use the multiplication rule to find an equivalent fraction with no decimals.

Step 3: Think of this equivalent fraction as a division problem.

Step 4: Solve the equivalent problem using partial-quotients division or another method.

The answer to the original problem is the same as the answer to the equivalent problem.

> **Note**
>
> Fractions can be used to show division problems. The fraction $\frac{a}{b}$ is another way of saying a divided by b, $a \div b$, or a / b.

Example $2.05 / 0.004 = ?$

Step 1: Think of the division problem as a fraction.

$$2.05 / 0.004 = \frac{2.05}{0.004}$$

Step 2: Find an equivalent fraction with no decimals.

$$\frac{2.05 * 1,000}{0.004 * 1,000} = \frac{2,050}{4}$$

Step 3: Think of this equivalent fraction as a division problem.

$$\frac{2,050}{4} = 2,050 / 4$$

Step 4: Solve the equivalent problem.

```
  4) 2,050
   - 2,000   | 500
       50
     - 48    |  12
        2      512
```

$$2,050 / 4 = 512\frac{2}{4} = 512\frac{1}{2} = 512.5$$

Since the two fractions, $\frac{2.05}{0.004}$ and $\frac{2,050}{4}$, are equivalent, the answer to $2.05 / 0.004$ is the same as the answer to $2,050 / 4$.

So, $2.05 / 0.004 = 512.5$.

Column Division with Decimal Quotients

Column division can be used to find quotients that have a decimal part. Think of sharing $50 equally among 8 people.

Example $8\overline{)50} = ?$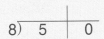

1. Set the problem up. Draw a line to separate the digits in the dividend. Work left to right. Think of the 5 in the tens column as 5 $10 bills.

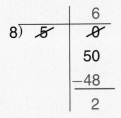

2. The 5 $10 bills can't be equally shared by 8 people. Trade them for 50 $1 bills. Think of the 0 in the ones column as 0 $1 bills. There are 50 + 0, or 50 $1 bills.

3. If 8 people share 50 $1 bills, each person gets 6 $1 bills. There are 2 $1 bills left over.

4. Draw a line and make decimal points to show amounts less than $1. Write 0 after the decimal point in the dividend to show there are 0 dimes. Then trade the 2 $1 bills for 20 dimes.

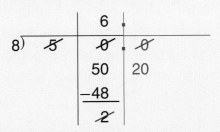

5. If 8 people share 20 dimes, each person gets 2 dimes. There are 4 dimes left over. Draw another line and write another 0 in the dividend to show pennies.

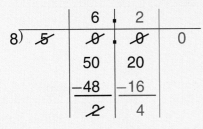

6. Trade the 4 dimes for 40 pennies.

7. If 8 people share 40 pennies, each person gets 5 pennies.

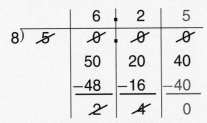

The column division shows that 50 / 8 = 6.25.

This means that $50 shared equally among 8 people is $6.25 each.

Rounding Decimals

Sometimes numbers have more digits than you need. Rounding is a way of adjusting numbers to get rid of unnecessary digits. Rounding also helps in making estimates because it makes numbers easier to use.

Here is one way to round decimals.

Step 1: Find the digit in the place you are rounding to.

Step 2: Rewrite the number, replacing all digits to the right of this digit with zeros. This is the **lower number.**

Step 3: Add 1 to the digit in the place you are rounding to. If the sum is 10, write 0 and add 1 to the digit to its left. This is the **higher number.**

Step 4: Ask, "Is the number I am rounding closer to the lower number or to the higher number?"

Step 5: Round to the closer of the two numbers. If it is halfway between the higher and the lower number, round to the higher number. Drop any 0s that are to the right of the decimal point AND to the right of the place you are rounding to.

Examples Round the decimals to the nearest place.

	2.851 (nearest tenth)	8.35 (nearest tenth)	2.891 (nearest hundredth)
Step 1: Find the place to which you are rounding.	2.**8**51	8.**3**5	2.8**9**1
Step 2: Find the lower number.	2.800	8.30	2.890
Step 3: Find the higher number.	2.900	8.40	2.900
Step 4: Is it closer to the lower or higher number?	higher	halfway	lower
Step 5: Round to the closer number.	2.900 = 2.9	8.40 = 8.4	2.890 = 2.89

Check Your Understanding

Round the numbers below to the nearest tenth, hundredth, and thousandth.

1. 4.6737   **2.** 8.3096   **3.** 0.0529   **4.** 0.0016

Check your answers on page 416.

Percents

A percent is another way to name a fraction or decimal. **Percent** means *per hundred* or *out of a hundred*. So 1% has the same meaning as the fraction $\frac{1}{100}$ and the decimal 0.01. And 40% has the same meaning as $\frac{40}{100}$ and 0.40.

The statement "40% of students were absent" means that 40 out of 100 students were absent. This does not mean that there were exactly 100 students and that 40 of them were absent. It does mean that for every 100 students, 40 students were absent.

A percent usually represents a percent of something. The "something" is the whole (or ONE, or 100%). In the statement, "40% of the students were absent," the whole is the total number of students enrolled in the school.

Note

The word *percent* comes from the Latin *per centum*: *Per* means *for* and *centum* means *one hundred*.

Example There are 350 students enrolled in Clissold School. One day, 40% of the students were absent. How many students were absent that day?

Think: 350 = 100 + 100 + 100 + 50

For every 100 students, 40 were absent.

So, for every 50 students ($\frac{1}{2}$ of 100), 20 were absent ($\frac{1}{2}$ of 40).

350 =	100	+	100	+	100	+	50
	↓		↓		↓		↓
	40 absent		40 absent		40 absent		20 absent

40 + 40 + 40 + 20 = 140 students were absent that day.

Percents are used in many ways in everyday life:

♦ *Business:* "75% off" means that the price of an item will be reduced by 75 cents for every 100 cents the item usually costs.

**SALE – 75% OFF
Everything Must Go**

♦ *Statistics:* "45% voter turnout" means that 45 out of every 100 registered voters actually voted.

Voter Turnout Pegged at 45% of Registered Voters

♦ *School:* A 90% score on a spelling test means that a student scored 90 out of 100 possible points for that test. One way to score 90% is to spell 90 words correctly out of 100. Another way to score 90% is to spell 9 words correctly out of 10.

♦ *Probability:* A "20% chance of snow" means that for every 100 days that have similar weather conditions, you would expect it to snow on 20 of the days.

For Tuesday, there is a 20% chance of snow.

Fractions, Decimals, and Percents

Percents may be used to rename both fractions and decimals.

Percents are another way of naming fractions with a denominator of 100.

> You can think of the fraction $\frac{50}{100}$ as 50 parts per hundred, or 50 out of 100, and write 50%.
>
> You can rename the fraction $\frac{1}{4}$ as $\frac{1 * 25}{4 * 25}$, or $\frac{25}{100}$, or 25%.
>
> 20% can be written as $\frac{20}{100}$, or $\frac{1}{5}$.

Percents are another way of naming decimals in terms of hundredths.

> Since 0.01 can be written as $\frac{1}{100}$, you can think of 0.01 as 1%.
>
> And you can think of 0.62 as $\frac{62}{100}$, or 62%.
>
> 47% means 47 hundredths, or 0.47.

Percents can also be used to name the whole.

> 100 out of 100 can be written as the fraction $\frac{100}{100}$, or 100 hundredths. This is the same as 1 whole, or 100%.

Examples

$65\% = \frac{65}{100} = 0.65$ $\qquad$ $90\% = \frac{90}{100} = 0.9$

$300\% = \frac{300}{100} = 3$ $\qquad$ $62.5\% = \frac{62.5}{100} = 0.625$

Example   The amounts shown in the pictures below can be written as $\frac{3}{4}$, or 75%, or 0.75.

$\frac{3}{4} = \frac{3 * 25}{4 * 25} = \frac{75}{100}$

But $\frac{75}{100} = 0.75$. And $\frac{75}{100} = 75\%$.

So, $\frac{3}{4}$ and $\frac{75}{100}$ and 0.75 and 75% all name the same amount.

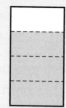

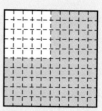

$\frac{3}{4} = 75\% = 0.75$

Finding a Percent of a Number

Finding a percent of a number is a basic problem that comes up over and over again. For example:

♦ A backpack that regularly sells for $70 is on sale for 30% off. What is the sale price?

♦ The sales tax on food is 5%. What is the tax on $90 worth of groceries?

♦ A borrower pays 10% interest on a car loan. If the loan is $8,000, how much is the interest?

There are many ways to find the percent of a number.

Use a Fraction

Some percents are equivalent to "easy" fractions. For example, 25% is the same as $\frac{25}{100}$, or $\frac{1}{4}$. It is usually easier to find 25% of a number by thinking of 25% as $\frac{1}{4}$.

Some "easy" fractions and percents		
$\frac{1}{2}$ =	$\frac{50}{100}$ =	50%
$\frac{1}{4}$ =	$\frac{25}{100}$ =	25%
$\frac{3}{4}$ =	$\frac{75}{100}$ =	75%
$\frac{1}{5}$ =	$\frac{20}{100}$ =	20%
$\frac{2}{5}$ =	$\frac{40}{100}$ =	40%
$\frac{3}{5}$ =	$\frac{60}{100}$ =	60%
$\frac{4}{5}$ =	$\frac{80}{100}$ =	80%
$\frac{1}{10}$ =	$\frac{10}{100}$ =	10%
$\frac{3}{10}$ =	$\frac{30}{100}$ =	30%
$\frac{7}{10}$ =	$\frac{70}{100}$ =	70%
$\frac{9}{10}$ =	$\frac{90}{100}$ =	90%

Example What is 25% of 64?

Think: 25% = $\frac{1}{4}$, so 25% of 64 is the same as $\frac{1}{4}$ of 64.

Divide 64 into 4 equal groups. Each group is $\frac{1}{4}$ of 64, and each group has 16.

So, 25% of 64 is 16.

Use Decimal Multiplication

Finding a percent of a number is the same as multiplying the number by the percent. Usually, it's easiest to change the percent to a decimal and use a calculator.

1% of 55 means $\frac{1}{100}$ * 55 or 0.01 * 55.

9% of 55 means $\frac{9}{100}$ * 55 or 0.09 * 55.

35% of 55 means $\frac{35}{100}$ * 55 or 0.35 * 55.

The word "of" in problems like these means multiplication.

Example What is 35% of 65?

35% of 65 is the same as 0.35 * 65.
Using a caluclator, we find 0.35 * 65 = 22.75.

So, 35% of 65 is 22.75.

Check Your Understanding

Solve.

1. The sales tax on food is 4%. What is the tax on $85 worth of groceries?

2. A backpack that regularly sells for $55 is on sale for 25% off. What is the savings? What is the sale price?

Check your answers on page 416.

Unit Percents

Unit percent is another name for 1%. And $1\% = \frac{1}{100} = 0.01$.

Finding a Percent of a Number

Example What is 7% of 400?

$1\% = \frac{1}{100}$, so 1% of 400 is the same as $\frac{1}{100}$ of 400.

If you divide 400 into 100 equal groups, there are 4 in each group.

So, 1% of 400 is 4. Then 7% of 400 is 7 * 4, or 28.

So, 7% of 400 = 28.

Sometimes it is helpful to find 10% first.

Example What is 30% of 70?

$10\% = \frac{10}{100} = \frac{1}{10}$. 10% of 70 is $\frac{1}{10}$ of 70.

If you divide 70 into 10 equal groups, each group has 7.
So, 10% of 70 is 7. Then 30% of 70 is 3 * 7, or 21.

So, 30% of 70 = 21.

Finding the Whole

Unit percents are used in solving problems in which part of the whole is given as a percent and you need to find the whole.

Example Ms. Partee spends $1,000 a month. $1,000 is 80% of her monthly earnings. How much does she earn per month?

Step 1: Find 1% of her monthly earnings.

$1,000 is 80% of her monthly earnings. Therefore, to find 1% of her earnings, divide $1,000 by 80. $1,000 / 80 = $12.50

So, $12.50 is 1% of her monthly earnings.

Step 2: Find her total monthly earnings. To find her total monthly earnings (or 100% of her monthly earnings), multiply $12.50 by 100. $12.50 * 100 = $1,250

So, Ms. Partee earns $1,250 per month.

Check Your Understanding

All bicycles at Art's Cycle Shop are on sale at 60% of the regular price. If the sale price of a bicycle is $150, how much did the bicycle cost before it was put on sale?

Check your answer on page 416.

Using Proportions to Solve Percent Problems

Many percent problems can be solved using proportions. Using proportions is not always the best approach, but it almost always works.

To solve a percent problem with proportions, start with this number model:

$$\frac{\text{part}}{\text{whole}} = \frac{\text{percent}}{100}$$

Often you can find two of the three unknowns in this number model right away, usually just by reading the problem. Then you can use what you know about solving proportions to find the third unknown.

Finding the Percent

> **Example** Jennifer Azzi made 30 of 58 three-point shots in the 1999 WNBA season. What was her three-point shooting percentage?
>
> **Step 1:** Write a proportion. Find two of the three unknowns by reading the problem.
>
> $$\frac{\text{part}}{\text{whole}} = \frac{\text{percent}}{100}$$
>
> $$\frac{\text{shots made}}{\text{shots attempted}} = \frac{\text{shooting percentage}}{100}$$
>
> $$\frac{30}{58} = \frac{\text{shooting percentage}}{100}$$
>
> **Step 2:** Find the cross products and set them equal.
>
> shooting percentage $* 58 = 30 * 100$
>
> **Step 3:** Solve the equation. Divide both sides by 58.
>
> shooting percentage $* 58 / 58 = 30 * 100 / 58$
>
> shooting percentage $= 30 * 100 / 58 = 51.7$
>
> So, Azzi made 51.7% of her three-point shots.

Note

If $\frac{x}{y} = \frac{a}{b}$ is a proportion, then the cross products are equal.

That is,
$x * b = a * y.$

You can also use the three steps shown above to solve percent problems in which you have to find the whole or the part. You will find examples of these problems on the following page.

Finding the Whole

Example Regina bought a CD on sale for $9. This was 25% off the everyday price. What was the everyday price for the CD?

Since Regina got 25% off, she paid only 75% of the everyday price.

$$\frac{\text{part}}{\text{whole}} = \frac{\text{percent}}{100}$$

$$\frac{\text{sale price}}{\text{everyday price}} = \frac{\text{percent paid}}{100}$$

$$\frac{\$9}{\text{everyday price}} = \frac{75}{100}$$

everyday price * 75 = $9 * 100   (Cross products are equal.)

everyday price * 75 / 75 = $9 * 100 / 75

everyday price = $9 * 100 / 75 = $12

So, the everyday price of the CD was $12.

Finding the Part

Example Dana Barros made 40% of his 160 three-point attempts in the 1998–1999 NBA season. How many three-point shots did he make?

$$\frac{\text{part}}{\text{whole}} = \frac{\text{percent}}{100}$$

$$\frac{\text{shots made}}{\text{shots attempted}} = \frac{\text{shooting percentage}}{100}$$

$$\frac{\text{shots made}}{160} = \frac{40}{100}$$

shots made * 100 = 160 * 40   (Cross products are equal.)

shots made * 100 / 100 = 160 * 40 / 100

shots made = 160 * 40 / 100 = 64

So, Barros made 64 three-point shots.

Check Your Understanding

Use proportions to solve these problems.

1. A dress is on sale for 20% off. The everyday price is $55. What is the sale price?

2. Ukari Figgs's free-throw shooting percentage for the 1999 WNBA season was 87.5%. She attempted 24 free throws. How many did she make?

Check your answers on page 416.

Renaming Fractions as Decimals

Any fraction can always be renamed as a decimal. Sometimes the decimal will end after a certain number of places. Decimals that end are called **terminating decimals.** Sometimes the decimal will have one or more digits that repeat in a pattern forever. Decimals that repeat in this way are called **repeating decimals.** The fraction $\frac{1}{2}$ is equal to the terminating decimal 0.5. The fraction $\frac{2}{3}$ is equal to the repeating decimal 0.6666....

One way to rename a fraction as a decimal is to remember the decimal name: $\frac{1}{2} = 0.5$, $\frac{3}{4} = 0.75$, $\frac{1}{8} = 0.125$, and so on. If you have memorized the decimal names for a few common fractions, then logical thinking can help you to rename many other common fractions as decimals. For example, if you know $\frac{1}{8} = 0.125$, then $\frac{3}{8} = 0.125 + 0.125 + 0.125 = 0.375$.

You can also find decimal names for fractions by using equivalent fractions, the Fraction-Stick Chart, division, or a calculator.

Using Equivalent Fractions

One way to rename a fraction as a decimal is to find an equivalent fraction with a denominator that is a power of 10, such as 10, 100, or 1,000. This method only works for some fractions.

Example   Rename $\frac{3}{5}$ as a decimal.

The solid lines divide the square into 5 equal parts. Each part is $\frac{1}{5}$ of the square. $\frac{3}{5}$ of the square is shaded.

The dashed lines divide each fifth into 2 equal parts. Each part is $\frac{1}{10}$, or 0.1, of the square. $\frac{6}{10}$, or 0.6, of the square is shaded.

$$\frac{3}{5} = \frac{6}{10} = 0.6$$

$\frac{1}{10} = 0.1$

Check Your Understanding

Rename these fractions as decimals.

1. $\frac{3}{4}$ **2.** $\frac{2}{5}$ **3.** $\frac{3}{2}$ **4.** $\frac{11}{20}$ **5.** $\frac{3}{25}$

Check your answers on page 416.

Using the Fraction-Stick Chart

The Fraction-Stick Chart below, and also on page 373, can be used to rename fractions as decimals. Note that the result is usually only an approximation. You can use division or a calculator to obtain better approximations.

Example Rename $\frac{2}{3}$ as a decimal.

1. Locate $\frac{2}{3}$ on the "thirds" stick.

2. Place one edge of a straightedge at $\frac{2}{3}$.

3. Find where the straightedge crosses the number line.

The straightedge crosses the number line between 0.66 and 0.67.

So, $\frac{2}{3}$ is equal to about 0.66 or 0.67.

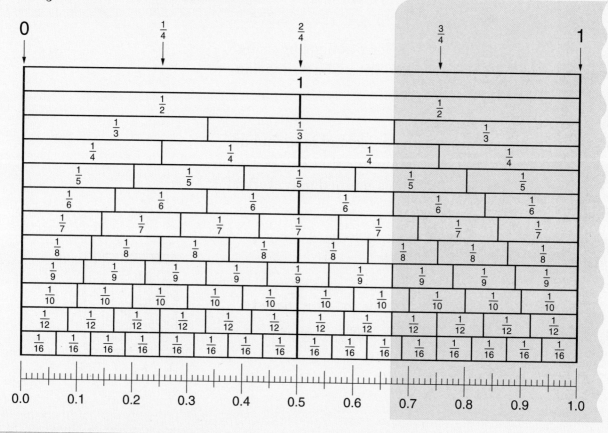

Check Your Understanding

Use the Fraction-Stick Chart to find an approximate decimal name for each fraction or mixed number.

1. $\frac{7}{10}$

2. $\frac{5}{8}$

3. $3\frac{1}{3}$

4. $\frac{9}{12}$

5. $1\frac{4}{9}$

6. $\frac{3}{7}$

Check your answers on page 416.

Using Division

The following examples illustrate how to rename a fraction as a decimal by dividing its numerator by its denominator.

Note

Fractions can be used to show division problems. For example, $\frac{7}{8}$ is another way to write $7 \div 8$. So, one way to rename $\frac{7}{8}$ as a decimal is to divide 7 by 8.

This method will *always* work. Any fraction can be renamed as a decimal by dividing its numerator by its denominator.

Example   Use partial-quotients division to rename $\frac{7}{8}$ as a decimal.

Step 1: Estimate the quotient. It will be less than 1 but greater than $\frac{1}{2}$.

Step 2: Decide how many digits you want to the right of the decimal point. For measuring or solving everyday problems, two or three digits are usually enough.

In this case, rename $\frac{7}{8}$ as a decimal with three digits to the right of the decimal point.

Step 3: Rewrite the numerator with a 0 for each decimal place you want. Rewrite the numerator, 7, as 7.000.

Step 4: Use partial-quotients division to divide 7.000 by 8. Ignore the decimal point for now, and divide 7000 by 8.

```
  8)7000
  - 6400    800
    600
  -  560     70
     40
  -   40      5
      0     875
```

Step 5: Use the estimate from Step 1 to place the decimal point in the quotient.

Since $\frac{7}{8}$ is between $\frac{1}{2}$ and 1, the decimal point should be placed before the 8, to give 0.875.

So, $\frac{7}{8} = 0.875$.

In this case, there was no remainder and the answer worked out to exactly three decimal places. Sometimes you will find a decimal name that is only approximately equal to the fraction.

Example Use partial-quotients division to rename $\frac{2}{3}$ as a decimal.

Step 1: Estimate the quotient. It will be less than 1 but greater than $\frac{1}{2}$.

Step 2: Decide how many digits you want to the right of the decimal point. In this example, rename $\frac{2}{3}$ as a decimal with four digits to the right of the decimal point.

Step 3: Rewrite the numerator with a 0 for each decimal place you want. Rewrite the numerator, 2, as 2.0000.

Step 4: Use partial-quotients division to divide 2.0000 by 3. Ignore the decimal point for now, and divide 20000 by 3.

```
3)20000
 - 18000    6000
   2000
 - 1800      600
    200
 - 180        60
     20
 - 18          6
      2     6666
```

Division shows that 20000 / 3 = $6666\frac{2}{3}$, which rounds to 6667.

Step 5: Use the estimate from Step 1 to place the decimal point in the quotient.

Since $\frac{2}{3}$ is between $\frac{1}{2}$ and 1, the decimal point should be placed before the first 6, to give 0.6667.

So, $\frac{2}{3} \approx 0.6667$. (The symbol $\approx$ means *is about equal to*.)

Check Your Understanding

Use partial-quotients division to find decimal equivalents for these fractions.

1. $\frac{3}{8}$ (to 3 decimal places)   2. $\frac{5}{6}$ (to 3 decimal places)   3. $\frac{7}{9}$ (to 4 decimal places)

Check your answers on page 416.

Example Use column division to rename $\frac{5}{8}$ as a decimal.

Step 1: Write $\frac{5}{8}$ as a division problem. Draw a line and make decimal points to show amounts smaller than 1. Write 0 in the first decimal place in the dividend to show there are 0 tenths.

8) 5 . 0

Step 2: Since 5 ones cannot be equally shared 8 ways, trade the 5 ones for 50 tenths. Share the 50 tenths 8 ways. Each share is 6 tenths. There are 2 tenths left over.

```
        0  |   6
8) 5  .      0
              50
            -48
              2
```

Step 3: Draw a line to show amounts smaller than 1 tenth. Write 0 to show there are no hundredths. Trade the 2 tenths for 20 hundredths. Share the 20 hundredths 8 ways. Each share is 2 hundredths. There are 4 hundredths left over.

```
      0  |   6  |   2
8) 5  .      0      0
            50     20
           -48    -16
             2      4
```

Step 4: Draw another line and write another 0. Trade the 4 hundredths for 40 thousandths. Share the 40 thousandths 8 ways. Each share is 5 thousandths.

```
      0  |   6  |   2  |   5
8) 5  .      0      0      0
            50     20     40
           -48    -16    -40
             2      4      0
```

The answer works out exactly.

$\frac{5}{8} = 0.625$

The digits in the decimal name for a fraction may repeat.

Example Rename $\frac{2}{11}$ as a decimal.

The column-division method keeps repeating.

The digits 1 and 8 will repeat forever.

```
        0 |  1 |  8 |  1 |  8 |  1 |  8
11) 2 .     0    0    0    0    0    0
            20   90   20   90   20   90
           -11  -88  -11  -88  -11  -88
            9    2    9    2    9    2
```

The decimal name for $\frac{2}{11}$ can be written 0.18181818..., or as $0.\overline{18}$.

Check Your Understanding

Use column division to find the decimal name for each fraction.

1. $\frac{3}{8}$

2. $\frac{5}{6}$

3. $\frac{5}{9}$

Check your answers on page 416.

Using a Calculator

You can also rename a fraction as a decimal by dividing the numerator by the denominator using a calculator.

Note

On some calculators, the final digit for a repeating decimal may not follow the pattern. For example, a calculator may show $\frac{2}{3}$ = 2 / 3 = 0.6666666667. The digit 6 really does repeat forever, but this calculator has rounded the final digit.

Examples Rename $\frac{3}{4}$ and $\frac{7}{8}$ as decimals.

Key in: 3 ÷ 4 =

Answer: 0.75

$\frac{3}{4} = 0.75$

Key in: 7 ÷ 8 =

Answer: 0.875

$\frac{7}{8} = 0.875$

In some cases, the decimal takes up the entire calculator display. If one or more digits repeat, the decimal can be written by writing the repeating digit or digits just once, and putting a bar above the digit or digits that repeat.

Examples

Fraction	Key in:	Calculator Display	Answer
$\frac{1}{3}$	1 ÷ 3 =	0.3333333	$0.\overline{3}$
$\frac{2}{3}$	2 ÷ 3 =	0.6666666 or 0.6666666667 (depending on the calculator)	$0.\overline{6}$
$\frac{1}{6}$	1 ÷ 6 =	0.1666666 or 0.1666666667 (depending on the calculator)	$0.1\overline{6}$
$\frac{4}{9}$	4 ÷ 9 =	0.4444444	$0.\overline{4}$
$\frac{6}{11}$	6 ÷ 11 =	0.5454545 or 0.5454545455 (depending on the calculator)	$0.\overline{54}$
$\frac{7}{12}$	7 ÷ 12 =	0.5833333	$0.58\overline{3}$

Some calculators have special keys for entering fractions and renaming them as decimals. For example, to rename $\frac{3}{5}$, you could key in 3 Ⓝ 5 Ⓓ (Enter) (F↔D) on Calculator A, or 3 (b/c) 5 = (F↔D) on Calculator B. The result would be the same as if you had divided the numerator by the denominator.

Check Your Understanding

Use a calculator to rename each fraction as a decimal.

1. $\frac{7}{8}$ 2. $\frac{4}{12}$ 3. $\frac{5}{12}$ 4. $\frac{5}{16}$ 5. $\frac{2}{9}$ 6. $\frac{7}{6}$

Check your answers on page 416.

Renaming Fractions, Decimals, and Percents

The previous pages describe how to rename fractions as decimals. Here we discuss how to change decimals to fractions, fractions to percents, and so on.

Renaming Decimals as Fractions

Any terminating decimal can be renamed as a fraction whose denominator is a power of 10. To change a terminating decimal to a fraction, use the place of the rightmost digit to help you write the denominator.

Examples

$0.4 = \frac{4}{10} = \frac{2}{5}$ $0.08 = \frac{8}{100} = \frac{4}{50} = \frac{2}{25}$ $0.25 = \frac{25}{100} = \frac{5}{20} = \frac{1}{4}$ $0.124 = \frac{124}{1,000} = \frac{62}{500} = \frac{31}{250}$

   ↓   ↓   ↓   ↓

tenths place   hundredths place   hundredths place   thousandths place

Some calculators have special keys for renaming decimals as fractions and for simplifying fractions.

Example

To rename 0.48 as a fraction, key in: .48 $=$ $F↔D$. Answer: $\frac{48}{100}$ (or $\frac{12}{25}$)

To simplify the fraction, key in: $SIMP$ $=$ $SIMP$ $=$. Answer: $\frac{12}{25}$

Renaming Fractions as Percents

One way to rename some fractions as percents is to find an equivalent fraction with 100 as the denominator first. Then write the fraction as a percent. For example, $\frac{3}{4} = \frac{75}{100} = 75\%$.

Another way is to divide the numerator by the denominator of the fraction. This renames the fraction as a decimal. Then multiply the result by 100 and write the % symbol. You can do this with paper and pencil or with a calculator.

Example   Use a calculator to rename $\frac{3}{8}$ as a percent.

- Divide 3 by 8. Key in: 3 $÷$ 8 $=$ Answer: 0.375
- Multiply 0.375 by 100. Key in: .375 $×$ 100 $=$ Answer: 37.5

So, $\frac{3}{8} = 37.5\%$.

Renaming Percents as Fractions

A percent can always be renamed as a fraction whose denominator is 100. Simply remove the % symbol and write a fraction bar and the denominator 100 below the number. The fraction can be renamed in simplest form if you want.

Examples

$40\% = \frac{40}{100} = \frac{2}{5}$ $85\% = \frac{85}{100} = \frac{17}{20}$

$150\% = \frac{150}{100} = \frac{3}{2} = 1\frac{1}{2}$

Renaming Percents as Decimals

A percent can be renamed as a decimal by first changing it to a fraction whose denominator is 100.

Examples

$75\% = \frac{75}{100} = 0.75$ $300\% = \frac{300}{100} = 3$

$37.5\% = \frac{37.5}{100} = \frac{375}{1,000} = 0.375$

Renaming Decimals as Percents

To rename a decimal as a percent, multiply it by 100 and write the % symbol.

Examples

$0.35 = (0.35 * 100)\% = 35\%$

$1.2 = (1.2 * 100)\% = 120\%$

$0.0675 = (0.0675 * 100)\% = 6.75\%$

Check Your Understanding

1. Rename each fraction or mixed number as a percent.
   a. $\frac{3}{5}$ b. $\frac{9}{10}$ c. $\frac{3}{8}$ d. $1\frac{1}{5}$ e. $1\frac{3}{4}$

2. Rename each percent as a fraction or mixed number.
   a. 80%   b. 15%   c. 125%

3. Rename each decimal as a fraction or mixed number. Check your answers on a calculator.
   a. 0.6   b. 0.45   c. 5.43   d. 1.019

Check your answers on page 416.

U.S. Traditional Addition: Decimals

You can use **U.S. traditional addition** to add decimals.

Did You Know?

When you add decimals, make sure you line up the places properly so that tenths are added to tenths, ones to ones, and so on. *For example:*

$$
\begin{array}{r}
27.4 \\
+\ 5.19 \\
\hline
\end{array}
$$

Example $9.23 + 4.29 = ?$

Step 1: Start with the 0.01s: $3 + 9 + 12$.
12 hundredths = 1 tenth + 2 hundredths

$$
\begin{array}{r}
1 \\
9\ .\ 2\ 3 \\
+\ 4\ .\ 2\ 9 \\
\hline
2
\end{array}
$$

Step 2: Add the 0.1s: $1 + 2 + 2 = 5$.
There are 5 tenths.

$$
\begin{array}{r}
1 \\
9\ .\ 2\ 3 \\
+\ 4\ .\ 2\ 9 \\
\hline
5\ 2
\end{array}
$$

Step 3: Add the 1s: $9 + 4 = 13$.
13 ones = 1 ten + 3 ones
Remember to include the decimal point in the answer.

$$
\begin{array}{r}
1 \\
9\ .\ 2\ 3 \\
+\ 4\ .\ 2\ 9 \\
\hline
1\ 3\ .\ 5\ 2
\end{array}
$$

$9.23 + 4.29 = 13.52$

Check Your Understanding

Add.

1. $9.56 + 4.28 = ?$ **2.** 37.89 **3.** $3.126 + 923.5 = ?$ **4.** $5.095 + 9.12 = ?$
                                   $+\ 6.37$

Check your answers on page 425.

U.S. Traditional Subtraction: Decimals

You can use **U.S. traditional subtraction** to subtract decimals.

Example $6.52 - 3.74 = ?$

Step 1: Start with the 0.01s.
Since 4 > 2, you need to regroup.
Trade 1 tenth for 10 hundredths:
$6.52 = 6$ ones $+ 4$ tenths $+ 12$ hundredths.
Subtract the 0.01s: $12 - 4 = 8$.

$$
\begin{array}{r}
{\scriptstyle 4\ \ 12} \\
6\ .\ \cancel{5}\ \cancel{2} \\
-\ 3\ .\ 7\ 4 \\
\hline
8
\end{array}
$$

Step 2: Go to the 0.1s.
Since 7 > 4, you need to regroup.
Trade 1 one for 10 tenths:
$6.52 = 5$ ones $+ 14$ tenths $+ 12$ hundredths.
Subtract the 0.1s: $14 - 7 = 7$.

$$
\begin{array}{r}
{\scriptstyle 14} \\
{\scriptstyle 5\ \ \cancel{4}\ \ 12} \\
\cancel{6}\ .\ \cancel{5}\ \cancel{2} \\
-\ 3\ .\ 7\ 4 \\
\hline
7\ 8
\end{array}
$$

Step 3: Go to the 1s.
You don't need to regroup.
Subtract the 1s: $5 - 3 = 2$.
Remember to include the decimal point
in the answer.

$$
\begin{array}{r}
{\scriptstyle 14} \\
{\scriptstyle 5\ \ \cancel{4}\ \ 12} \\
\cancel{6}\ .\ \cancel{5}\ \cancel{2} \\
-\ 3\ .\ 7\ 4 \\
\hline
2\ .\ 7\ 8
\end{array}
$$

$6.52 - 3.74 = 2.78$

Check Your Understanding

Subtract.

1. $97.56 - 23.76 = ?$

2.
$$
\begin{array}{r}
3.70 \\
-\ 2.73 \\
\hline
\end{array}
$$

3.
$$
\begin{array}{r}
58.04 \\
-\ 4.82 \\
\hline
\end{array}
$$

4. $5 - 3.75 = ?$

Check your answers on page 425.

U.S. Traditional Multiplication: Decimals

You can also use **U.S. traditional multiplication** with decimals. Here are two ways to use this method for multiplying decimals.

Method 1

One way to use U.S. traditional multiplication to multiply decimals is to multiply them as though they were whole numbers and then use estimation to place the decimal point.

Step 1: Multiply as though both factors were whole numbers.

Step 2: Estimate the product.

Step 3: Use your estimate to place the decimal point in the answer.

Example $6 * 4.79 = ?$

Step 1: Multiply as though the factors were whole numbers.

Step 2: Estimate the product.

$6 * 4.79 \approx 6 * 5 = 30$

$$\begin{array}{r} {\scriptstyle 4\ \ 5} \\ 4\ \ 7\ \ 9 \\ *\ \ \ \ \ \ \ 6 \\ \hline 2\ \ 8\ \ 7\ \ 4 \end{array}$$

Step 3: Use the estimate to place the decimal point in the answer. The estimate is 30, so place the decimal point to make the answer close to 30: 28.74 is close to 30.

$6 * 4.79 = 28.74$

Method 2

The second method for multiplying decimals is useful when there are many decimal places in the factors and it becomes harder to estimate the answer.

Step 1: Multiply as though both factors were whole numbers.

Step 2: Count the total number of places to the RIGHT of the decimal point for both factors.

Step 3: Place the decimal point so that you have the same number of decimal places as the total in Step 2.

Example $0.065 * 0.032$

Step 1: Multiply as though the factors were whole numbers.

Step 2: Count the total number of decimal places in both factors. 0.065 has 3 decimal places; 0.032 has 3 decimal places. There are 6 decimal places in all.

Step 3: Place the decimal point 6 places from the right. Insert 0s as needed.

$$0.002080.$$

$$0.065 * 0.032 = 0.002080$$

```
        1
        1
        3 2
    *   6 5
      1 6 0
  + 1 9 2 0
    2 0 8 0
```

Check Your Understanding

Multiply.

1. $74.5 * 3 = ?$ 2. $4.86 * 0.36 = ?$ 3. $0.073 * 2.07 = ?$ 4. $0.05 * 6.3 = ?$

Check your answers on page 425.

U.S. Traditional Long Division: Decimal Dividends

You can use **U.S. traditional long division** to divide money in dollars-and-cents notation.

Example   Share $5.29 equally among 3 people.

Step 1: Share the dollars.

```
     1          ← Each person gets 1 dollar.
  3)5.29
    −3          ← 1 dollar each for 3 people
     2          ← 2 dollars are left.
```

Step 2: Trade the dollars for dimes. Share the dimes.

```
     1.7        ← Each person gets 7 dimes. Write a decimal point
  3)5.29            to show amounts less than a dollar.
    −3
     2 2        ← 20 dimes + 2 dimes
    −2 1        ← 7 dimes each for 3 people
       1        ← 1 dime is left.
```

Step 3: Trade the dime for pennies. Share the pennies.

```
     1.76       ← Each person gets 6 pennies.
  3)5.29
    −3
     2 2
    −2 1
      19        ← 10 pennies + 9 pennies
     −18        ← 6 pennies each for 3 people
       1        ← 1 penny is left.
```

Each person gets $1.76. There is 1¢ left.

$5.29 / 3 → $1.76 R1¢

Check Your Understanding

Divide.

1. $7.26 / 6   **2.** 7)$8.61   **3.** 7)$5.62   **4.** $8.04 / 3

Check your answers on page 425.

SRB
60E

You can use U.S. long division to divide decimals that do not represent money.

Example 3.97 / 5

Step 1: Trade the ones for tenths and share the tenths.

```
   .7
5)3.97
 − 3 5
    4
```

← Each share gets 7 tenths. Write a decimal point in the quotient.
← 3 ones + 9 tenths = 39 tenths
← 7 tenths * 5 = 35 tenths
← 4 tenths are left.

Step 2: Trade the remaining tenths for hundredths. Share the hundredths.

```
   .79
5)3.97
 − 3 5
    47
  − 45
     2
```

← Each share gets 9 hundredths.

← 4 tenths + 7 hundredths = 47 hundredths
← 9 hundredths * 5 = 45 hundredths
← 2 hundredths are left.

At this point, you can either round 0.79 to 0.8 and write 3.97 / 5 ≈ 0.8, or you can continue dividing into the thousandths.

Step 3: Continue dividing into the thousandths. Add a 0 to the end of 3.97. (Adding 0s or "padding" a decimal with 0s doesn't change its value.)

```
   .794
5)3.970
 − 3 5
    47
  − 45
    20
  − 20
     0
```

← Each share gets 4 thousandths.
← 3.97 = 3.970

← 2 hundredths + 0 thousandths = 20 thousandths
← 4 thousandths * 5 = 20 thousandths
← No thousandths are left.

3.97 / 5 = 0.794

Check Your Understanding

Divide.

1. 8.28 / 4   **2.** 4)9.64   **3.** 6)8.67   **4.** 38.65 / 5

Check your answers on page 425.

SRB
60F

U.S. Traditional Long Division: Decimal Divisors

To use **U.S. traditional long division** to divide by a decimal number, such as 0.6 or 3.5, you can find an equivalent problem that has no decimal in the divisor. The answer to the equivalent problem is the same as the answer to your original problem.

Step 1: Think of the division problem as a fraction.

Step 2: Use the multiplication rule to find an equivalent fraction that has no decimal in the denominator.

Step 3: Think of the equivalent fraction as a division problem.

Step 4: Solve the division problem. The answer to the equivalent problem is the same as the answer to the original problem.

Example $194 / 0.4 = ?$

Step 1: Think of the division problem as a fraction.

$$194 / 0.4 = \frac{194}{0.4}$$

Step 2: Find an equivalent fraction with no decimal in the denominator.

$$\frac{194 * 10}{0.4 * 10} = \frac{1940}{4}$$

Step 3: Think of the equivalent fraction as a division problem.

$$\frac{1940}{4} = 1940 / 4$$

Step 4: Solve the equivalent division problem.

$$
\begin{array}{r}
485 \\
4\overline{)1940} \\
-\,16 \\
\hline
34 \\
-\,32 \\
\hline
20 \\
-\,20 \\
\hline
0
\end{array}
$$

Because $\frac{1940}{4}$ and $\frac{194}{0.4}$ are equivalent fractions, the division problems 1940 / 4 and 194 / 0.4 are equivalent. So the answer to 1940 / 4 is the same as the answer to 194 / 0.4.

$194 / 0.4 = 485$

U.S. Traditional Long Division: Decimal Divisors and Dividends

Sometimes *both* the divisor (the number you are dividing by) and the dividend (the number being divided) are decimal numbers. To use **U.S. traditional long division** in such cases, you can first find an equivalent problem that has no decimal in the divisor. (Having a decimal part in the dividend is okay.) The answer to the equivalent problem is the same as the answer to your original problem.

Example $3.78 / 0.7 = ?$

Step 1: Think of the division problem as a fraction.

$$3.78 / 0.7 = \frac{3.78}{0.7}$$

Step 2: Find an equivalent fraction with no decimal in the denominator.

$$\frac{3.78 * 10}{0.7 * 10} = \frac{37.8}{7}$$

Step 3: Think of the equivalent fraction as a division problem.

$$\frac{37.8}{7} = 37.8 / 7$$

Step 4: Solve the division problem.

$$
\begin{array}{r}
5.4 \\
7\overline{)37.8} \\
-\,35 \\
\hline
2\,8 \\
-\,2\,8 \\
\hline
0
\end{array}
$$

Because $\frac{37.8}{7}$ and $\frac{3.78}{0.7}$ are equivalent fractions, the division problems $37.8 / 7$ and $3.78 / 0.7$ are equivalent. So the answer to $37.8 / 7$ is the same as the answer to $3.78 / 0.7$.

$3.78 / 0.7 = 5.4$

Check Your Understanding

Divide.

1. $784 / 0.7$ **2.** $36.9 / 1.5$ **3.** $4.68 / 0.03$ **4.** $3.05 / 0.005$

Check your answers on page 425.

U.S. Traditional Long Division: Renaming Fractions as Decimals

U.S. traditional long division can be used to rename fractions as decimals.

Example   Use U.S. traditional long division to rename $\frac{3}{8}$ as a decimal.

Step 1: Write $\frac{3}{8}$ as a division problem. Write 3 with several 0s after the decimal point: 3.000. (You can always add more 0s if you need them.)

$$8\overline{)3.000}$$

Step 2: Solve the division problem. Stop when the remainder is 0, or when you have enough precision for your purposes, or when you notice a repeating pattern.

This division problem divided evenly in three decimal places.

$\frac{3}{8} = 0.375$

```
    .375
8)3.000
  -2 4
    60
   -56
    40
   -40
     0
```

Example   Use U.S. traditional long division to rename $\frac{9}{11}$ as a decimal.

Step 1: Write $\frac{9}{11}$ as a division problem. Write 9 with several 0s after the decimal point: 9.000. (You can always add more 0s if you need them.)

$$11\overline{)9.000}$$

Step 2: Solve the division problem. Stop when the remainder is 0, or when you have enough precision for your purposes, or when you notice a repeating pattern.

The digits 8 and 1 in the quotient appear to repeat forever.

$\frac{9}{11} = 0.818181... = 0.\overline{81}$

```
     .818181
11)9.000000
  -8 8
    20
   -11
    90
   -88
    20
   -11
    90
   -88
    20
   -11
     9
```

Check Your Understanding

Use long division to rename these fractions as decimals.

1. $\frac{2}{3}$ 2. $\frac{3}{11}$ 3. $\frac{8}{9}$ 4. $\frac{5}{6}$

Check your answers on page 425.

The U.S. Census

Since 1790, the monumental task of counting and gathering information about every person living in the United States has been undertaken by the government once every ten years. With the growth of the U.S. population and advances in technology, this process, called the **decennial census,** has changed a great deal.

A Constitutional Mandate

The authors of the Constitution wanted each state's number of members in the House of Representatives to be based on its total population. For this reason, a census, or count of the people, was required by the Constitution.

Article 1, Section 2, calls for an "enumeration" of the entire U.S. population every ten years. ➤

We the People of the Un...

The Return for SOUTH CAROLINA having been made since the foregoing Schedule was originally printed, the whole Enumeration is here given complete, except for the N. Western Territory, of which no Return has yet been published.

DISTICTS	Free white Males of 16 years and upwards, including heads of families.	Free white Males under sixteen years.	Free white Females, including heads of families.	All other free persons.	Slaves.	Total.
Vermont	22435	22328	40505	255	16	85539
N. Hampshire	36086	34851	70160	630	158	141885
Maine	24384	24748	46870	538	NONE	96540
Maſsachuſetts	95453	87289	190582	5403	NONE	378787
Rhode Iſland	16019	15799	32652	3407	948	68825
Connecticut	60523	54403	117448	2808	2764	237946
New York	83700	78122	152320	4654	21324	340120
New Jerſey	45251	41416	83287	2762	11423	184139
Pennſylvania	110788	106948	206363	6537	3737	434373
Delaware	11783	12143	22384	3899	8887	59094
Maryland	55915	51339	101395	8043	103036	319728
Virginia	110936	116135	215046	12866	292627	747610
Kentucky	15154	17057	28922	114	12430	73677
N. Carolina	69988	77506	140710	4975	100572	393751
S. Carolina	35576	37722	66880	1801	107094	249073
Georgia	13103	14044	25739	398	29264	82548
	807094	791850	1541263	59150	694280	3893635

Total number of Inhabitants of the United States exclufive of S. Weſtern and N. Territory.	Free white Males of 21 years and up.	Free Males under 21 years of age.	Free white Females.	All other Perſons.	Slaves.	Total
S.W. territory N. Ditto	6271	10277	15365	361	3417	35691

◀ The 1790 census was conducted by 17 U.S. Marshals and hundreds of assistants. They recorded information by hand on paper. It took 18 months to gather data and make this report, which shows a total of about 3.9 million people.

Dealing with Data

From the beginning, the census counted people and gathered basic information, such as names, ages, and genders. In 1820, the government added questions in order to learn more about manufacturing, agriculture, and social issues, such as illiteracy and crime. The task of processing the data became much more complex.

◄ As in this 1870 scene, census-takers, or **enumerators**, hand-wrote answers to the census questions during home visits. This face-to-face method of data collection was not changed until 1960.

In 1880, the population was 50 million, and processing the information took seven years! With the invention of a punchcard tabulating system, the 1890 census was processed in just two and a half years.

▲ A punch-card machine made holes in specific locations on a card to represent each person's information.

A tabulating machine pushed metal pins through the holes in each card, which completed an electrical circuit and caused the dial counters to move. ➤

◄ In 1940, in addition to the decennial census, the U.S. Census Bureau began tracking unemployment through monthly surveys of a representative part of the population. Statisticians used the data to draw conclusions about the whole country. This method, called **sampling,** saved time and money, and led to the development of surveys on over 100 topics.

DEPARTMENT OF COMMERCE—BUREAU OF THE CENSUS

SIXTEENTH CENSUS OF THE UNITED STATES: 1940—HOUSING
OCCUPIED-DWELLING SCHEDULE
(To be used for dwelling units occupied by households enumerated on the Population Schedule)

III. CHARACTERISTICS OF DWELLING UNIT

13	14	15	16	17	18	19	20	21	22
Num-ber of rooms	Water supply	Toilet facilities	Bathtub or shower with running water in structure	Principal lighting equip't	Principal refrig-eration equip't	Radio in dwell-ing unit?	Heating equipment	Principal fuel used for heating	Principal fuel used for cooking

◄ In 1940, a detailed set of questions about housing was permanently added to the census. The government hoped to use this information to improve the standard of living for the population of 132 million Americans.

The Census Bureau helped pay for the development of the first electronic computer for processing data. Starting with the 1950 Census, data from punch cards were transferred to magnetic tape, which enabled the room-sized UNIVAC (Universal Automatic Computer) to tabulate 4,000 items per minute. ➤

More Advances

To take advantage of the ability of computers to count data quickly, there also had to be advances in collecting data and preparing it for processing.

◄ Here, high-speed cameras photograph questionnaires as part of a new system developed in the 1950s. Dots were filled in with pencil to indicate answers, then photographed onto microfilm and "read" by computers.

In 1960, the Census Bureau began using self-enumeration forms, which were mailed to households for people to fill out themselves. Enumerators collected the competed forms. ➤

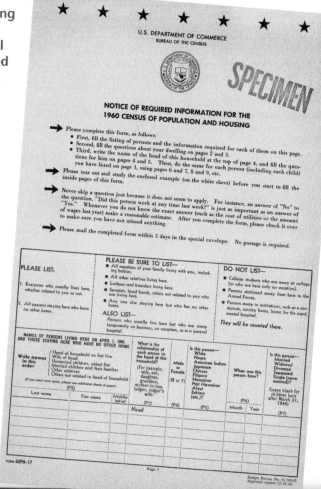

▲ Sampling techniques were also used to improve the efficiency and value of the decennial census. Most of the population received a "short form" questionnaire with 6 or 7 basic questions. About 17% of the population received the more detailed and involved "long form."

◄ Since 1970, the Census Bureau has asked people to return their questionnaires by mail. However, enumerators are still needed to help the Bureau reach as close to 100% of the population as possible.

Enumerators now use handheld and laptop computers to locate addresses and enter data. ➤

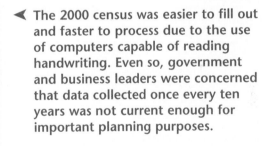

◄ The 2000 census was easier to fill out and faster to process due to the use of computers capable of reading handwriting. Even so, government and business leaders were concerned that data collected once every ten years was not current enough for important planning purposes.

In an effort to have reliable data on a yearly basis, the Census Bureau began testing the new American Community Survey (ACS) in 1996. The ACS uses monthly surveys to provide estimates of the population and its characteristics. If approved by Congress, the ACS will replace the census long form in 2010. ➤

Putting the Data to Use

The U.S. Census Bureau publishes thousands of reports every year, which are used by governments, businesses, journalists, social scientists, non-profit organizations, and ordinary people. Because of the huge amount of information it collects and shares, the Bureau is often called "The Fact-Finder for the Nation."

The number of representatives each state sends to Congress is still determined by census results. Population data are also used to redraw voting district lines every decade so that each person's vote carries equal weight. ➤

◄ Government leaders use census data to plan how to distribute money for schools, libraries, healthcare clinics, and senior centers.
▼

Business planners use census data to decide which communities are good markets for things such as stores, movie theaters, and restaurants.
▼

You can access most of the Census Bureau data on the Internet or at your library. What do you want to know about your community?

Fractions

Fractions

Fractions were invented thousands of years ago to name numbers between whole numbers. People probably first used these in-between numbers for making more exact measurements.

Today most rulers and other measuring tools have marks to name numbers that are between whole measures. Learning how to read these in-between marks is an important part of learning to use these tools. Here are some examples of measurements that use fractions: $\frac{1}{3}$ cup, $\frac{3}{4}$ hour, $\frac{7}{10}$ km, and $24\frac{1}{2}$ lb.

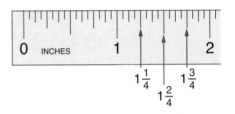

The $\frac{1}{4}$-in. marks between 1 and 2 are labeled.

Fractions are also used to name parts of wholes. The whole might be one single thing, like a stick of butter. Or, the whole might be a collection of things, like a box of crayons. The whole is sometimes called the ONE. In measurements, the whole is called the *unit*.

| Whole |
| 16 Crayons |

The "whole" box names the ONE being considered.

To understand fractions you need to know what the ONE is, or what the unit is. Half a box of crayons might be many crayons or a few crayons, depending on the number of crayons in the box. Half a yard is much less than half a mile.

Fractions are also used to show division, in rates and ratios, and in many other ways.

Naming Fractions

A fraction is written as $\frac{a}{b}$, with two whole numbers that are separated by a fraction bar. The top number is called the **numerator.** The bottom number is called the **denominator.** (The denominator cannot be 0.) In fractions that name parts of wholes, the denominator names the number of equal parts into which the whole is divided. The numerator names the number of parts being considered.

$\frac{a}{b}$ $\leftarrow$ numerator
   $\leftarrow$ denominator

$(b \neq 0)$

A number can be written as a fraction in many ways. Different fractions that name the same number are called **equivalent.** Multiplying or dividing a fraction's numerator and denominator by the same number (except 0) produces an equivalent fraction. All fractions also have decimal and percent names, which can be found from any of its fraction names by dividing the numerator by the denominator.

$\frac{1}{5} = \frac{2}{10} = \frac{3}{15} = 0.2 = 20\%$

$\frac{1}{3} = \frac{2}{6} = \frac{3}{9} = 0.\overline{3} = 33\frac{1}{3}\%$

Fractions have many equivalent names.

Fraction Uses

Parts of Wholes

Fractions are used to name a part of a whole object or a part of a collection of objects.

$\frac{5}{6}$ of the hexagon is blue.

$\frac{6}{10}$ of the dimes are circled.

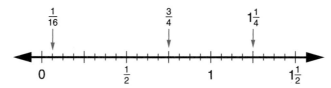

Points on Number Lines

Fractions can name points on a number line that are between points named by whole numbers.

"In-Between" Measures

Fractions can name measures that are between whole-number measures.

Division Notation

The fraction $\frac{a}{b}$ is another way of saying a divided by b.

The division problem 24 divided by 3 can be written in any of these ways: $24 \div 3$, or $3\overline{)24}$, or $24 / 3$, or $\frac{24}{3}$.

$$\frac{a}{b} \qquad a/b$$
$$a \div b \qquad b\overline{)a}$$

Ratios

Fractions are used to compare quantities with the *same unit*.

Gage Park won 8 games and lost 10 games during last year's basketball season. The fraction $\frac{\text{games won}}{\text{games lost}} = \frac{8}{10}$ compares quantities with the same unit (games and games).

We say that the ratio of wins to losses is 8 to 10.

PUBLIC-RED CENTRAL		
	Conf.	Overall
Dunbar	4−0	9−6
King	4−1	14−4
Robeson	3−2	8−9
Gage Park	2−3	8−10
Harper	2−3	8−7
Curie	1−3	7−10
Hubbard	1−4	8−9

Rates

Fractions are used to compare quantities with *different units*.

Luke's car can travel about 36 miles on 1 gallon of gasoline. The fraction $\frac{36 \text{ miles}}{1 \text{ gallon}}$ compares quantities with different units (miles and gallons). At this rate, the car can travel about 324 miles on 9 gallons of gasoline.

$$\frac{36 \text{ miles}}{1 \text{ gallon}} = \frac{324 \text{ miles}}{9 \text{ gallons}}$$

Scales

Fractions are used to compare the size of a drawing or a model to the size of the actual object.

A scale on a map given as 1:10,000 means that every distance on the map is $\frac{1}{10,000}$ of the real-world distance. A 1-inch distance on the map stands for a real-world distance of 10,000 inches, or about 830 feet.

$$\frac{\text{map distance}}{\text{real distance}} = \frac{1 \text{ inch}}{10,000 \text{ inches}}$$

Probabilities

Fractions are a way to describe the chance that an event will happen.

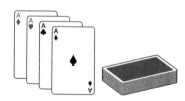

If one card is drawn from a well-shuffled deck of 52 playing cards, the chance of selecting the ace of spades is $\frac{1}{52}$, or about 2%. The chance of drawing any ace is $\frac{4}{52}$, or about 8%.

Other Uses

People use fractions in different ways every day.

A film critic gave the movie $4\frac{1}{2}$ stars out of 5.

A *half-baked idea* is an idea that is not practical or has not been thought out properly.

A shoe size of $6\frac{1}{2}$ is recommended for women whose feet are $9\frac{1}{2}$ inches long.

Renaming a Mixed or Whole Number as a Fraction

Numbers like $2\frac{3}{4}$, $8\frac{4}{5}$, and $1\frac{1}{16}$ are called **mixed numbers.** A mixed number has a whole-number part and a fraction part. In the mixed number $2\frac{3}{4}$, the whole-number part is 2 and the fraction part is $\frac{3}{4}$. A mixed number is equal to the sum of the whole-number part and the fraction part: $2\frac{3}{4} = 2 + \frac{3}{4}$.

$$2\frac{3}{4} \;=\; 2 \text{ wholes} \;+\; \frac{3}{4}$$

A mixed number can be renamed as a fraction. Study the following example and its shortcut.

Did You Know?

If the inside band of a man's hat is rearranged to form a perfect circle, the circle's diameter (in inches) is the hat size. Hat sizes are given as mixed numbers, where the fractions can be renamed as *eighths*. Most men's hat sizes are between $6\frac{1}{2}$ and $7\frac{7}{8}$. The most common size is $7\frac{1}{8}$.

Example Rename $2\frac{3}{4}$ as a fraction. A circle is the ONE.

$$1 \;+\; 1 \;+\; \frac{3}{4} \;=\; \frac{4}{4} \;+\; \frac{4}{4} \;+\; \frac{3}{4}$$

$$2\frac{3}{4} \;=\; \frac{11}{4}$$

Shortcut:

- Multiply the whole-number part, 2, by the denominator of the fraction part, 4.
  $2 * 4 = 8$. This is the number of fourths in 2 wholes: $2 = \frac{8}{4}$.

- Add the numerator of the fraction part, 3, to the result, 8.
  $8 + 3 = 11$. This is the number of fourths in the mixed number $2\frac{3}{4}$.

So, $2\frac{3}{4} = \frac{8}{4} + \frac{3}{4} = \frac{11}{4}$.

Check Your Understanding

Rename as a fraction.

1. $3\frac{1}{3}$ 2. $1\frac{1}{2}$ 3. $2\frac{3}{5}$ 4. $1\frac{5}{6}$ 5. $3\frac{3}{4}$ 6. 5

Check your answers on page 416.

Renaming a Fraction as a Mixed or Whole Number

An **improper fraction** is a fraction that is greater than or equal to 1. Fractions like $\frac{4}{3}$, $\frac{5}{5}$, and $\frac{125}{10}$ are improper fractions. In an improper fraction, the numerator is greater than or equal to the denominator. A **proper fraction** is a fraction that is less than 1. In a proper fraction, the numerator is less than the denominator.

An improper fraction can be renamed as a mixed number or a whole number.

Note

Even though they are called *improper*, there is nothing wrong about improper fractions. Do not avoid them.

Example Rename $\frac{23}{6}$ as a mixed number. A circle is the ONE.

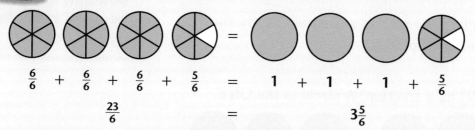

$$\frac{6}{6} + \frac{6}{6} + \frac{6}{6} + \frac{5}{6} = 1 + 1 + 1 + \frac{5}{6}$$

$$\frac{23}{6} = 3\frac{5}{6}$$

Shortcut: Divide the numerator, 23, by the denominator, 6.

- The quotient, 3, is the whole-number part of the mixed number. It tells how many wholes there are in $\frac{23}{6}$.
- The remainder, 5, is the numerator of the fraction part. It tells how many sixths there are left that cannot be made into wholes.

$$\frac{23}{6} = 3\frac{5}{6}$$

$$6)\overline{23}$$
$$\underline{-\ 18}\ \ \big\rfloor\ 3$$
$$5\ \ \ \ 3$$
$$23\ /\ 6 \rightarrow 3\ R5$$

Some calculators have special keys for renaming fractions as whole numbers or mixed numbers.

Example Use a calculator to rename $\frac{23}{6}$ as a mixed number.

On Calculator A: Key in: 23 ⒩ 6 ⒟ Enter U$\frac{n}{d}$↔$\frac{n}{d}$ U$\frac{n}{d}$↔$\frac{n}{d}$ U$\frac{n}{d}$↔$\frac{n}{d}$

On Calculator B: Key in: 23 b/c 6 = $\frac{a\,b/c}{↔d/c}$ $\frac{a\,b/c}{↔d/c}$ $\frac{a\,b/c}{↔d/c}$

$$\frac{23}{6} = 3\frac{5}{6}$$

Check Your Understanding

Rename each improper fraction as a mixed number or a whole number.

1. $\frac{6}{5}$
2. $\frac{21}{8}$
3. $\frac{24}{6}$
4. $\frac{11}{2}$
5. $\frac{15}{4}$
6. $\frac{20}{3}$

Check your answers on page 416.

Equivalent Fractions

Two or more fractions that name the same number are called **equivalent fractions.** Equivalent fractions are equal.

One way to rename a fraction as an equivalent fraction is to *multiply* the numerator and denominator by the same number.

Example Rename $\frac{3}{4}$ as an equivalent fraction using multiplication.

The rectangle is divided into 4 equal parts. 3 of the parts are blue. $\frac{3}{4}$ of the rectangle is blue.

If each of the 4 parts is split into 2 equal parts, there are now 8 equal parts. 6 of them are blue. $\frac{6}{8}$ of the rectangle is blue.

$$\frac{3}{4} = \frac{6}{8}$$

$\frac{3}{4}$ and $\frac{6}{8}$ both name the same amount of the rectangle that is blue.

The number of parts in the rectangle was doubled. You can show this by multiplying the numerator and the denominator of $\frac{3}{4}$ by 2.

$$\frac{3 * 2}{4 * 2} = \frac{6}{8}$$

$\frac{3}{4}$ is equivalent to $\frac{6}{8}$. $\frac{3}{4} = \frac{6}{8}$

If each part in the rectangle is divided into 3 equal parts, the number of parts is tripled. You can show this by multiplying the numerator and the denominator of $\frac{3}{4}$ by 3.

$$\frac{3 * 3}{4 * 3} = \frac{9}{12}$$

$\frac{9}{12}$ is equivalent to $\frac{3}{4}$. $\frac{9}{12} = \frac{3}{4}$

Another way to rename a fraction as an equivalent fraction is to *divide* the numerator and the denominator by the same number.

Example Rename $\frac{6}{12}$ as an equivalent fraction using division.

$\frac{6}{12}$ of the region is green.

Divide the region into groups of 3.

$\frac{2}{4}$ is green.

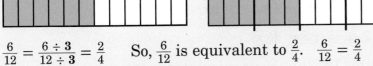

$$\frac{6}{12} = \frac{6 \div 3}{12 \div 3} = \frac{2}{4} \qquad \text{So, } \frac{6}{12} \text{ is equivalent to } \frac{2}{4}. \quad \frac{6}{12} = \frac{2}{4}$$

Check Your Understanding

Rename each fraction as an equivalent fraction. Use multiplication for Problems 1–3 and division for Problems 4–6.

1. $\frac{1}{4}$ 2. $\frac{5}{10}$ 3. $\frac{5}{4}$ 4. $\frac{8}{12}$ 5. $\frac{6}{10}$ 6. $\frac{9}{15}$

Check your answers on page 417.

Simplifying Fractions

When a fraction is renamed as an equivalent fraction with a smaller numerator and denominator, the new fraction is in **simpler form.** You can simplify a fraction by dividing its numerator and denominator by a common factor greater than 1.

Example   Rename $\frac{8}{12}$ in simpler form.

First, find common factors of the numerator and the denominator. 2 and 4 are common factors of 8 and 12.

Then, divide the numerator and the denominator of $\frac{8}{12}$ by either 2 or 4.

$\frac{8 \div 2}{12 \div 2} = \frac{4}{6}$ $\frac{4}{6}$ is equivalent to $\frac{8}{12}$ and is in simpler form.

$\frac{8 \div 4}{12 \div 4} = \frac{2}{3}$ $\frac{2}{3}$ is equivalent to $\frac{8}{12}$ and is in simpler form.

$\frac{4}{6}$ and $\frac{2}{3}$ are simpler forms of $\frac{8}{12}$.

Note

The expression *lowest terms* means the same as *simplest form.*

In the example at the left, 4 is the greatest common factor of 8 and 12. Therefore, $\frac{2}{3}$ is equivalent to $\frac{8}{12}$ and is in simplest form.

See page 80 for more on greatest common factors.

A proper fraction is in **simplest form** if it cannot be renamed in simpler form. You can rename a fraction in simplest form by dividing its numerator and denominator by the **greatest common factor** of both the numerator and the denominator.

A fraction is in simplest form when 1 is the only common factor of both the numerator and the denominator. Some calculators have a special key for renaming fractions in simplest form.

Example   Rename $\frac{8}{12}$ in simplest form.

On Calculator A: Key in: 8 $\boxed{n}$ 12 $\boxed{d}$ $\boxed{Simp}$ $\boxed{Enter}$

The display shows $\frac{4}{6}$. If, without clearing the display, you press $\boxed{Simp}$ $\boxed{Enter}$ again, the display will show $\frac{2}{3}$. If you press $\boxed{Simp}$ $\boxed{Enter}$ one more time, the display will show $\frac{2}{3}$ again.

On Calculator B: Key in: 8 $\boxed{b/c}$ 12 $\boxed{Simp}$ $\boxed{Simp}$ $\boxed{Simp}$.

Therefore, $\frac{2}{3}$ is in simplest form. Try it on your calculator.

Check Your Understanding

Write the fractions in Problems 1–3 in simpler form and those in Problems 4–6 in simplest form.

1. $\frac{6}{12}$ 2. $\frac{16}{20}$ 3. $\frac{16}{28}$ 4. $\frac{9}{15}$ 5. $\frac{12}{28}$ 6. $\frac{20}{24}$

Check your answers on page 417.

Comparing Fractions

There are several strategies that can help you compare fractions.

Use a Common Numerator	$\frac{5}{7} > \frac{5}{8}$ because sevenths are larger than eighths and there are 5 of each.	 $\frac{5}{7} > \frac{5}{8}$

Compare to $\frac{1}{2}$	$\frac{4}{9}$ is less than $\frac{1}{2}$, and $\frac{3}{5}$ is more than $\frac{1}{2}$. So, $\frac{3}{5} > \frac{4}{9}$.	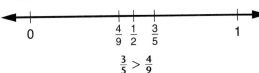 $\frac{3}{5} > \frac{4}{9}$

Compare to 1	Both $\frac{3}{4}$ and $\frac{2}{3}$ are less than 1. But $\frac{3}{4}$ is $\frac{1}{4}$ away from 1, and $\frac{2}{3}$ is $\frac{1}{3}$ away from 1. So, $\frac{3}{4}$ is closer to 1 than $\frac{2}{3}$ is. This means $\frac{3}{4} > \frac{2}{3}$.	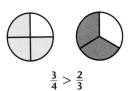 $\frac{3}{4} > \frac{2}{3}$

Use an Equivalent for One of the Fractions	To compare $\frac{2}{7}$ and $\frac{1}{4}$, change $\frac{1}{4}$ to $\frac{2}{8}$. Since $\frac{2}{8} < \frac{2}{7}$ (because eighths are smaller than sevenths), you know that $\frac{1}{4} < \frac{2}{7}$.	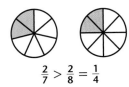 $\frac{2}{7} > \frac{2}{8} = \frac{1}{4}$

Use a Common Denominator	To compare $\frac{3}{5}$ and $\frac{5}{8}$, rename both fractions with a common denominator. $5 * 8 = 40$ is a common denominator. Since $\frac{24}{40} < \frac{25}{40}$, you know that $\frac{3}{5} < \frac{5}{8}$.	$\frac{3}{5} = \frac{3*8}{5*8} = \frac{24}{40}$ $\frac{5}{8} = \frac{5*5}{8*5} = \frac{25}{40}$

Convert to Decimals	To compare $\frac{13}{17}$ and $\frac{32}{41}$, use a calculator to convert both to decimals. Since $0.76 \ldots < 0.78 \ldots$, you know that $\frac{13}{17} < \frac{32}{41}$.	$\frac{13}{17} = 13 \boxed{\div} 17 \boxed{=} 0.7647058$ $\frac{32}{41} = 32 \boxed{\div} 41 \boxed{=} 0.7804878$

Check Your Understanding

Compare. Use > or <.

1. $\frac{5}{9} \square \frac{2}{5}$ **2.** $\frac{14}{15} \square \frac{10}{11}$ **3.** $\frac{12}{6} \square \frac{12}{5}$ **4.** $\frac{4}{5} \square \frac{5}{7}$ **5.** $\frac{17}{23} \square \frac{67}{93}$

Check your answers on page 417.

Fraction-Stick Chart

Each stick on the Fraction-Stick Chart represents 1 whole.
Each stick (except the 1-stick) is divided into equal pieces.
Each piece represents a fraction of 1 whole.

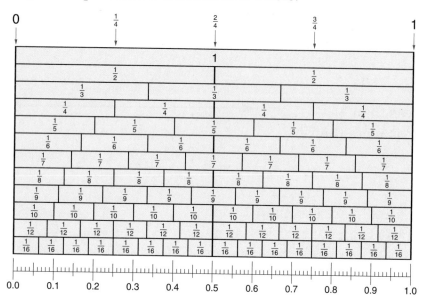

Locating a Fraction on the Fraction-Stick Chart

1. Select the stick shown by the denominator of the fraction.

2. Count the number of pieces shown by the numerator, starting at the left edge of the chart.

Example   Find $\frac{3}{4}$ on the Fraction-Stick Chart.

The "fourths" stick is equally divided into 4 pieces, each labeled $\frac{1}{4}$.
This stick can be used to locate fractions whose denominators are 4.
To locate the fraction $\frac{3}{4}$, count 3 pieces, starting at the left.

0   1

| $\frac{1}{4}$ | $\frac{1}{4}$ | $\frac{1}{4}$ | $\frac{1}{4}$ |

$\frac{0}{4}$ $\frac{1}{4}$ $\frac{2}{4}$ $\frac{3}{4}$ $\frac{4}{4}$

$\frac{3}{4}$ is located at the right edge of the third piece.

Check Your Understanding

Locate each fraction on the Fraction-Stick Chart.

1. $\frac{3}{8}$ **2.** $\frac{8}{12}$ **3.** $\frac{0}{4}$ **4.** $\frac{5}{9}$ **5.** $\frac{6}{6}$ **6.** $\frac{2}{3}$

Check your answers on page 417.

Finding Equivalent Fractions

Example Find fractions that are equivalent to $\frac{2}{3}$.

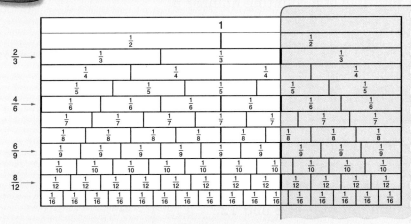

Step 1: Place one edge of a straightedge at $\frac{2}{3}$.

Step 2: Find all the pieces whose right edges touch the edge of the straightedge. The right edge of each piece shows the location of a fraction that is equivalent to $\frac{2}{3}$.

$\frac{4}{6}$, $\frac{6}{9}$, and $\frac{8}{12}$ are equivalent to $\frac{2}{3}$.

Comparing Two Fractions

Example Compare $\frac{4}{9}$ and $\frac{3}{8}$. Which is less?

Step 1: Place one edge of a straightedge at $\frac{4}{9}$.

Step 2: Locate $\frac{3}{8}$ on the "eighths" stick.

Step 3: $\frac{3}{8}$ is to the left of $\frac{4}{9}$.

$\frac{3}{8}$ is less than $\frac{4}{9}$.

Check Your Understanding

Use the Fraction-Stick Chart to find an equivalent fraction for each fraction.

1. $\frac{3}{9}$ **2.** $\frac{2}{4}$ **3.** $\frac{8}{12}$ **4.** $\frac{6}{16}$ **5.** $\frac{3}{5}$

Which fraction is less? Use the Fraction-Stick Chart to decide.

6. $\frac{1}{9}$ or $\frac{1}{8}$ **7.** $\frac{7}{9}$ or $\frac{8}{10}$ **8.** $\frac{3}{7}$ or $\frac{3}{6}$ **9.** $\frac{6}{7}$ or $\frac{13}{16}$ **10.** $\frac{11}{12}$ or $\frac{15}{16}$

Check your answers on page 417.

Least Common Multiples

A **multiple of a number *n*** is the product of any counting number and the number *n*. A multiple of a counting number *n* is always divisible by *n*.

Examples   6 is a multiple of 3 because 3 * 2 = 6; 6 is divisible by 3.

24 is a multiple of 4 because 4 * 6 = 24; 24 is divisible by 4.

The **least common multiple** of two numbers is the smallest number that is a multiple of both numbers.

Example   Find the least common multiple of 8 and 12.

Step 1: List multiples of 8: 8, 16, **24**, 32, 40, **48**, ...

Step 2: List multiples of 12: 12, **24**, 36, **48**, 60, ...

24 and 48 are common multiples of 8 and 12.

24 is smaller, so it is the least common multiple of 8 and 12.

Another way to find the least common multiple of two counting numbers is to use **prime factorization.**

Example   Find the least common multiple of 8 and 12.

Step 1:	Step 2:	Step 3:	Step 4:
Write the prime factorization of each number.	Circle pairs of common factors.	Cross out one factor in each circled pair.	Multiply the factors that have not been crossed out.
8 = 2 * 2 * 2 12 = 2 * 2 * 3	8 = 2 * 2 * 2 12 = 2 * 2 * 3	8 = 2 * 2 * 2 12 = 2 * 2 * 3	2 * 2 * 2 * 3 = 24

The least common multiple of 8 and 12 is 24.

The least common multiple of the denominators of two fractions is the **least common denominator** of the fractions. For example, 12 is the least common denominator of $\frac{3}{4}$ and $\frac{5}{6}$ because 12 is the least common multiple of 4 and 6.

Check Your Understanding

Find the least common multiple of each pair of numbers.

1. 6 and 12   **2.** 4 and 10   **3.** 3 and 4   **4.** 6 and 8   **5.** 6 and 9   **6.** 9 and 15

Check your answers on page 417.

Common Denominators

When solving problems that involve fractions with different denominators, rename the fractions so they have the same denominator. If two fractions have the same denominator, that denominator is called a **common denominator.**

There are several methods for renaming fractions so they have a common denominator.

Examples   Rename $\frac{3}{4}$ and $\frac{1}{6}$ with a common denominator.

Equivalent Fractions Method

List equivalent fractions for $\frac{3}{4}$ and $\frac{1}{6}$.

$$\frac{3}{4} = \frac{6}{8} = \frac{9}{12} = \frac{12}{16} = \dots$$

$$\frac{1}{6} = \frac{2}{12} = \frac{3}{18} = \frac{4}{24} = \dots$$

Both $\frac{3}{4}$ and $\frac{1}{6}$ can be renamed as fractions with the common denominator 12.

$$\frac{3}{4} = \frac{9}{12} \text{ and } \frac{1}{6} = \frac{2}{12}$$

The Multiplication Method

Multiply the numerator and the denominator of each fraction by the denominator of the other fraction.

$$\frac{3}{4} = \frac{3*\mathbf{6}}{4*\mathbf{6}} = \frac{18}{24} \qquad \frac{1}{6} = \frac{1*\mathbf{4}}{6*\mathbf{4}} = \frac{4}{24}$$

Least Common Multiple Method

Find the least common multiple of the denominators.

Multiples of 4: 4, 8, **12**, 16, 20, …

Multiples of 6: 6, **12**, 18, 24, …

The least common multiple of 4 and 6 is 12.

Rename the fractions so that their denominator is the least common multiple.

$$\frac{3}{4} = \frac{3*3}{4*3} = \frac{9}{12} \text{ and } \frac{1}{6} = \frac{1*2}{6*2} = \frac{2}{12}$$

This method gives what is known as the **least common denominator.**

> **Note**
>
> The Multiplication Method gives what *Everyday Mathematics* calls the **quick common denominator.** The quick common denominator can be used with variables, so it is common in algebra.

> **Note**
>
> The least common denominator is usually easier to use in complicated calculations, although finding it can often take more time.

Check Your Understanding

Rename each pair of fractions as fractions with a common denominator.

1. $\frac{1}{3}$ and $\frac{5}{6}$ 2. $\frac{3}{4}$ and $\frac{3}{5}$ 3. $\frac{7}{10}$ and $\frac{3}{2}$ 4. $\frac{1}{4}$ and $\frac{3}{10}$ 5. $\frac{4}{6}$ and $\frac{7}{8}$

Check your answers on page 417.

Greatest Common Factors

The **greatest common factor** of two counting numbers is the largest counting number that is a factor of both numbers.

> **Example**   Find the greatest common factor of 20 and 24.
>
> **Step 1:** List all the factors of 20: **1, 2, 4,** 5, 10 and 20.
>
> **Step 2:** List all the factors of 24: **1, 2,** 3, **4,** 6, 8, 12, and 24.
>   1, 2, and 4 are common factors of 20 and 24.
>
> 4 is the **greatest common factor** of 20 and 24.

Another way to find the greatest common factor of two counting numbers is to use prime factorization.

> **Example**   Find the greatest common factor of 20 and 24.
>
> | **Step 1:** | **Step 2:** | **Step 3:** | **Step 4:** |
> | --- | --- | --- | --- |
> | Write the prime factorization of each. | Circle pairs of common prime factors. | Cross out the factors that are not circled. | Multiply one factor from each pair of circled factors. |
> | $20 = 2 * 2 * 5$ | $20 = $ ②$*$② $* 5$ | $20 = $ ②$*$② $* \cancel{5}$ | $2 * 2 = 4$ |
> | $24 = 2 * 2 * 2 * 3$ | $24 = $ ②$*$② $* 2 * 3$ | $24 = $ ②$*$② $* \cancel{2} * \cancel{3}$ | |
>
> The greatest common factor of 20 and 24 is 4.

An Interesting Fact

The product of the least common multiple and the greatest common factor of two counting numbers is the same as the product of the two numbers themselves.

> **Example**   The least common multiple of 4 and 6 is 12.
>
>   The greatest common factor of 4 and 6 is 2.
>
> The product of the least common multiple and the greatest common factor of 4 and 6 is 12 * 2, or 24, and the product of 4 and 6 is also 24.

Check Your Understanding

Find the greatest common factor of each pair of numbers.

1. 3 and 5   **2.** 4 and 10   **3.** 8 and 24   **4.** 35 and 28 **5.** 18 and 12   **6.** 9 and 15

Check your answers on page 417.

Using Unit Fractions

A **unit fraction** is a fraction with 1 in its numerator, such as $\frac{1}{2}, \frac{1}{10}$, and $\frac{1}{25}$. If you know the part of a whole represented by a unit fraction, you can find the whole by multiplying the part of the whole by the denominator of the unit fraction.

Example   6 is $\frac{1}{4}$ of what number?

Since $\frac{1}{4}$ is 6, $\frac{2}{4}$ must be 2 * 6, or 12;

$\frac{3}{4}$ must be 3 * 6, or 18;

and $\frac{4}{4}$ must be 4 * 6, or 24.

So, 6 is $\frac{1}{4}$ of 24.

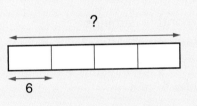

Example   Mark owns 2 white shirts. This is $\frac{1}{4}$ of the total number of shirts he owns. How many shirts does he own?

If 2 shirts are $\frac{1}{4}$ of the total number of shirts, then the total number of shirts is 4 times that number.

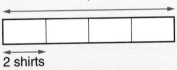

Since 4 * 2 = 8, Mark owns a total of 8 shirts.

Unit fractions can be used even when a part is given as a fraction that is not a unit fraction.

Example   Sara lives 8 blocks from the library. This is $\frac{2}{3}$ of the distance from her home to school. How many blocks is it from Sara's home to school?

Step 1: Find $\frac{1}{3}$ of the distance to school. 8 blocks is $\frac{2}{3}$ of the distance to school. Therefore, to find $\frac{1}{3}$ of the distance, divide 8 blocks by 2.   8 / 2 = 4

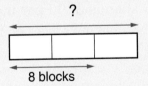

Step 2: Find the total distance to school. 4 blocks is $\frac{1}{3}$ of the distance to school. Therefore, to find the total distance to school (or $\frac{3}{3}$ of the distance), multiply 4 blocks by 3.   3 * 4 = 12

Sara lives 12 blocks from school.

Example Ms. Partee spends $1,000 a month. This amount is $\frac{4}{5}$ of her monthly earnings. How much does she earn per month?

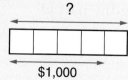

?

$1,000

Step 1: Find $\frac{1}{5}$ of her monthly earnings.

$1,000 is $\frac{4}{5}$ of her monthly earnings. Therefore, to find $\frac{1}{5}$ of her earnings, divide $1,000 by 4.

$1,000 / 4 = $250

Step 2: Find her total monthly earnings.

$250 is $\frac{1}{5}$ of her monthly earnings. Therefore, to find her total earnings (or $\frac{5}{5}$ of her earnings), multiply $250 by 5.

5 * $250 = $1,250

Ms. Partee earns $1,250 each month.

Check Your Understanding

Solve each problem.

1. Natalie collects movie posters. Her 4 posters from *Star Wars Episode III* are $\frac{1}{8}$ of her collection. How many posters does Natalie have in her whole collection?

2. If 12 counters are $\frac{3}{5}$ of a set, how many counters are in the whole set?

3. Mr. Hart spends $800 per month on rent. This is $\frac{3}{15}$ of his monthly earnings. How much does he earn per month?

Check your answers on page 417.

Addition and Subtraction of Fractions

To find the sum of fractions that have the same denominator, you add just the numerators. The denominator does not change. Subtraction of fractions with like denominators is done the same way.

Examples Find $\frac{1}{4} + \frac{2}{4}$.

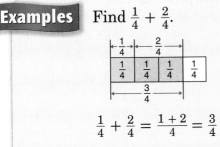

$$\frac{1}{4} + \frac{2}{4} = \frac{1+2}{4} = \frac{3}{4}$$

Find $\frac{4}{5} - \frac{3}{5}$.

$$\frac{4}{5} - \frac{3}{5} = \frac{4-3}{5} = \frac{1}{5}$$

To find the sum of fractions that do not have the same denominator, first rename the fractions as fractions with a common denominator. Then proceed as above. Subtraction of fractions with unlike denominators is done the same way.

Examples Find $\frac{3}{4} + \frac{1}{8}$.

$$\frac{3}{4} + \frac{1}{8} = \frac{6}{8} + \frac{1}{8} = \frac{7}{8} \quad \left(\frac{3}{4} = \frac{6}{8}\right)$$

So, $\frac{3}{4} + \frac{1}{8} = \frac{7}{8}$.

Find $\frac{5}{6} - \frac{1}{4}$.

$$\frac{5}{6} - \frac{1}{4} = \frac{10}{12} - \frac{3}{12} = \frac{7}{12} \quad \left(\frac{5}{6} = \frac{10}{12}, \frac{1}{4} = \frac{2}{8} = \frac{3}{12}\right)$$

So, $\frac{5}{6} - \frac{1}{4} = \frac{7}{12}$.

It is possible to add and subtract fractions on some calculators. See if you can do this on your calculator.

Examples Find $\frac{3}{8} + \frac{1}{3}$.

On Calculator A:

Key in 3 ⓝ 8 ⓓ ⊕ 1 ⓝ 3 ⓓ (Enter)

Answer: $\frac{17}{24}$

Find $\frac{7}{8} - \frac{3}{5}$.

On Calculator B:

Key in 7 (b/c) 8 ⊖ 3 (b/c) 5 (=)

Answer: $\frac{11}{40}$.

Check Your Understanding

Solve each problem. Check the answers on a calculator.

1. $\frac{5}{9} + \frac{3}{9}$ **2.** $\frac{7}{8} - \frac{1}{4}$ **3.** $\frac{2}{3} + \frac{1}{4}$ **4.** $\frac{13}{6} - \frac{7}{4}$ **5.** $\frac{3}{8} + \frac{5}{6}$

Check your answers on page 417.

Addition of Mixed Numbers

One way to add mixed numbers is to add the fractions and the
whole numbers separately. This may require renaming the sum.

Example Find $4\frac{5}{8} + 2\frac{7}{8}$.

Step 1: Add the fractions.

$$4\frac{5}{8}$$
$$+\,2\frac{7}{8}$$
$$\frac{12}{8}$$

$$4\frac{5}{8} + 2\frac{7}{8} = 7\frac{1}{2}$$

Step 2: Add the whole numbers.

$$4\frac{5}{8}$$
$$+\,2\frac{7}{8}$$
$$6\frac{12}{8}$$

Step 3: Rename the sum.

$$6\frac{12}{8} = 6 + \frac{8}{8} + \frac{4}{8}$$
$$= 6 + 1 + \frac{4}{8}$$
$$= 7\frac{4}{8}$$
$$= 7\frac{1}{2}$$

If the fractions do not have the same denominator, first rename
the fractions so they have a common denominator.

Example Find $3\frac{3}{4} + 5\frac{2}{3}$.

Step 1: Rename and add the fractions.

$$3\frac{3}{4} = 3\frac{9}{12}$$
$$+\,5\frac{2}{3} = +\,5\frac{8}{12}$$
$$\frac{17}{12}$$

$$3\frac{3}{4} + 5\frac{2}{3} = 9\frac{5}{12}$$

Step 2: Add the whole numbers.

$$3\frac{9}{12}$$
$$+\,5\frac{8}{12}$$
$$8\frac{17}{12}$$

Step 3: Rename the sum.

$$8\frac{17}{12} = 8 + \frac{12}{12} + \frac{5}{12}$$
$$= 8 + 1 + \frac{5}{12}$$
$$= 9\frac{5}{12}$$

Some calculators have special keys for entering mixed numbers.

Example Solve $3\frac{3}{4} + 5\frac{2}{3}$ on a calculator.

On Calculator A: Key in 3 (Unit) 3 (n) 4 (d) (+) 5 (Unit) 2 (n) 3 (d) (Enter)

On Calculator B: Key in 3 (a) 3 (b/c) 4 (+) 5 (a) 2 (b/c) 3 (=)

Check Your Understanding

Solve Problems 1–3 without a calculator. Solve Problem 4 with a calculator.

1. $2\frac{1}{8} + 7\frac{7}{8}$ 2. $3\frac{4}{5} + 2\frac{1}{2}$ 3. $6\frac{2}{3} + 3\frac{3}{4}$ 4. $14\frac{4}{9} + 8\frac{6}{7}$

Check your answers on page 417.

Subtraction of Mixed Numbers

If the fractions do not have the same denominator, first rename them as fractions with a common denominator.

Example   Find $3\frac{7}{8} - 1\frac{3}{4}$.

Step 1: Rename the fractions.	**Step 2:** Subtract the fractions.	**Step 3:** Subtract the whole numbers.
$3\frac{7}{8} = 3\frac{7}{8}$ $-1\frac{3}{4} = -1\frac{6}{8}$	$3\frac{7}{8}$ $-1\frac{6}{8}$ $\frac{1}{8}$	$3\frac{7}{8}$ $-1\frac{6}{8}$ $2\frac{1}{8}$

$3\frac{7}{8} - 1\frac{3}{4} = 2\frac{1}{8}$

To subtract a mixed number from a whole number, first rename the whole number as the sum of a whole number and a fraction that is equivalent to 1.

Example   Find $5 - 2\frac{2}{3}$.

Step 1: Rename the whole number.	**Step 2:** Subtract the fractions.	**Step 3:** Subtract the whole numbers.
$5 = 4\frac{3}{3}$ $-2\frac{2}{3} = -2\frac{2}{3}$	$4\frac{3}{3}$ $-2\frac{2}{3}$ $\frac{1}{3}$	$4\frac{3}{3}$ $-2\frac{2}{3}$ $2\frac{1}{3}$

$5 - 2\frac{2}{3} = 2\frac{1}{3}$

When subtracting mixed numbers, rename the larger mixed number if it contains a fraction that is less than the fraction in the smaller mixed number.

Example   Find $7\frac{1}{5} - 3\frac{3}{5}$.

Step 1: Rename the larger mixed number.	**Step 2:** Subtract the fractions.	**Step 3:** Subtract the whole numbers.
$7\frac{1}{5} = 6\frac{6}{5}$ $-3\frac{3}{5} = -3\frac{3}{5}$	$6\frac{6}{5}$ $-3\frac{3}{5}$ $\frac{3}{5}$	$6\frac{6}{5}$ $-3\frac{3}{5}$ $3\frac{3}{5}$

$7\frac{1}{5} - 3\frac{3}{5} = 3\frac{3}{5}$

The example below shows three methods of solving $4\frac{1}{6} - 2\frac{2}{3}$.

Example Find $4\frac{1}{6} - 2\frac{2}{3}$.

Method 1: This is the method shown on page 85.

Step 1: Rename the fractions.

$$4\frac{1}{6} = 4\frac{1}{6}$$
$$-2\frac{2}{3} = -2\frac{4}{6}$$

Step 2: Rename the larger mixed number.

$$4\frac{1}{6} = 3\frac{7}{6}$$
$$-2\frac{4}{6} = -2\frac{4}{6}$$

Step 3: Subtract.

$$3\frac{7}{6}$$
$$-2\frac{4}{6}$$
$$1\frac{3}{6} = 1\frac{1}{2}$$

Method 2: Work with the fraction names for the mixed numbers.

Step 1: Rename the mixed numbers.

$$4\frac{1}{6} = \frac{25}{6}$$
$$-2\frac{2}{3} = -\frac{8}{3}$$

Step 2: Rename the fractions. Subtract.

$$\frac{25}{6} = \frac{25}{6}$$
$$-\frac{8}{3} = -\frac{16}{6}$$
$$\frac{9}{6}$$

Step 3: Rename the result as a mixed number.

$$\frac{9}{6} = 1\frac{3}{6} = 1\frac{1}{2}$$

Method 3: Use the partial-differences method.

$$4\frac{1}{6}$$
$$-2\frac{2}{3}$$

Subtract the whole numbers. $4 - 2 \rightarrow 2$

Subtract the fractions. $\frac{1}{6} - \frac{2}{3} = \frac{1}{6} - \frac{4}{6} \rightarrow -\frac{3}{6}$

Find the total. $2 - \frac{3}{6} \rightarrow 1\frac{3}{6} = 1\frac{1}{2}$

$$4\frac{1}{6} - 2\frac{2}{3} = 1\frac{1}{2}$$

You can also subtract mixed numbers on calculators that have keys for entering mixed numbers. Try it.

Check Your Understanding

Subtract. Check your answers on a calculator.

1. $5\frac{2}{5} - 1\frac{1}{3}$

2. $6 - 2\frac{7}{8}$

3. $4\frac{5}{9} - \frac{7}{9}$

4. $8\frac{1}{2} - 3\frac{2}{3}$

Check your answers on page 417.

Finding a Fraction of a Number

Many problems with fractions involve finding a fraction of a number.

Example   Find $\frac{2}{3}$ of 24.

Model the problem by using 24 pennies. Divide the pennies into 3 equal groups.

Each group has $\frac{1}{3}$ of the pennies. So, $\frac{1}{3}$ of 24 pennies is 8 pennies.

$\frac{1}{3}$ of 24 = 8, so $\frac{2}{3}$ of 24 = 16.

$\frac{2}{3}$ of 24 = $\frac{2}{3}$ * 24 = 16

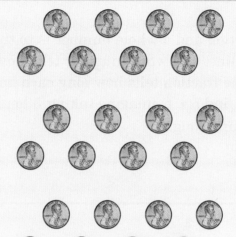

Note

"$\frac{2}{3}$ of 24" has the same meaning as "$\frac{2}{3}$ * 24." When you find the fraction of a number, you can replace the word "of" by the multiplication symbol.

Other examples:

$\frac{1}{6}$ of 18 means $\frac{1}{6}$ * 18.

$\frac{3}{4}$ of 40 means $\frac{3}{4}$ * 40.

$\frac{3}{4}$ of 2.8 means $\frac{3}{4}$ * 2.8.

Example   A jacket that sells for \$45 is on sale for $\frac{2}{3}$ of the regular price. What is the sale price?

To find the sale price, you have to find $\frac{2}{3}$ of \$45.

Step 1:   Find $\frac{1}{3}$ of 45. 45 ÷ 3 = 15, so $\frac{1}{3}$ of 45 is 15.

Step 2:   Use the answer from Step 1 to find $\frac{2}{3}$ of 45.

Since $\frac{1}{3}$ of 45 is 15, then $\frac{2}{3}$ of 45 is 2 * 15 = 30.

The sale price is \$30.

$\frac{2}{3}$ of \$45 = $\frac{2}{3}$ * \$45 = \$30

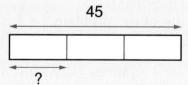

45 ÷ 3 = 15, so $\frac{1}{3}$ of 45 is 15.

Check Your Understanding

Solve each problem.

1. $\frac{1}{4}$ of 32   2. $\frac{3}{4}$ of 32   3. $\frac{3}{5}$ of 30   4. $\frac{5}{6}$ of 72

5. Gina and Robert earned \$56 raking lawns, but Gina did most of the work. They decided that Gina should get $\frac{3}{4}$ of the money. How much does each person get?

Check your answers on page 417.

Multiplying Fractions and Whole Numbers

People usually think that "multiplication makes things bigger." But multiplication involving fractions can lead to products that are smaller than at least one of the factors. For example, $10 * \frac{1}{2} = 5$.

Number-Line Model

One way to multiply a fraction and a whole number is to think about "hops" on a number line. The whole number tells how many hops to make, and the fraction tells how long each hop should be. For example, to find $5 * \frac{2}{3}$, imagine taking 5 hops on a number line, each $\frac{2}{3}$ of a unit long.

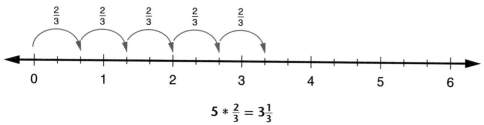

$$5 * \frac{2}{3} = 3\frac{1}{3}$$

Addition Model

You can use addition to multiply a fraction and a whole number. For example, to find $4 * \frac{2}{3}$, draw 4 models of $\frac{2}{3}$. Then add up all of the fractions.

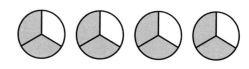

$$4 * \frac{2}{3} = \frac{2}{3} + \frac{2}{3} + \frac{2}{3} + \frac{2}{3} = \frac{8}{3}$$

Area Model

Think of the problem $\frac{2}{3} * 4$ as "What is $\frac{2}{3}$ of an area that has 4 square units?"

Draw 4 squares, each with an area of 1 square unit. The rectangle has an area of 4 square units.

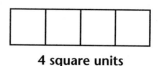

4 square units

Divide the rectangle into 3 equal strips and shade 2 strips ($\frac{2}{3}$ of the area). The shaded area equals $\frac{8}{3}$ (8 small rectangles, each with an area of $\frac{1}{3}$).

So, $\frac{2}{3}$ of 4 square units equals $\frac{8}{3}$ square units.

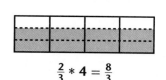

$$\frac{2}{3} * 4 = \frac{8}{3}$$

$$\frac{2}{3} \text{ of } 4 = \frac{2}{3} * 4 = \frac{8}{3}$$

Check Your Understanding

Multiply.

1. $5 * \frac{3}{4}$ **2.** $\frac{2}{3} * 6$ **3.** $4 * \frac{4}{5}$ **4.** $6 * \frac{2}{5}$ **5.** $\frac{3}{4} * 6$

Check your answers on page 417.

Multiplying Fractions

When both numbers to be multiplied are fractions, addition and number-line hopping are not helpful in finding the answer. Fortunately, the **area model** does help.

Example $\frac{3}{4} * \frac{2}{3} = ?$

$\frac{2}{3}$ of the rectangular region is shaded this way:

$\frac{3}{4}$ of the region is shaded this way:

$\frac{3}{4}$ of $\frac{2}{3}$ of the region is shaded both ways:

That's $\frac{6}{12}$, or $\frac{1}{2}$, of the whole region.

$\frac{3}{4} * \frac{2}{3} = \frac{6}{12} = \frac{1}{2}$

Multiplication of Fractions Property

The problem above is an example of this general pattern: To multiply fractions, simply multiply the numerators and multiply the denominators.

This pattern can be expressed as follows:

$\frac{a}{b} * \frac{c}{d} = \frac{a * c}{b * d}$ (b and d may not be 0.)

Examples $\quad \frac{3}{4} * \frac{2}{3} = \frac{3 * 2}{4 * 3} = \frac{6}{12} = \frac{1}{2}$ $\qquad\qquad \frac{5}{4} * \frac{6}{7} = \frac{5 * 6}{4 * 7} = \frac{30}{28} = \frac{15}{14} = 1\frac{1}{14}$

The Multiplication of Fractions Property can be used to multiply a whole number and a fraction. First, rename the whole number as a fraction with 1 in the denominator.

Examples $\quad 5 * \frac{2}{3} = \frac{5}{1} * \frac{2}{3} = \frac{5 * 2}{1 * 3} = \frac{10}{3} = 3\frac{1}{3}$ $\qquad \frac{2}{11} * 4 = \frac{2}{11} * \frac{4}{1} = \frac{2 * 4}{11 * 1} = \frac{8}{11}$

Check Your Understanding

Multiply.

1. $\frac{1}{2} * \frac{3}{5}$ **2.** $\frac{3}{4} * \frac{5}{6}$ **3.** $\frac{4}{5} * \frac{20}{3}$ **4.** $\frac{6}{7} * \frac{0}{7}$ **5.** $\frac{10}{3} * \frac{6}{25}$

Check your answers on page 417.

Multiplying Mixed Numbers

One way to multiply two mixed numbers is to rename each mixed number as an improper fraction and multiply the fractions. Then rename the product as a mixed number.

Example Find $3\frac{1}{4} * 1\frac{5}{6}$.

Rename the mixed numbers as fractions and multiply.

$$3\frac{1}{4} * 1\frac{5}{6} = \frac{13}{4} * \frac{11}{6}$$

$$= \frac{13 * 11}{4 * 6} = \frac{143}{24}$$

So, $3\frac{1}{4} * 1\frac{5}{6} = 5\frac{23}{24}$.

Rename the product as a mixed number.

$$24\overline{)143}$$
$$\underline{-120}\ \Big|\ 5$$
$$23\ \ \ 5$$

$$\frac{143}{24} = 5\frac{23}{24}$$

Note

The remainder in a division problem can be rewritten as a fraction. The remainder is the numerator and the divisor is the denominator. For example,

$143 \div 24 \rightarrow 5\ R23$

so $143 \div 24 = 5\frac{23}{24}$.

Another way to multiply mixed numbers is to find partial products and add them.

Example Find $6\frac{1}{2} * 3\frac{3}{5}$.

Step 1: Find all the partial products.

$6 * 3 = 18$

$6 * \frac{3}{5} = \frac{6}{1} * \frac{3}{5} = \frac{18}{5} = 3\frac{3}{5}$

$\frac{1}{2} * 3 = \frac{1}{2} * \frac{3}{1} = \frac{3}{2} = 1\frac{1}{2}$

$\frac{1}{2} * \frac{3}{5} = \frac{3}{10}$

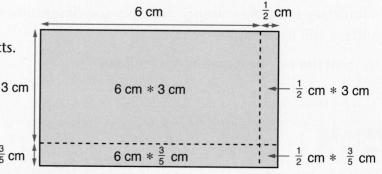

Step 2: Add the partial products.

$$18 + 3\frac{3}{5} + 1\frac{1}{2} + \frac{3}{10} = 22 + \frac{3}{5} + \frac{1}{2} + \frac{3}{10}$$

$$= 22 + \frac{6}{10} + \frac{5}{10} + \frac{3}{10}$$

$$= 22 + \frac{14}{10} = 23\frac{4}{10} = 23\frac{2}{5}$$

So, $6\frac{1}{2} * 3\frac{3}{5} = 23\frac{2}{5}$.

Check Your Understanding

Multiply.

1. $\frac{1}{4} * 1\frac{1}{2}$

2. $2\frac{2}{3} * 5$

3. $3\frac{2}{5} * 2\frac{1}{2}$

Check your answers on page 417.

Division of Fractions

Dividing a number by a fraction often gives a quotient that is larger than the dividend. For example, $4 \div \frac{1}{2} = 8$. To understand why this is, it's helpful to think about what division means.

Equal Groups

A division problem like $a \div b = ?$ is asking "How many bs are there in a?" For example, the problem $6 \div 3 = ?$ asks, "How many 3s are there in 6?" The figure at the right shows that there are two 3s in 6, so $6 \div 3 = 2$.

$6 \div 3 = 2$

$6 \div \frac{1}{3} = 18$

A division problem like $6 \div \frac{1}{3} = ?$ is asking, "How many $\frac{1}{3}$s are there in 6?" The figure at the right shows that there are 18 thirds in 6, so $6 \div \frac{1}{3} = 18$.

Example Scott has 5 pounds of rice. A cup of rice is about $\frac{1}{2}$ pound. How many cups of rice does Scott have?

$\frac{1}{2}$ lb $+$ $\frac{1}{2}$ lb $+$ $\frac{1}{2}$ lb $+$ $\frac{1}{2}$ lb $+$ $\frac{1}{2}$ lb $+$ $\frac{1}{2}$ lb $+$ $\frac{1}{2}$ lb $+$ $\frac{1}{2}$ lb $+$ $\frac{1}{2}$ lb $+$ $\frac{1}{2}$ lb $=$ 5 lb

This problem is solved by finding how many $\frac{1}{2}$s are in 5, which is the same as $5 \div \frac{1}{2}$.

So, Scott has about 10 cups of rice.

Missing Factors

A division problem is equivalent to a multiplication problem with a missing factor.

A problem like $6 \div \frac{1}{2} = \square$ is equivalent to $\frac{1}{2} * \square = 6$.

$\frac{1}{2} * \square = 6$ is the same as asking "$\frac{1}{2}$ of what number equals 6?"

Since $\frac{1}{2} * 12 = 6$, you know that $6 \div \frac{1}{2} = 12$.

Example Find $10 \div \frac{2}{3}$. Write $10 \div \frac{2}{3} = \square$.

This problem is equivalent to $\frac{2}{3} * \square = 10$, which means "$\frac{2}{3}$ of what number is 10?"

The diagram shows that $\frac{2}{3}$ of the missing number is 10. Since $\frac{2}{3}$ of the missing number is 10, $\frac{1}{3}$ must be 5. Since $\frac{1}{3}$ of the missing number is 5, the missing number must be $3 * 5 = 15$.

So, $\frac{2}{3}$ of 15 = 10, which means that $\frac{2}{3} * 15 = 10$.

$10 \div \frac{2}{3} = 15$

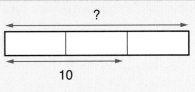

$\frac{2}{3}$ of ? = 10

Common Denominators

One way to solve a fraction division problem is to rename both the dividend and the divisor as fractions with a common denominator. Then divide the numerators and the denominators.

Example Find $6 \div \frac{2}{3}$.

Rename 6 as $\frac{18}{3}$.

$$6 \div \frac{2}{3} = \frac{18}{3} \div \frac{2}{3}$$

Divide the numerators and the denominators.

$$= \frac{18 \div 2}{3 \div 3}$$

$$= \frac{9}{1}, \text{ or } 9$$

So, $6 \div \frac{2}{3} = 9$.

To see why this method works, imagine putting the 18 thirds into equal groups that have $\frac{2}{3}$ in each group. There would be 9 groups. There are 9 two-thirds in $\frac{18}{3}$.

Example Julia has 6 pounds of modeling clay. She wants to put it in packages that hold $\frac{3}{4}$ of a pound each. How many packages can she make?

To solve $6 \div \frac{3}{4}$, rename 6 as $\frac{24}{4}$. Then divide.

$$6 \div \frac{3}{4} = \frac{24}{4} \div \frac{3}{4}$$

The 24 fourths can be put into 8 groups of $\frac{3}{4}$ each.

$$= \frac{24 \div 3}{4 \div 4}$$

$$= \frac{8}{1}, \text{ or } 8$$

So, Julia can make 8 packages of modeling clay.

Check Your Understanding

Solve. Then write a division number model for each problem.

1. Regina has 9 pizzas. If each person can eat $\frac{1}{2}$ of a pizza, how many people can Regina serve?

2. Selena has 10 yards of plastic strips for making bracelets. She needs $\frac{1}{3}$ yard for each bracelet. How many bracelets can she make?

3. 7 is $\frac{1}{4}$ of a number. What is the number?

Check your answers on page 417.

Division of Fractions and Mixed Numbers

The **reciprocal** of any fraction $\frac{a}{b}$ is the fraction $\frac{b}{a}$. For example, the reciprocal of $\frac{3}{4}$ is $\frac{4}{3}$.

If n is a whole number, $n = \frac{n}{1}$. So the reciprocal of n is $\frac{1}{n}$. For example, the reciprocal of 9 is $\frac{1}{9}$. 0 can be written as $\frac{0}{1}$, but 0 does not have a reciprocal, because $\frac{1}{0}$ is not defined. (Division by 0 is *never* allowed.)

The product of a number and its reciprocal is always 1.

$$\frac{a}{b} * \frac{b}{a} = 1$$

Examples $\quad \frac{4}{7} * \frac{7}{4} = \frac{4 * 7}{7 * 4} = \frac{28}{28} = 1 \qquad \frac{120}{3} * \frac{3}{120} = \frac{360}{360} = 1 \qquad 29 * \frac{1}{29} = \frac{29}{1} * \frac{1}{29} = \frac{29}{29} = 1$

Reciprocals are useful when dividing fractions.

Division of Fractions Property
To find the quotient of two fractions, multiply the first fraction by the reciprocal of the second fraction.

$$\frac{a}{b} \div \frac{c}{d} = \frac{a}{b} * \frac{d}{c}$$

Examples

$$\begin{aligned}\frac{4}{5} \div \frac{2}{3} &= \frac{4}{5} * \frac{3}{2} \\ &= \frac{12}{10} \\ &= 1\tfrac{2}{10}, \text{ or } 1\tfrac{1}{5}\end{aligned} \qquad \begin{aligned}2\tfrac{3}{4} \div 1\tfrac{1}{3} &= \frac{11}{4} \div \frac{4}{3} \\ &= \frac{11}{4} * \frac{3}{4} \\ &= \frac{33}{16}, \text{ or } 2\tfrac{1}{16}\end{aligned}$$

Example Roger has $6\tfrac{1}{4}$ pounds of ground hamburger. He wants to make 5-ounce hamburger patties. How many can he make?

1 pound = 16 ounces. So, 1 ounce = $\frac{1}{16}$ pound, and 5 ounces = $\frac{5}{16}$ pound.

Think: How many $\frac{5}{16}$s are in $6\tfrac{1}{4}$? This is a division problem: $6\tfrac{1}{4} \div \frac{5}{16}$.

Rename the mixed number $6\tfrac{1}{4}$ as $\frac{25}{4}$ and use the Division of Fractions Property:

$$\frac{25}{4} \div \frac{5}{16} = \frac{25}{4} * \frac{16}{5} = \frac{400}{20} = 20$$

Roger can make 20 hamburger patties.

Check Your Understanding

Divide.

1. $\frac{3}{5} \div \frac{1}{4}$ **2.** $5 \div \frac{5}{3}$ **3.** $\frac{1}{7} \div \frac{3}{7}$ **4.** $2\tfrac{2}{3} \div 4$ **5.** $3\tfrac{1}{2} \div 1\tfrac{1}{4}$

Check your answers on page 417.

Positive and Negative Numbers

Positive and negative numbers are used on temperature scales and thermometers. They express temperatures with reference to a zero point (0 degrees).

Many other real-world situations have zero as a starting point. Numbers go in opposite directions from zero. The numbers greater than zero are called **positive numbers;** the numbers less than zero are called **negative numbers.**

Situation	Negative (−)	Zero (0)	Positive (+)
bank account	withdrawal	no transaction	deposit
weight	loss	no change	gain
time	past	present	future
games	behind	even	ahead
elevation	below sea level	at sea level	above sea level

Did You Know?

In golf, the *par* score for a course is the average number of strokes that an expert golfer should take to complete the course. A score of −5 indicates that the player took 5 fewer strokes than par. A score of +11 indicates that the player took 11 more strokes than par. The winners of most professional tournaments have negative scores.

Notation

A positive number may be written using the "+" symbol, but is usually written without it. For example, $+10 = 10$ and $\pi = +\pi$. The symbol "−" is written before a positive number to show that it represents a negative number. For example, −5 is read as "negative 5," −12.3 as "negative 12.3," and $-\frac{5}{8}$ as "negative $\frac{5}{8}$."

Note

Sometimes people say "minus" instead of "negative," reading −5 as "minus 5." This is less correct than saying "negative 5," but it is usually acceptable.

Relations

Two numbers are **opposite numbers** if they are the same distance from 0 on the number line, but on opposite sides of 0. For example, the opposite of 1.5 is −1.5 and the opposite of −1.5 is 1.5. The opposite of 0 is 0 itself.

When two numbers are shown on a number line, the number that is farther to the right is the larger number. Thus, $3 > -1$ and $-2.435 > -3.627$.

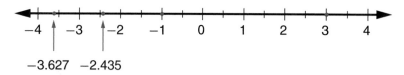

−3.627   −2.435

Absolute Value

The **absolute value** of a number is the distance between the number and 0 on the number line. The absolute value of any number is either positive or 0. The absolute value of a positive number is the number itself; for example, the absolute value of 4.5 is 4.5. The absolute value of a negative number is the opposite of the number; for example, the absolute value of −6 is 6. The absolute value of 0 is 0.

Note

Absolute value is shown by vertical lines before and after a number.

absolute value of 4.5:
$|4.5| = 4.5$

absolute value of −6:
$|-6| = 6$

Addition and Subtraction of Positive and Negative Numbers

Note

When a negative number follows an operation symbol, the negative number appears in parentheses. For example, $8 + (-7)$ tells you to add 8 and negative 7.

One way to add positive and negative numbers is to imagine walking along a number line.

◆ The first number tells you where to start.

◆ The operation sign ($+$ or $-$) tells you which way to face:
   $+$ means face the positive end of the number line.
   $-$ means face the negative end of the number line.

◆ If the second number is negative (has a $-$ sign), walk backward. Otherwise, walk forward.

◆ The second number tells you how many steps to take.

◆ The number where you end is the answer.

Example   Find $-3 + 5$.

Start at -3.
The operation sign is $+$, so face the positive direction.
The second number is positive, so walk forward 5 steps.
You will end at 2.

So, $-3 + 5 = 2$.

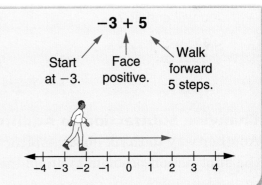

$-3 + 5$

Start at -3.   Face positive.   Walk forward 5 steps.

Example   Find $-1 + (-4)$.

Start at -1.
The operation sign is $+$, so face the positive direction.
The second number is negative, so walk backward 4 steps.
You will end at -5.

So, $-1 + (-4) = -5$.

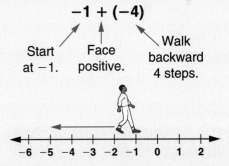

$-1 + (-4)$

Start at -1.   Face positive.   Walk backward 4 steps.

Check Your Understanding

Add.

1. $4 + (-5)$ **2.** $-4 + (-5)$ **3.** $-3 + 8$ **4.** $-3 + \frac{3}{2}$

Check your answers on page 417.

Example Find $1 - 4$.

Start at 1.
The operation sign is −, so face the negative direction.
The second number is positive, so walk forward 4 steps.
You will end at −3.

So, $1 - 4 = -3$.

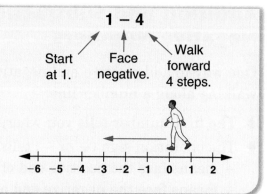

Example Find $-2 - (-4)$.

Start at −2.
The operation sign is −, so face the negative direction.
The second number is negative, so walk backward 4 steps.
You will end at 2.

So, $-2 - (-4) = 2$.

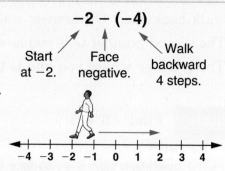

Changing Subtraction to Addition

Another way to think about subtraction of numbers is to change the subtraction problem to an addition problem.

For any numbers a and b, $a - b = a +$ (the opposite of b),
or $a - b = a + (-b)$.

Examples $1 - 4 = 1 + (-4) = -3$ $3 - (-4) = 3 + 4 = 7$

   $-2 - (-4) = -2 + 4 = 2$ $-3 - 4 = -3 + (-4) = -7$

Check Your Understanding

Subtract.

1. $-4 - 2$ 2. $-5 - (-8)$ 3. $4 - (-3)$ 4. $2 - 6$

5. $8 - (-25)$ 6. $5.3 - 5.7$ 7. $-14 - (-14)$ 8. $-\frac{1}{2} - \frac{3}{2}$

Check your answers on page 417.

Multiplication and Division of Positive and Negative Numbers

To multiply or divide two numbers, calculate the answer as if both numbers were positive. Then use one of the following rules to decide whether the answer is a positive or a negative number.

♦ If the signs on both numbers match, the result is a positive number.

♦ If the signs on both numbers do not match, the result is a negative number.

Study the examples below.

	Multiplication	**Division**	**Result**
Signs match • both positive • both negative	$8 * 3 = 24$ $-2 * (-3) = 6$	$51 \div 17 = 3$ $-81 \div (-3) = 27$	positive
Signs don't match • one positive and one negative	$12 * (-6) = -72$ $-25 * 8 = -200$	$-75 \div 25 = -3$ $10 \div (-2) = -5$	negative

One way to understand why the above rules are correct is to look at patterns in products of positive and negative numbers.

The first table begins with a positive number times a positive number. Each time the second factor gets smaller, the product is reduced by 5. When the second factor is negative, the product must be negative to keep the pattern going.

The second table begins with a negative number times a positive number. Each time the second factor gets smaller, the product is increased by 5. When both factors are negative, the product must be positive to keep the pattern going.

a	b	$a * b$
5	2	10
5	1	5
5	0	0
5	−1	−5
5	−2	−10

a	b	$a * b$
−5	2	−10
−5	1	−5
−5	0	0
−5	−1	5
−5	−2	10

Check Your Understanding

Multiply or divide.

1. $-7 * 9$
2. $10 * (-32)$
3. $9 * (-7)$
4. $48 \div (-12)$
5. $-42 \div 6$
6. $-100 \div (-25)$

Check your answers on page 417.

Multiplication and Division with Zero

Multiplication with Zero: $a * 0 = 0$

If one (or more) of the factors in a multiplication problem is 0, the product is 0.

Examples $-38 * 0 = 0$ $5 * 12 * 0 * 4 = 0$ $0 * 0 = 0$

Division of Zero: $0 / a = 0$

When 0 is divided by any number (except 0), the answer is 0.

To understand this, think about multiplication. Any division problem can be rewritten as a multiplication problem with a missing factor. And here the missing factor must be 0.

Division Problem	Multiplication Problem	Missing Factor
$0 / 8 = \square$	$8 * \square = 0$	0
$0 / 75 = \square$	$75 * \square = 0$	0
$0 / a = \square$ (if $a \neq 0$)	$a * \square = 0$	0

Division by Zero: $a / 0 = ?$

Division by 0 is not allowed.

Again, thinking about multiplication can help you understand this fact. Any division problem can be rewritten as a multiplication problem with a missing factor.

Division Problem	Multiplication Problem	Missing Factor
$8 / 0 = \square$	$0 * \square = 8$	no solution
$75 / 0 = \square$	$0 * \square = 75$	no solution
$a / 0 = \square$ (if $a \neq 0$)	$0 * \square = a$	no solution

Any number multiplied by 0 is 0, so there are no solutions to the multiplication problems above. This means *none* of the division problems have answers either. This is why we say division by 0 is not allowed: There is no answer.

Did You Know?

The problem $0 / 0 = \square$ is interesting. It can be rewritten as the multiplication problem $0 * \square = 0$.

Since any number at all (5, 73, -245, $\frac{1}{2}$, and so on) will make $0 * \square = 0$ true, there are too many answers. So, people usually say $0 / 0$ is not allowed.

If you study calculus in high school or college, you may learn about special cases when $0 / 0$ is allowed, but for elementary mathematics, division of 0 by 0 is not allowed.

Check Your Understanding

Multiply or divide.

1. $0 * (-1,234)$ **2.** $4 * 12 * 10 * 0$ **3.** $0 / 100$ **4.** $32 / 0$

Check your answers on page 418.

Different Types of Numbers

Counting is almost as old as the human race and has been used in some form by every human society. Long ago, people found that the **counting numbers** (1, 2, 3, and so on) did not meet all their needs.

♦ Counting numbers cannot be used to express measures between two consecutive whole numbers, such as $2\frac{1}{2}$ inches and 1.6 kilometers.

♦ With the counting numbers, division problems such as 8/5 and 3/7 do not have an answer.

Fractions were invented to meet these needs. Fractions can also be renamed as decimals and percents. And most of the numbers you have seen, such as $\frac{1}{2}$, $5\frac{1}{6}$, 1.23, and 25%, are either fractions or can be renamed as fractions. With the invention of fractions, it became possible to express rates and ratios, to name many more points on the number line, and to solve any division problem involving whole numbers (except division by 0).

> **Note**
>
> Every whole number can be renamed as a fraction. For example, 0 can be written as $\frac{0}{1}$ and 8 can be written as $\frac{8}{1}$.

However, even fractions did not meet every need. For example, problems such as $5 - 7$ and $2\frac{3}{4} - 5\frac{1}{4}$ have answers that are less than 0 and cannot be named as fractions. (Fractions, by the way they are defined, can never be less than 0). This led to the invention of **negative numbers.** Negative numbers are numbers that are less than 0. The numbers $-\frac{1}{4}$, -3.25, and -100 are negative numbers. The number -3 is read as "negative 3."

> **Note**
>
> Since every whole number can be renamed as a fraction, every negative whole number can be renamed as a negative fraction. For example, $-7 = -\frac{7}{1}$.

Negative numbers serve several purposes:

♦ to name locations such as temperatures below zero on a thermometer and depths below sea level

♦ to show changes such as yards lost in a football game and decreases in weight

♦ to extend the number line to the left of zero

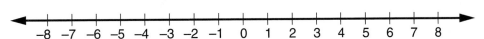

♦ to calculate answers to many subtraction problems

The **opposite** of every positive number is a negative number. And the opposite of every negative number is a positive number. The diagram below shows this relationship. The number 0 is neither positive nor negative. 0 is its own opposite.

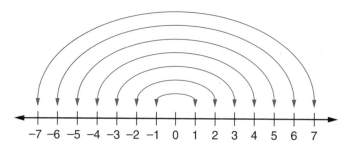

A Summary of Different Types of Numbers

The **counting numbers** are 1, 2, 3, and so on.

The **whole numbers** are 0, 1, 2, 3, and so on. The whole numbers include all the counting numbers, together with 0.

The **integers** are 0, 1, −1, 2, −2, 3, −3, and so on. So the integers include all the whole numbers, together with the opposites of all the whole numbers.

The **fractions** are all numbers written as $\frac{a}{b}$, where a and b can be any whole numbers ($b \neq 0$).

♦ Every whole number can be renamed as a fraction. For example, 5 can be renamed as $\frac{5}{1}$ and 0 can be renamed as $\frac{0}{1}$.

♦ Every fraction can be renamed as a decimal and as a percent. For example, $\frac{3}{4}$ can be renamed as 0.75 and as 75%, and $\frac{9}{8}$ can be renamed as 1.125 and as 112.5%.

The **rational numbers** are all numbers that are written or can be renamed as fractions or their opposites.

♦ Counting numbers, whole numbers, integers, fractions, and the opposites of these numbers are all rational numbers.

♦ Every number shown on the Probability Meter at the right is a rational number.

There are other numbers that are called **irrational numbers.** Some of these, like the number π, you have used before. An irrational number cannot be renamed as a fraction or its opposite. You will learn more about irrational numbers when you study algebra.

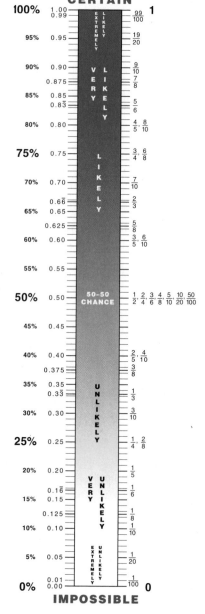

Notation

Any fraction can always be renamed as a decimal. Sometimes the decimal will end after a certain number of places. Decimals that end are called **terminating decimals.** Other times the decimal will have one or more digits that repeat in a pattern forever. These are called **repeating decimals.** If the decimal name for a fraction is not a terminating decimal, then it is a repeating decimal.

The *rational numbers* are all numbers that are written or can be renamed as fractions or their opposites. Just as for fractions, the decimal name for a rational number is either terminating or repeating.

> Every rational number (including every fraction) can be renamed as either a terminating or a repeating decimal.

Examples $\frac{5}{8} = 0.625$ $\frac{1}{3} = 0.\overline{3} = 0.3333...$

$\frac{1}{7} = 0.\overline{142857} = 0.142857142857142857142857142857142857...$

An *irrational number* can be written as a decimal, but not as a terminating or repeating decimal. An irrational number cannot be renamed as a fraction or the opposite of a fraction.

> An irrational number can't be renamed as a terminating or a repeating decimal.

Examples $\pi = 3.141592653...$ The decimal continues without a repeating pattern.

$\sqrt{2} = 1.414213562...$ The decimal continues without a repeating pattern.

$1.010010001...$ There is a pattern in the decimal, but it does not repeat.

Check Your Understanding

Refer to The Real Number Line on page 102 to answer these questions.

1. Which counting numbers are labeled on the number line?
2. Which integers are labeled on the number line?
3. Can 0.2 be written as a fraction?
4. Is 1.333... a rational number?
5. Is 10^{-1} a positive rational number?
6. Is $\sqrt{16}$ a rational number?
7. Which rational numbers between 1 and 2 are labeled?
8. Can 1.1666... be written as a fraction?
9. Which numbers labeled on the number line cannot be written as either terminating or repeating decimals?

Check your answers on page 418.

The Real Number Line

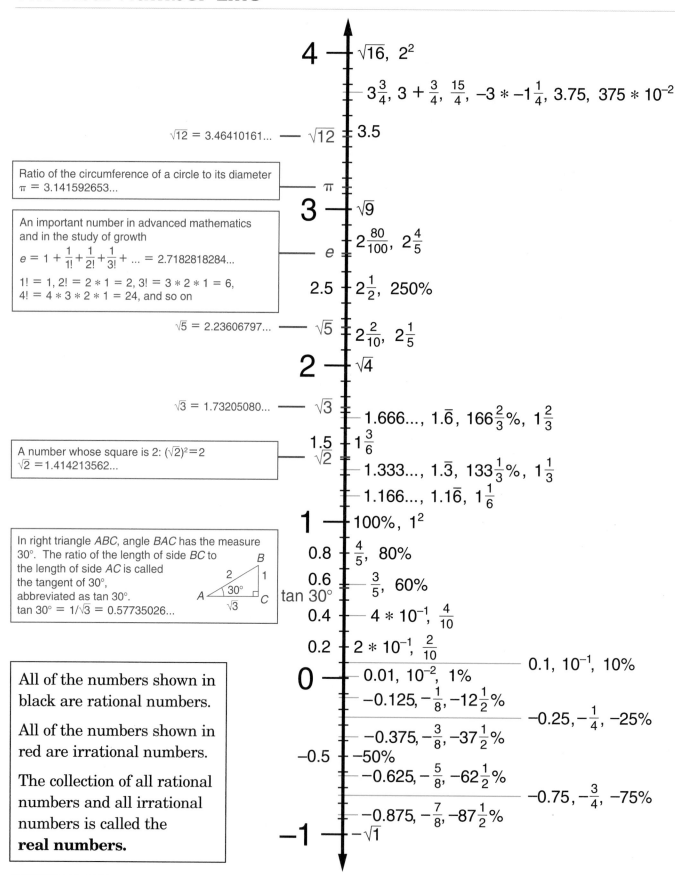

$\sqrt{12} = 3.46410161\ldots$ — $\sqrt{12}$

4 — $\sqrt{16}$, 2^2

$3\frac{3}{4}$, $3 + \frac{3}{4}$, $\frac{15}{4}$, $-3 * -1\frac{1}{4}$, 3.75, $375 * 10^{-2}$

3.5

Ratio of the circumference of a circle to its diameter
$\pi = 3.141592653\ldots$ — π

3 — $\sqrt{9}$

An important number in advanced mathematics and in the study of growth
$e = 1 + \frac{1}{1!} + \frac{1}{2!} + \frac{1}{3!} + \ldots = 2.7182818284\ldots$ — e

$1! = 1$, $2! = 2 * 1 = 2$, $3! = 3 * 2 * 1 = 6$,
$4! = 4 * 3 * 2 * 1 = 24$, and so on

$2\frac{80}{100}$, $2\frac{4}{5}$

2.5 — $2\frac{1}{2}$, 250%

$\sqrt{5} = 2.23606797\ldots$ — $\sqrt{5}$ — $2\frac{2}{10}$, $2\frac{1}{5}$

2 — $\sqrt{4}$

$\sqrt{3} = 1.73205080\ldots$ — $\sqrt{3}$

1.666..., $1.\overline{6}$, $166\frac{2}{3}\%$, $1\frac{2}{3}$

1.5 — $1\frac{3}{6}$
$\sqrt{2}$

A number whose square is 2: $(\sqrt{2})^2 = 2$
$\sqrt{2} = 1.414213562\ldots$

1.333..., $1.\overline{3}$, $133\frac{1}{3}\%$, $1\frac{1}{3}$

1.166..., $1.1\overline{6}$, $1\frac{1}{6}$

1 — 100%, 1^2

In right triangle ABC, angle BAC has the measure 30°. The ratio of the length of side BC to the length of side AC is called the tangent of 30°, abbreviated as tan 30°.
tan 30° = $1/\sqrt{3}$ = 0.57735026...

B
2   1
A 30° C
$\sqrt{3}$
tan 30°

0.8 — $\frac{4}{5}$, 80%

0.6 — $\frac{3}{5}$, 60%

0.4 — $4 * 10^{-1}$, $\frac{4}{10}$

0.2 — $2 * 10^{-1}$, $\frac{2}{10}$

0.1, 10^{-1}, 10%

0 — 0.01, 10^{-2}, 1%

-0.125, $-\frac{1}{8}$, $-12\frac{1}{2}\%$

-0.25, $-\frac{1}{4}$, -25%

-0.375, $-\frac{3}{8}$, $-37\frac{1}{2}\%$

−0.5 — −50%

-0.625, $-\frac{5}{8}$, $-62\frac{1}{2}\%$

-0.75, $-\frac{3}{4}$, -75%

-0.875, $-\frac{7}{8}$, $-87\frac{1}{2}\%$

−1 — $-\sqrt{1}$

All of the numbers shown in black are rational numbers.

All of the numbers shown in red are irrational numbers.

The collection of all rational numbers and all irrational numbers is called the **real numbers.**

General Patterns and Special Cases

Many rules involving numbers can be described with the help of variables, such as n or $\square$. For example, the rule "The sum of a number and its opposite is 0" can be expressed as $a + (-a) = 0$, where a stands for any number. This is sometimes called a **general pattern**. A **special case** of this general pattern can be given by replacing the variable a with any number. For example, $5 + (-5) = 0$ and $27\frac{1}{2} + (-27\frac{1}{2}) = 0$ are special cases of the general pattern $a + (-a) = 0$.

Did You Know?

The general pattern below may be used to add all of the counting numbers from 1 up to, and including, n.

$1 + 2 + 3 + ... + n = n * (n + 1) / 2$

In the special case where $n = 100$, the sum of the numbers 1 through 100 equals $100 * 101 / 2$, or 5,050.

Example   Express the following rule with variables:

"The square of a number is the number multiplied by itself."
Give three special cases of the general pattern.

General pattern: $n^2 = n * n$

Special cases: For $n = 3$: $3^2 = 3 * 3$

For $n = -7$: $(-7)^2 = -7 * (-7)$

For $n = \frac{2}{3}$: $\left(\frac{2}{3}\right)^2 = \frac{2}{3} * \frac{2}{3}$

Example   Use variables to describe the general pattern for the following special cases.

Special cases: $8 / 8 = 1$ $0.5 / 0.5 = 1$ $2\frac{3}{4} / 2\frac{3}{4} = 1$

Step 1: Write everything that is the same for all the special cases. Use blanks for the parts that change.   _____ / _____ $= 1$

Step 2: Fill in the blanks. Use a variable for the number that varies. (Use 2 different variables if there are 2 different numbers that vary.) $\underline{b} / \underline{b} = 1$

The general pattern is $b / b = 1$. (Division by 0 is not allowed, so b may not equal 0.)

Check Your Understanding

Write three special cases for each general pattern.

1. $2 * r + r = 3 * r$

2. $b + (b + 1) + (b + 2) = 3 * (b + 1)$

Use variables to express the general pattern for these special cases.

3. $3 + 6 = 6 + 3$

$0.25 + 0.75 = 0.75 + 0.25$

$\frac{2}{4} + \frac{1}{5} = \frac{1}{5} + \frac{2}{4}$

4. $\frac{3}{2} * \frac{2}{3} = 1$

$\frac{7}{1} * \frac{1}{7} = 1$

$\frac{3}{5} * \frac{5}{3} = 1$

Check your answers on page 418.

Properties of Numbers

The following properties are true for all numbers. The variables a, b, c, and d stand for any numbers (except 0 if the variable stands for a divisor).

Properties	Examples
Binary Operations Property When any two numbers are added, subtracted, multiplied, or divided, the result is a single number. $a + b$, $a - b$, $a * b$, and $a \div b$ are equal to single numbers.	$5 + 7 = 12$ $-3 - \frac{8}{3} = -5\frac{2}{3}$ $0.5 * (-4) = -2$ $2\frac{3}{5} \div \frac{8}{3} = \frac{39}{40}$
Commutative Property The sum or product of two numbers is the same, regardless of the order of the numbers. $a + b = b + a$ $a * b = b * a$	$7 + 8 = 8 + 7 = 15$ $-5 * (-6) = -6 * (-5) = 30$ $\frac{3}{4} * (-\frac{4}{5}) = -\frac{4}{5} * \frac{3}{4} = -\frac{3}{5}$
Associative Property The sum or product of three or more numbers is the same, regardless of how the numbers are grouped. $a + (b + c) = (a + b) + c$ $a * (b * c) = (a * b) * c$	$(7 + 5) + 8 = 7 + (5 + 8)$ $12 + 8 = 7 + 13$ $20 = 20$ $2\frac{1}{2} * (2 * 3) = (2\frac{1}{2} * 2) * 3$ $2\frac{1}{2} * 6 = 5 * 3$ $15 = 15$
Distributive Property When a number a is multiplied by the sum or difference of two other numbers, the number a is "distributed" to each of these numbers. $a * (b + c) = (a * b) + (a * c)$ $a * (b - c) = (a * b) - (a * c)$	$5 * (8 + 2) = (5 * 8) + (5 * 2)$ $5 * 10 = 40 + 10$ $50 = 50$ $-2 * (8 - 3) = (-2 * 8) - (-2 * 3)$ $-2 * 5 = -16 - (-6)$ $-10 = -10$
Addition Property of Zero The sum of any number and 0 is equal to the original number. $a + 0 = 0 + a = a$	$5.37 + 0 = 5.37$ $0 + (-6) = -6$
Multiplication Property of One The product of any number and 1 is equal to the original number. $a * 1 = 1 * a = a$	$\frac{2}{3} * 1 = \frac{2}{3}$ $1 * 19 = 19$

Properties	**Examples**
Opposites Property The opposite of a number, a, is usually written $-a$. In *Everyday Mathematics,* we sometimes write OPP(a) for the opposite of a. If a is a positive number, then OPP(a) is a negative number. If a is a negative number, then OPP(a) is a positive number. If $a = 0$, then OPP(a) = 0. Zero is the only number that is its own opposite.	OPP(8) $= -8$ OPP($-\frac{3}{4}$) $= \frac{3}{4}$ OPP(-7) $= 7$ OPP(0) $= 0$
Opposite of Opposites Property The opposite of the opposite of a number is equal to the original number. OPP(OPP(a)) = OPP($-a$) = a	OPP(OPP($\frac{2}{3}$)) = OPP($-\frac{2}{3}$) = $\frac{2}{3}$ OPP(OPP(-9)) = OPP(9) = -9
Sum of Opposites Property The sum of any number and its opposite is 0. $a + (-a) = (-a) + a = 0$	$15 + (-15) = 0$ $-2 + 2 = 0$
Multiplication of Reciprocals Property The product of any number and its reciprocal is 1. $a * \frac{1}{a} = \frac{1}{a} * a = 1$ $\frac{a}{b} * \frac{b}{a} = \frac{b}{a} * \frac{a}{b} = 1$	$20 * \frac{1}{20} = 1$ $\frac{2}{5} * \frac{5}{2} = 1$
Addition Property of Positive and Negative Numbers The sum of two positive numbers is a positive number. The sum of two negative numbers is the opposite of the sum of the "number parts" of the addends. The sum of a positive number and a negative number may be positive or negative or zero.	$7 + 8 = 15$ $-7 + (-8) = (OPP)(7 + 8) = -15$ $\frac{3}{4} + (-\frac{1}{4}) = \frac{1}{2}$ $-2.5 + 1.5 = -1$
Multiplication Property of Positive and Negative Numbers The product of two positive numbers or two negative numbers is a positive number. The product of a positive number and a negative number is a negative number.	$6 * 3 = 18$ $-6 * (-3) = 18$ $\frac{1}{3} * (-\frac{4}{5}) = -\frac{4}{15}$
Subtraction and Division Properties All subtraction problems can be solved by addition. All division problems can be solved by multiplication. $a - b = a + (-b)$ $a \div b = a * \frac{1}{b}$ $\frac{a}{b} = a * \frac{1}{b}$	$15 - 7 = 15 + (-7) = 8$ $-9 - (-6) = -9 + 6 = -3$ $25 \div (-5) = 25 * (-\frac{1}{5}) = -\frac{25}{5} = -5$ $\frac{12}{4} = 12 * \frac{1}{4} = 3$

Properties	Examples

Equivalent Fractions Property

If the numerator and denominator of a fraction are multiplied or divided by the same number, the resulting fraction is equal to the original fraction.

$$\frac{a}{b} = \frac{a * c}{b * c} \qquad\qquad \frac{a}{b} = \frac{a \div c}{b \div c}$$

$$\frac{2}{3} = \frac{2 * 5}{3 * 5} = \frac{10}{15}$$

$$\frac{6}{8} = \frac{6 \div 2}{8 \div 2} = \frac{3}{4}$$

Addition and Subtraction of Fractions Properties

The sum or difference of fractions with like denominators is the sum or difference of the numerators over the denominator.

$$\frac{a}{c} + \frac{b}{c} = \frac{a + b}{c}$$

$$\frac{a}{c} - \frac{b}{c} = \frac{a - b}{c}$$

To add or subtract fractions with unlike denominators, rename the fractions so that they have a common denominator.

$$\frac{a}{b} + \frac{c}{d} = \frac{ad + bc}{bd}$$

$$\frac{a}{b} - \frac{c}{d} = \frac{ad - bc}{bd}$$

$$\frac{3}{5} + \frac{1}{5} = \frac{3 + 1}{5} = \frac{4}{5}$$

$$\frac{5}{6} - \frac{1}{6} = \frac{5 - 1}{6} = \frac{4}{6} = \frac{2}{3}$$

$$\frac{2}{3} + \frac{1}{5} = \frac{10}{15} + \frac{3}{15} = \frac{10 + 3}{15} = \frac{13}{15}$$

$$\frac{2}{3} - \frac{1}{4} = \frac{8}{12} - \frac{3}{12} = \frac{8 - 3}{12} = \frac{5}{12}$$

Multiplication of Fractions Property

The product of two fractions is the product of the numerators over the product of the denominators.

$$\frac{a}{b} * \frac{c}{d} = \frac{a * c}{b * d}$$

$$\frac{5}{8} * \frac{3}{4} = \frac{5 * 3}{8 * 4} = \frac{15}{32}$$

Division of Fractions Property

The quotient of two fractions is the product of the dividend and the reciprocal of the divisor.

$$\frac{a}{b} \div \frac{c}{d} = \frac{a}{b} * \frac{d}{c} = \frac{a * d}{b * c}$$

$$9 \div \frac{2}{3} = 9 * \frac{3}{2} = \frac{27}{2}, \text{ or } 13\frac{1}{2}$$

$$\frac{5}{6} \div \frac{1}{4} = \frac{5}{6} * \frac{4}{1} = \frac{20}{6}, \text{ or } 3\frac{1}{3}$$

Powers of a Number Property

If a is any number and b is a positive whole number, then a^b is the product of a used as a factor b times.

$$a^b = \underbrace{a * a * a * \ldots * a}_{b \text{ factors}}$$

a^0 is equal to 1.

If a is any nonzero number and b is a positive whole number, then a^{-b} is 1 divided by the product of a used as a factor b times.

$$a^{-b} = \frac{1}{a^b} = \underbrace{\frac{1}{a * a * a * \ldots * a}}_{b \text{ factors}}$$

$$5^2 = 5 * 5 = 25$$

$$\left(\frac{2}{3}\right)^4 = \frac{2}{3} * \frac{2}{3} * \frac{2}{3} * \frac{2}{3} = \frac{16}{81}$$

$$4^0 = 1$$

$$3^{-2} = \frac{1}{3^2} = \frac{1}{3 * 3} = \frac{1}{9}$$

Rates, Ratios, and Proportions

Rates, Ratios, and Proportions

Many fractions, such as $\frac{3}{8}$ of a pizza, name parts of wholes. Other fractions, such as $\frac{1}{2}$ inch, are used in measurement. In working with fractions, it's important to keep in mind what the ONE, or whole, is; $\frac{3}{4}$ of a mile is much longer than $\frac{3}{4}$ of an inch.

Some fractions compare two different amounts, where one amount is not part of the other. For example, a store might sell apples at 3 apples for 89 cents. Or a car's gas mileage might be 147 miles per 7 gallons. These facts can be written as fractions: $\frac{3 \text{ apples}}{89¢}$ and $\frac{147 \text{ miles}}{7 \text{ gallons}}$. These fractions do not name parts of wholes. The apples are *not* part of the money; the miles are *not* part of the gallons of gasoline.

Fractions like $\frac{147 \text{ miles}}{7 \text{ gallons}}$ show rates. A **rate** tells how many of one thing there are for a certain number of another thing. Rates often contain the word **per,** meaning *for each* or *for every*.

Example   Alan rode his bicycle 12 miles in 1 hour. His rate was 12 miles per hour. This rate describes the distance he traveled and the time it took. The rate *12 miles per hour* is written as "12 mph." The fraction for this rate is $\frac{12 \text{ miles}}{1 \text{ hour}}$.

Ratios are like rates, but they compare two amounts that have the same unit.

Example   The ratio of the length of the side of a square to the perimeter of the square is 1 to 4, or $\frac{1}{4}$. A square with a side 1 inch long has a perimeter of 4 inches. A square with a side 1 yard long has a perimeter of 4 yards.

A **proportion** is a number model which states that two fractions are equal.

Examples   Write a proportion for each situation.

1. Alan's speed is 12 miles per hour. At the same speed, he can travel 36 miles in 3 hours.

$$\frac{12 \text{ miles}}{1 \text{ hour}} = \frac{36 \text{ miles}}{3 \text{ hours}}$$

2. Since the ratio of the length of a side of a square to the perimeter is 1 to 4, a square with a side 6 inches long has a perimeter of 24 inches.

$$\frac{1 \text{ inch}}{4 \text{ inches}} = \frac{6 \text{ inches}}{24 \text{ inches}}$$

Rates and Rate Tables

A rate tells how many of one thing there are for a certain number of another thing. Rates often contain the word **per** meaning *for each,* or *for every.*

Some rates use special abbreviations. For example, *miles per hour* can be written as "mph" and *miles per gallon* can be written as "mpg."

Examples

Cynthia jogged 5 miles in 1 hour. She traveled at a rate of 5 miles per hour, or 5 mph. This rate describes the distance Cynthia traveled and the time it took her.

The table shows other rates.

typing speed	50 words per minute	$\frac{50 \text{ words}}{1 \text{ minute}}$
price	$14\frac{1}{2}$ cents per ounce	$\frac{14\frac{1}{2} \text{¢}}{1 \text{ ounce}}$
scoring average	17 points per game	$\frac{17 \text{ points}}{1 \text{ game}}$
exchange rate	0.78 Euros for each U.S. dollar	$\frac{0.78 \text{ Euros}}{1 \text{ U.S. dollar}}$

Rate information can be organized in a **rate table.**

Example   Make a rate table for the statement, "A computer printer prints 4 pages per minute."

The table shows that if a printer prints 4 pages per minute, it will print 8 pages in 2 minutes, 12 pages in 3 minutes, and so on.

pages	4	8	12	16	20	24	28
minutes	1	2	3	4	5	6	7

Rates are often written with a slash (/) or as a fraction. The slash and fraction bar can be read as "per" or "for each."

Rate	Slash	Fraction
per-hour rate: 65 miles per hour	65 miles/hour	$\frac{65 \text{ miles}}{1 \text{ hour}}$
per-candy rate: $\frac{1}{2}$ cent per candy	$\frac{1}{2}$ cent/candy	$\frac{\frac{1}{2} \text{ cent}}{1 \text{ candy}}$

Check Your Understanding

Write each rate with a slash and as a fraction. Make a rate table for each rate.

1. Rebecca earns $5 per hour helping her neighbor in the garden.

2. Todd can type 45 words per minute.

Check your answers on page 418.

Solving Rate Problems

The units in a rate table are written to the left of each row.
The units are "pages" and "minutes" in the table shown here.

The fractions formed by the two numbers in each column
are equivalent fractions. In the table here, $\frac{4}{1} = \frac{8}{2} = \frac{12}{3}$,
and so on. This is why the rates in a rate table are called
equivalent rates.

pages	4	8	12	16
minutes	1	2	3	4

In most cases, the units are included when rates are written.
Sometimes rates are written as fractions without units, to save
time in working problems. For example, you may write $\frac{16}{4}$
instead of $\frac{16 \text{ pages}}{4 \text{ minutes}}$. But remember to include the units when
reporting the answer to a problem.

In many rate problems, one rate is given and you need to find an
equivalent rate. These problems can be solved in several ways.

Example   Beth's car can travel 35 miles on 1 gallon of
gasoline. How far can it travel on 7 gallons?

First, set up a rate table and enter what you know and what
you want to find.

miles	35					?
gallons	1					7

Next, work from what you know to what you need to find.
In this case, by doubling, you can find how far the car could
travel on 2 gallons, 4 gallons, and 8 gallons of gasoline.

miles	35	70	140	280			?
gallons	1	2	4	8			7

There are two different ways to use the rate table.

• Add the distances for 1 gallon, 2 gallons, and 4 gallons
  (a total of 7 gallons):
  35 miles + 70 miles + 140 miles = 245 miles.

• Subtract the distance for 1 gallon from the distance for
  8 gallons: 280 miles − 35 miles = 245 miles.

So, Beth's car can travel 245 miles on 7 gallons of gas.

Example Krystal receives an allowance of $20 for 4 weeks. At this rate, how much does she receive for 10 weeks?

First, set up a table. Enter what you know and what you want to find.

allowance	$20					?
weeks	4					10

Next, work from what you know to what you want to know. By halving $20, you can find how much Krystal gets for 2 weeks.

allowance	$20	$10	$5	$40			?
weeks	4	2	1	8			10

By halving again you can find what she gets for 1 week. Then, by doubling $20 for 4 weeks, you find that Krystal will get $40 for 8 weeks.

Since 10 weeks is 8 weeks plus 2 weeks, Krystal will get $40 + $10, or $50, for 10 weeks.

Using Per-Unit Rates

A **per-unit rate** is a rate that tells how many of something there are for a single one of another thing. The fraction for a per-unit rate has a 1 in the denominator.

Examples
$$\frac{\$2.00}{1 \text{ gallon}} \qquad \frac{50 \text{ miles}}{1 \text{ hour}} \qquad \frac{36 \text{ inches}}{1 \text{ yard}}$$

A rate problem may be easier to solve if you can find a per-unit rate that is equivalent to the rate given in the problem. Any rate can be renamed as a per-unit rate by dividing the numerator and the denominator by the denominator.

Examples Find equivalent per-unit rates.

$$\frac{60 \text{ pages}}{3 \text{ hours}} = \frac{60 \text{ pages} \div 3}{3 \text{ hours} \div 3} \qquad \frac{300 \text{ miles}}{5 \text{ hours}} = \frac{300 \text{ miles} \div 5}{5 \text{ hours} \div 5}$$
$$= \frac{20 \text{ pages}}{1 \text{ hour}} \qquad\qquad\qquad = \frac{60 \text{ miles}}{1 \text{ hour}}$$

Did You Know?

The heartbeat rate of a normal human being changes from birth to old age.

Age	Heartbeats per Minute
newborn	120 to 150
7-year-old	about 90
adult	60 to 80

A rate problem can be solved in two steps using a per-unit rate:

Step 1: Find a per-unit rate equivalent to the rate given in the problem.

Step 2: Use the per-unit rate to solve the problem.

one hundred eleven

The following example was also solved using a rate table on page 111. Here it is solved using a per-unit rate.

Example Krystal receives an allowance of $20 for 4 weeks. At this rate, how much does she receive for 10 weeks?

Step 1: Find a per-unit rate equivalent to the given rate.

$$\frac{\$20}{4 \text{ weeks}} = \frac{\$20 \div 4}{4 \text{ weeks} \div 4} = \frac{\$5}{1 \text{ week}}$$

Step 2: Use the per-unit rate to solve the problem. If Krystal gets $5 for 1 week, then she gets 10 * $5, or $50, for 10 weeks.

Krystal receives $50 for 10 weeks.

Example A carton of 12 eggs costs $1.44. At this rate, how much do 8 eggs cost?

Step 1: Find a per-unit rate equivalent to the given rate.

$$\frac{\$1.44}{12 \text{ eggs}} = \frac{\$1.44 \div 12}{12 \text{ eggs} \div 12} = \frac{12¢}{1 \text{ egg}}$$

Step 2: Use the per-unit rate to solve the problem. If 1 egg costs 12 cents, then 8 eggs cost 8 * 12 cents, or 96 cents.

So, 8 eggs cost 96 cents.

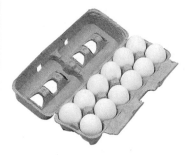

Check Your Understanding

Solve.

1. There are 4 quarts in 1 gallon. How many quarts are there in 6 gallons?

2. Andy baby-sat for 6 hours. He was paid $42. How much did he earn per hour?

3. Judy saved $360 last year. How much did she save per month?

4. Jean swam 15 laps in 9 minutes. At this rate, how many laps would she swim in 15 minutes?

Check your answers on page 418.

Proportions

A **proportion** is a number sentence which states that two fractions are equivalent (equal).

Examples $\frac{1}{2} = \frac{3}{6}$ $\frac{3}{4} = \frac{9}{12}$ $\frac{5}{8} = \frac{10}{16}$ $\frac{21}{9} = \frac{7}{3}$

If you know any three numbers in a proportion, you can find the fourth number. Finding a missing number in a proportion is called **solving the proportion.**

Solving Proportions Using Rules for Equivalent Fractions

One way to solve a proportion is to use the multiplication and division rules for equivalent fractions.

Example   Solve: $\frac{2}{3} = \frac{n}{15}$

To rename $\frac{2}{3}$ as an equivalent fraction with 15 in the denominator, multiply the numerator and the denominator of $\frac{2}{3}$ by 5.

$\frac{2}{3} = \frac{2 * 5}{3 * 5} = \frac{10}{15}$

So, $n = 10$.

Multiplication and division rules for equivalent fractions:

$$\frac{a}{b} = \frac{a * c}{b * c}$$

$$\frac{a}{b} = \frac{a \div c}{b \div c}$$

Example   Solve: $\frac{6}{15} = \frac{x}{5}$

To rename $\frac{6}{15}$ as an equivalent fraction with 5 in the denominator, divide the numerator and the denominator of $\frac{6}{15}$ by 3.

$\frac{6}{15} = \frac{6 \div 3}{15 \div 3} = \frac{2}{5}$

So, $x = 2$.

Example   Solve: $\frac{8}{b} = \frac{32}{20}$

To rename $\frac{32}{20}$ as an equivalent fraction with 8 in the numerator, divide the numerator and the denominator of $\frac{32}{20}$ by 4.

$\frac{32}{20} = \frac{32 \div 4}{20 \div 4} = \frac{8}{5}$

So, $b = 5$.

Solving Proportions by Cross Multiplication

Another way to solve proportions is by using cross multiplication.

In a proportion $\frac{a}{b} = \frac{c}{d}$, the products $a * d$ and $b * c$ are called the **cross products** of the proportion. For example, if $\frac{2}{3} = \frac{6}{9}$, then $2 * 9$ and $3 * 6$ are the cross products.

The cross products are found by multiplying the numerator of each fraction by the denominator of the other fraction.

The process of finding the cross products is called **cross multiplication.** The diagram below shows why this is a good name to describe finding the cross products.

$$2 * 9 = 18 \qquad\qquad 3 * 6 = 18$$

$$\frac{2}{3} = \frac{6}{9}$$

One arrow passes through 2 and 9: $2 * 9$ is one of the cross products.

One arrow passes through 3 and 6: $3 * 6$ is the other cross product.

Solving proportions by cross multiplication depends on the following key idea:

In a proportion, the cross products are equal.

We can use an example to show why the cross products in a proportion are equal.

Example   Show that the cross products in $\frac{2}{3} = \frac{8}{12}$ are equal.

Rename each fraction using a common denominator of $3 * 12 = 36$.

$$\frac{2}{3} = \frac{2 * 12}{3 * 12} = \frac{2 * 12}{36} \text{ and } \frac{8}{12} = \frac{8 * 3}{12 * 3} = \frac{8 * 3}{36}$$

Since $\frac{2}{3} = \frac{8}{12}$, the renamed fractions must also be equal.

So, $\frac{2 * 12}{36} = \frac{8 * 3}{36}$.

These fractions are equal and their denominators are equal. Therefore, the numerators must also be equal: $2 * 12 = 8 * 3$.

So, the cross products $2 * 12$ and $8 * 3$ are equal.

To solve a proportion using cross multiplication:

♦ Find the cross products and write an equation to show that they are equal.

♦ Solve the equation.

Example Solve $\frac{3}{4} = \frac{z}{20}$ by cross multiplication.

One cross product is 3 * 20, which equals 60.

The other cross product is 4 * z, which equals $z + z + z + z$ and can also be written as 4z.

$$3 * 20 = 60 \qquad\qquad 4 * z = 4z$$

$$\frac{3}{4} = \frac{z}{20}$$

Since the cross products are equal, 60 = 4z.

To solve the equation, divide both sides by 4: 60 / 4 = 4z / 4

$$15 = z$$

So, z = 15 and $\frac{3}{4} = \frac{15}{20}$.

Note

To solve an equation, you may add, subtract, multiply, or divide by any amount.

But you must always perform the same operation on both sides of the equal sign. (Multiplying by 0 and dividing by 0 are not allowed.)

Example Solve $\frac{60}{a} = \frac{5}{3}$ by cross multiplication.

One cross product is 60 * 3, which equals 180.

The other cross product is 5 * a, or $a + a + a + a + a$, or 5a.

The cross products are equal: 180 = 5a.

$$60 * 3 = 180 \qquad\qquad 5 * a = 5a$$

$$\frac{60}{a} = \frac{5}{3}$$

To solve the equation, divide both sides by 5: 180 / 5 = 5a / 5

$$36 = a$$

So, a = 36 and $\frac{60}{36} = \frac{5}{3}$.

Check Your Understanding

Solve the proportion. Use cross multiplication for problems 4–6.

1. $\frac{3}{2} = \frac{a}{12}$ 2. $\frac{2}{3} = \frac{t}{60}$ 3. $\frac{10}{3} = \frac{5}{m}$ 4. $\frac{4}{b} = \frac{3}{6}$ 5. $\frac{15}{20} = \frac{6}{x}$ 6. $\frac{8}{c} = \frac{12}{9}$

Check your answers on page 418.

Using Proportions to Solve Rate Problems

Proportions can be used to solve many rate problems. The first step is to write a proportion that fits the problem you want to solve. The second step is to solve the proportion.

Example Bill's new car can travel 35 miles on 1 gallon of gasoline. At this rate, how far can the car travel on 7 gallons?

Set up a simple rate table. This will help you write a correct proportion.

miles	35	n
gallons	1	7

The rates in a rate table are equivalent. So, you can write $\frac{35 \text{ miles}}{1 \text{ gallon}} = \frac{n \text{ miles}}{7 \text{ gallons}}$.

To save time, write this proportion without the units: $\frac{35}{1} = \frac{n}{7}$.

Solve for n.

Method 1: Find an equivalent fraction: $\frac{35}{1} = \frac{35 * 7}{1 * 7} = \frac{245}{7}$. So, $n = 245$.

Method 2: Use cross multiplication: The cross products are equal.
$35 * 7 = 1 * n$, or $245 = n$.

Bill's car can travel 245 miles on 7 gallons.

Example Henry typed 192 words in 4 minutes. At this rate, how long would it take him to type 288 words?

Set up a simple rate table. This will help you write a correct proportion.

words	192	288
minutes	4	n

The rates in a rate table are equivalent.

So, you can write the proportion $\frac{192 \text{ words}}{4 \text{ minutes}} = \frac{288 \text{ words}}{n \text{ minutes}}$. Or, simply write $\frac{192}{4} = \frac{288}{n}$.

The cross products are equal: $192 * n = 4 * 288$. So, $192n = 1,152$.

Divide both sides of the equation by 192: $192n / 192 = 1,152 / 192$. So, $n = 6$.

Henry would take 6 minutes to type 288 words.

Check Your Understanding

A marathon champion ran 26 miles in 2 hours and 15 minutes (135 minutes). About how long did it take him to cover 10 miles?

Check your answer on page 418.

Ratios

A **ratio** is a comparison of two counts or measures that have the same unit.

Ratios can be described in words, with a fraction, with a percent, and in other ways.

Examples   Susan weighs 90 pounds and her brother John weighs 120 pounds. Compare their weights.

In words: The ratio of Susan's weight to John's weight is 90 to 120.

With a fraction: This ratio may be written as the fraction $\frac{90}{120}$.
$\frac{90}{120} = \frac{9}{12} = \frac{3}{4}$ So, Susan weighs $\frac{3}{4}$ as much as John weighs.

With a percent: $\frac{90}{120} = \frac{3}{4} = 75\%$ So, Susan's weight is 75% of John's weight.

With a colon between the two amounts being compared: The ratio of Susan's weight to John's weight is 90 : 120, 9 : 12, or 3 : 4.

In a proportion: $\dfrac{\text{Susan's weight}}{\text{John's weight}} = \dfrac{90 \text{ pounds}}{120 \text{ pounds}}$

A ratio can compare amounts in either order. For the example here:

The ratio of John's weight to Susan's weight is 120 to 90.

This ratio may be written as a fraction and as a percent:
$\frac{120}{90} = \frac{4}{3} = 1\frac{1}{3} = 133\frac{1}{3}\%$.

John weighs $1\frac{1}{3}$ as much as Susan. John's weight is $133\frac{1}{3}\%$ of Susan's weight.

The ratio of John's weight to Susan's weight is 120 : 90, or 4 : 3.

Written as a proportion, $\dfrac{\text{John's weight}}{\text{Susan's weight}} = \dfrac{120 \text{ pounds}}{90 \text{ pounds}}$.

Ratios that can be named by equivalent fractions are called **equivalent ratios.**

The ratios 90 to 120, 9 to 12, and 3 to 4 are equivalent because $\frac{90}{120}$, $\frac{9}{12}$, and $\frac{3}{4}$ are equivalent fractions.

The ratios 120 to 90, 12 to 9, and 4 to 3 are equivalent because $\frac{120}{90}$, $\frac{12}{9}$, and $\frac{4}{3}$ are equivalent fractions.

Some ratios compare quantities that involve a whole and its parts.

> **Examples** In a class of 20 students, there are 12 girls and 8 boys.
>
> You can think of the 20 students as the **whole** and the 12 girls and 8 boys as **parts of the whole.**
>
> A **part-to-whole** ratio compares a part of the whole to the whole. Each of the statements "8 out of 20 students are boys" and "12 out of 20 students are girls," expresses a part-to-whole ratio.
>
> A **part-to-part** ratio compares a part of the whole to another part of the whole. The statement, "There are 12 girls for every 8 boys," expresses a part-to-part ratio.

Note

For the counts in this example, "girls" and "boys" and "students" are not different units.

There is only one unit here — students.

Girls are one type of student. Boys are another type of student.

Ratios can be expressed in a number of ways. For the example above, the ratio of girls to the total number of students can be written—

♦ In *words:* Twelve out of 20 students are girls.
  Twelve in 20 students are girls.
  There are 12 girls for every 20 students.
  The ratio of girls to all students is 12 to 20.

♦ With a *fraction:* $\frac{12}{20}$, or $\frac{3}{5}$, of the students are girls.

♦ With a *percent:* 60% of the students are girls.

♦ With a *colon* between the two numbers being compared:
  The ratio of girls to all students is 12 : 20 (12 to 20).

♦ In a *proportion:* $\frac{\text{number of girls}}{\text{number of students}} = \frac{12 \text{ girls}}{20 \text{ students}}$

For the example above, the part-to-part ratio is 12 girls to 8 boys. This ratio may be written as the fraction $\frac{12}{8}$.

Since $\frac{12}{8} = \frac{6}{4} = \frac{3}{2}$ are equivalent fractions, the ratios 12 to 8, 6 to 4, and 3 to 2 are equivalent ratios. If there are 12 girls for every 8 boys, you can also say that there are 6 girls for every 4 boys, or 3 girls for every 2 boys.

The ratio of girls to boys is 3 to 2.

Comparing Ratios

Writing ratios as percents can help you compare the ratios. For example, suppose that Esther got 14 correct answers out of 25 on her homework and 18 correct out of 30 on a test. Each ratio of correct answers to total answers can be written as a percent.

$$\text{homework: } \frac{14}{25} = \frac{56}{100} = 56\% \qquad \text{test: } \frac{18}{30} = \frac{6}{10} = 60\%$$

Once each ratio is renamed in percent form, it is easy to see that Esther did better on her test.

Every ratio can be converted to a ratio of some number to 1. These are called **n-to-1** ratios. One way to convert a ratio to an n-to-1 ratio is to divide both numbers in the ratio by the second number. For example, to convert the ratio of 3 girls to 2 boys to an n-to-1 ratio, divide both 3 and 2 by 2. The answer is an equivalent ratio: 1.5 girls to 1 boy. This means that there are 1.5 girls for every boy, or that there are 1.5 times as many girls as boys. When comparing ratios, n-to-1 ratios are useful.

Example A book is in a hardcover version and a paperback version. The hardcover version costs $25. The paperback version costs $10. Compare the costs.

The ratio of the hardcover cost to the paperback cost is 25 to 10, or $\frac{25}{10}$, or $\frac{5}{2}$.

To find the n-to-1 ratio, divide. $\frac{5 \div 2}{2 \div 2} = \frac{2.5}{1}$ $\frac{\text{hardcover cost}}{\text{paperback cost}} = \frac{25}{10} = \frac{5}{2} = \frac{2.5}{1}$

The hardcover version costs 2.5 times as much as the paperback version.

Rates versus Ratios

Ratios and rates are each used to compare two different amounts.

♦ Ratios and rates can each be written as fractions.

♦ Rates compare amounts that have *different units,* while ratios compare amounts that have the *same unit.*

♦ You must always mention both units when you name a rate. But a ratio is a "pure number," and there are no units to mention when you name a ratio.

Check Your Understanding

Last month, Mae Li received an allowance of $20. She spent $12 and saved the rest.
1. What is the ratio of the money she spent to her total allowance?
2. What is the ratio of the money she saved to the money she spent?
3. The money she spent is how many times the money she saved?
4. What percent of her allowance did she save?

Check your answers on page 418.

Using Proportions to Solve Ratio Problems

Proportions can be used to solve many ratio problems. The first step is to write a proportion that fits the problem you want to solve. The second step is to solve the proportion.

Example Jack is 70 inches tall. Alice is $\frac{3}{5}$ as tall as Jack. How tall is Alice?

Write a proportion using the ratio of Alice's height to Jack's height.

$\frac{\text{Alice's height (in.)}}{\text{Jack's height (in.)}} = \frac{3}{5}$

Substitute 70 for Jack's height in the proportion.
Let the variable a stand for Alice's height, in inches.

Then solve $\frac{a}{70} = \frac{3}{5}$ using cross multiplication.

$5 * a = 5a \qquad 70 * 3 = 210$

$$\frac{a}{70} = \frac{3}{5}$$

The cross products are equal: $5a = 210$.
To solve the equation, divide both sides by 5.
$5a / 5 = 210 / 5$ So, $a = 42$.
Alice is 42 inches tall.

Example In the 6th grade of the Ames School, the ratio of boys to girls is $4 : 7$. There are 32 boys. How many girls are there?

Write a proportion, using the ratio of boys to girls.

$\frac{\text{number of boys}}{\text{number of girls}} = \frac{4}{7}$

Substitute 32 for the number of boys in the proportion.
Let the variable g stand for the number of girls.

Then solve $\frac{32}{g} = \frac{4}{7}$ by using cross multiplication.

$32 * 7 = 224 \qquad 4 * g = 4g$

$$\frac{32}{g} = \frac{4}{7}$$

The cross products are equal: $4g = 224$.
To solve the equation, divide both sides by 4.
$4g / 4 = 224 / 4$ So, $g = 56$.
There are 56 girls in the 6th grade of the Ames School.

Check Your Understanding

Of Sarah's cousins, $\frac{2}{5}$ are girls. Sarah has 8 girl cousins. How many cousins does Sarah have in all?

Check your answer on page 418.

Using Ratios to Describe Size Changes

Many situations produce a **size change.** For example, a magnifying glass, a microscope, and an overhead projector all produce size changes that enlarge the original images. Most copying machines can create a variety of size changes—both enlargements and reductions of the original document.

Similar figures are figures that have the same shape but not necessarily the same size. Enlargements or reductions are **similar** to the originals; that is, they have the same shape as the originals.

The **size-change factor** is a number that tells the amount of enlargement or reduction that takes place. For example, if you use a copy machine to make a 2X change in size, then every length in the copy is twice the size of the original. The size-change factor is 2, or 200%. If you make a 0.5X change in size, then every length in the copy is half the size of the original. The size-change factor is $\frac{1}{2}$, or 0.5, or 50%.

original

You can think of the size-change factor as a ratio. For a 2X size change, the ratio of a length in the copy to the corresponding length in the original is 2 to 1.

$$\text{size-change factor 2: } \frac{\text{copy size}}{\text{original size}} = \frac{2}{1}$$

size-change factor 2

For a 0.5X size change, the ratio of a length in the copy to a corresponding length in the original is 0.5 to 1.

$$\text{size-change factor 0.5: } \frac{\text{copy size}}{\text{original size}} = \frac{0.5}{1}$$

size-change factor 0.5

If the size-change factor is greater than 1, then the copy is an **enlargement** of the original. If it is less than 1, then the copy is a **reduction** of the original.

A photographer uses an enlarger to make prints from negatives. If the size of the image on the negative is 2" by 2" and the size of the image on the print is 6" by 6", then the size-change factor is 3. Binoculars that are 8X, or "8 power," magnify all the lengths you see with the naked eye to 8 times their actual size.

Scale Models

A model that is a careful, reduced copy of an actual object is called a **scale model.** You have probably seen scale models of cars, trains, and airplanes. The size-change factor in scale models is usually called the **scale factor.**

Dollhouses often have a scale factor of $\frac{1}{12}$. You can write this as "$\frac{1}{12}$ of actual size," "scale 1 : 12," "$\frac{1}{12}$ scale," or as a proportion:

$$\frac{\text{dollhouse length}}{\text{real house length}} = \frac{1 \text{ inch}}{12 \text{ inches}}$$

All the dimensions of an E-scale model railroad are $\frac{1}{96}$ of the actual size. The scale factor is $\frac{1}{96}$. We can write this as "scale 1 : 96," or as a proportion:

$$\frac{\text{model railroad length}}{\text{real railroad length}} = \frac{1 \text{ inch}}{96 \text{ inches}}$$

We can also write this as "scale: $\frac{1}{8}$ inch represents 1 foot," or as "scale: 0.125 inch represents 12 inches." To see this, write

$$\frac{\frac{1}{8}}{12} = \frac{\frac{1}{8} * 8}{12 * 8} = \frac{1}{96}$$

$\frac{1}{8}$ inch : 12 inches is the same as 1 inch : 96 inches.

Scale Drawings

The size-change factor for scale drawings is usually called the **scale.** If an architect's scale drawing shows "scale $\frac{1}{4}$ inch : 1 foot" or "scale: $\frac{1}{4}$ inch represents 1 foot," then $\frac{1}{4}$ inch on the drawing represents 1 foot of actual length.

$$\frac{\text{drawing length}}{\text{real length}} = \frac{\frac{1}{4} \text{ inch}}{1 \text{ foot}}$$

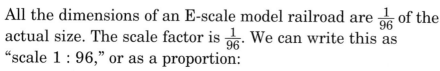

Since 1 foot = 12 inches, we can rename

$$\frac{\frac{1}{4} \text{ inch}}{1 \text{ foot}} \text{ as } \frac{\frac{1}{4} \text{ inch}}{12 \text{ inches}}.$$

Multiply by 4 to change this to an easier fraction:

$$\frac{\frac{1}{4} \text{ inch} * 4}{12 \text{ inches} * 4} = \frac{1 \text{ inch}}{48 \text{ inches}}.$$

The drawing is $\frac{1}{48}$ of the actual size.

Map Scales

Cartographers (mapmakers) show large areas of land and water in small areas on paper. The size-change factor for a map is usually called the **map scale.** Using a map and a map scale, you can estimate actual distances. Different maps use different scales.

If a map scale is 1 : 24,000, then every length on the map is $\frac{1}{24,000}$ of the actual length, and any real distance is 24,000 times the distance shown on the map.

$$\frac{\text{map distance}}{\text{real distance}} = \frac{1}{24,000}$$

On the map scale below, the length of the bar stands for 10 actual miles. Half the length of the bar stands for 5 actual miles.

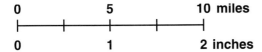

Since the bar is 2 inches long and stands for 10 actual miles, the map scale can also be written:

$$\frac{\text{map distance}}{\text{real distance}} = \frac{2 \text{ inches}}{10 \text{ inches}}.$$

You can find the exact size-change factor for a map with this scale by converting miles to inches in the proportion.

1 mile = 5,280 feet and 1 foot = 12 inches.
So, 10 miles = 52,800 feet = 52,800 * 12 inches = 633,600 inches.

$$\frac{\text{map distance}}{\text{real distance}} = \frac{2 \text{ inches}}{633,600 \text{ inches}} = \frac{1 \text{ inch}}{316,800 \text{ inches}}$$

The size-change factor is 1: 316,800, or $\frac{1}{316,800}$.

Caution: You may see scales written with an equal sign, such as "$\frac{1}{4}$ inch = 1 foot." But $\frac{1}{4}$ inch is certainly not equal to 1 foot, so "$\frac{1}{4}$ inch = 1 foot" is not mathematically correct. What is meant is that $\frac{1}{4}$ inch on the map or scale drawing stands for 1 foot in the real world.

Did You Know?

The U.S. Geological Survey (USGS) has made a detailed set of maps that covers the entire area of the U.S.

Their best known maps have a scale of 1:24,000 (1 inch represents 24,000 inches = 2,000 feet). Each of these maps covers an area of between 49 and 65 square miles.

Check Your Understanding

1. The side of a square is 2.5 cm. A copier is used to make an enlargement of the square. The size-change factor is 3.

   a. What is the side of the enlarged square?

   b. What is the perimeter of the enlarged square?

   Check your answers on page 418.

Finding Distances Using a Map Scale

There are many ways to find actual distances using a map. This method requires a ruler and string.

Step 1: Measure the map distance.

If the distance is along a straight path, such as the distance from Home to School, use the ruler to measure the map distance directly.

If the distance is along a curved path, such as the distance from Home to the Park:

♦ Lay the string along the path. Mark the beginning and ending points on the string.

♦ Straighten out the string. Use a ruler to measure between the beginning and ending points.

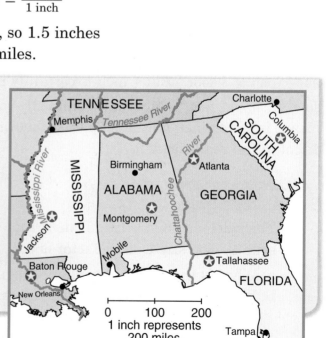

Scale: 2 in. = 10 mi

Step 2: Use the map scale to find the real distance for the map distance you measured. For example, the map distance from Home to School is 1.5 inches and the scale is 10 miles to 2 inches. To find the real distance we can solve this number model:

$$\frac{\text{real distance}}{1.5 \text{ inches}} = \frac{10 \text{ miles}}{2 \text{ inches}}$$

One way to solve this number model is to change the ratio to an equivalent n-to-1 ratio.

$$\frac{\text{real distance}}{1.5 \text{ inches}} = \frac{10 \text{ miles}}{2 \text{ inches}} = \frac{10 \text{ miles} \div 2}{2 \text{ inches} \div 2} = \frac{5 \text{ miles}}{1 \text{ inch}}$$

1 inch on the map stands for 5 miles, so 1.5 inches must stand for 1.5 * 5 miles, or 7.5 miles.

Check Your Understanding

Look at the map at the right.
The map scale is $\frac{\text{map distance}}{\text{real distance}} = \frac{1 \text{ inch}}{200 \text{ miles}}$

1. What is the actual distance between
   a. New Orleans and Charlotte?
   b. Mobile and Birmingham?
   c. Charlotte and Tampa?
   d. Tampa and Memphis?

Check your answers on page 418.

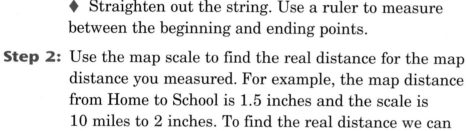

Scale Models

A scale model is a three-dimensional representation of a real object that is larger or smaller than the real thing. All of the parts in a model are proportional to the size of the real object. Scale models are used in fields such as architecture, design, science, engineering, education, and entertainment.

This is a model of a microscopic DNA molecule. The actual molecule is so tiny it cannot be seen with the naked eye. In science and industry, a large-scale model helps people view and study small parts. ➤

▲ Many scale models are smaller versions of the object they represent. A globe is a scale model of the earth. It shows features that can only be seen from locations hundreds of miles above the planet's surface.

Understanding Scale

Models are built according to different scales. The scale of a model is the ratio of the size of the model to the size of the actual object

◀ This space shuttle model is displayed at a NASA museum. Its scale is 1:15.

The scale of 1:15 means the dimensions of the actual space shuttle, shown here at liftoff, are 15 times as large as the dimensions of the model shown above. ➤

◀ Model railways are very common throughout the world. A popular scale is known as HO. The HO scale is 1:87.

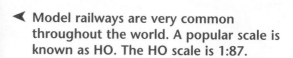

The scale of 1:87 means, for example, that the length of the real train engine is 87 times the length of the model train engine. ➤

Models in Architecture and Design

Architects and designers use scale models when planning structures and developing products. This enables them to make modifications to their plans before the actual structures or products are created.

These architects use a scale model of a building design to solve problems before construction begins. It is easier and cheaper to work with a model to make sure the plan for a structure is correct than to fix the structure after it has been built. ➤

◄ The sun simulator above this architectural model helps the architect see what the lighting conditions will be like at different times of day.

This scale model for a new housing development shows homes, other structures, roads, and landscaping. It offers city planners and potential buyers a view of the completed project. ➤

Models in Science and Engineering

Scientists and engineers create scale models when they design new equipment and machines, and when they test existing ones. They test function and safety by simulating the environment in which the machines will operate.

In the early 1900s, the Wright Brothers built and tested hundreds of scale models before finding a successful airplane design. ➤

◀ This is a replica of a wind tunnel the Wright Brothers used to test their designs. To simulate flight conditions, a fan blew air over models placed in the tunnel. From these tests, they could evaluate the performance and safety of their designs.

◀ Aircraft designers still use scale models and wind tunnels to collect data about their designs. They use wind tunnels to test safety and reliability during takeoffs and landings and at a variety of speeds.

In this photo from 1931, ship builders are testing an 18-foot model of an ocean liner by simulating heavy weather conditions in a 300-foot tank. Today, ship builders still use water tanks to test ship performance.

These scientists are trying to develop a bracing system to protect adobe houses during earthquakes. Here, they examine a scale model that was shaken by an earthquake simulator. ➤

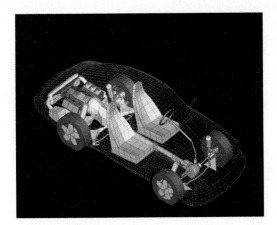

⋀ Scientists and engineers also create scale models using computers. This is a computer model of an automobile.

This scientist is looking at a computer model of a DNA molecule. ➤

Models in Education and Entertainment

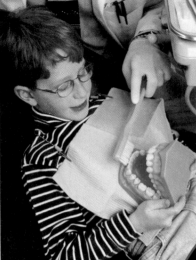

◄ With the help of this oversized model, a young boy learns how to correctly brush his teeth.

Scale models are often used in movie-making to film scenes that would be difficult, dangerous, or impossible to enact in real life. ➤

◄ Models are also created for the sets for operas, plays, ballets, and musicals. This painting shows a model of a stage design used in the opera "Boris Godunov" by Modest Musorgsky.

Where have you seen scale models?

Data and Probability

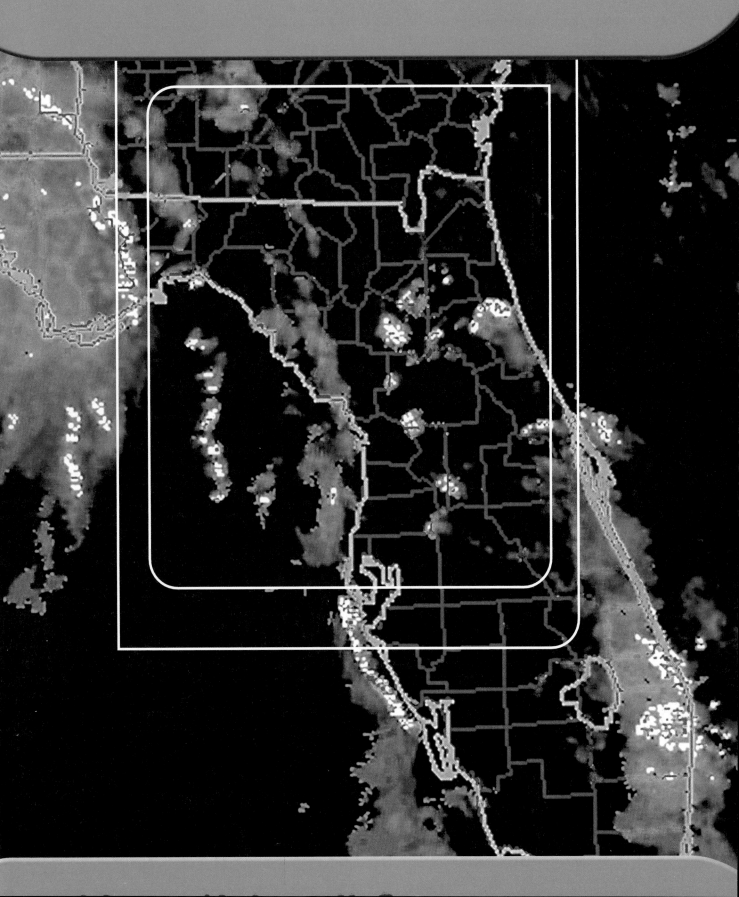

Collecting Data

There are different ways to collect information about something. You can count, measure, ask questions, or observe and describe what you see. The information you collect is called **data.**

Surveys

A **survey** is a study that collects data.

Much of the information used to make decisions comes from surveys. Many surveys collect data about people. Stores survey their customers to find out which products they should carry. Television stations survey viewers to learn which programs are popular. Politicians survey people to learn how they plan to vote in elections.

A face-to-face interview

Survey data are collected in several ways. These include face-to-face interviews, telephone interviews, printed questionnaires that are returned by mail, and group discussions (often called *focus groups*).

Not all surveys gather information about people. For example, there are surveys about cars, buildings, and animal groups. These surveys often collect data in ways other than through interviews or questionnaires.

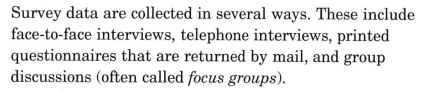 Highway engineers sometimes make videotapes of vehicles and drivers along a street or highway. They use the video data to analyze vehicle speeds and driving patterns.

 A bird survey is conducted during December and January each year in the Chicago area. Bird watchers list the different bird species observed. Then they count the number of each species observed. The lists are combined to create a final data set.

From a recent Chicago bird survey:

Species	Number of birds seen
Canada goose	3,768
house finch	304
snowy owl	1
starling	1,573

Samples

The **population** for a survey is the group of people or things that is being studied. Because the population may be very large, it may not be possible to collect data from every member of the population. Therefore, data are collected only from a sample group to provide information for the population. A **sample** is a part of the population that is chosen to represent the whole population.

Large samples usually give more dependable estimates than small ones. For example, if you want to estimate the percentage of adults who drive to work, a sample of 100 adults provides a better estimate than a sample of 10.

Example   A survey of teenagers collects data about people aged 13 to 19. There are about 28 million teenagers in the United States. It is not possible to collect data from every teenager. Instead, data are collected from a sample of teenagers.

Results from a recent survey of teens are shown in the table.

How Teens Divide Their Media Time	
Watching TV/videos/movies	51%
Listening to music	18%
Reading	9%
Playing video games	10%
Using computers	12%

Source: Henry J. Kaiser Foundation

Note

A *census* collects data from every member of the population. In a census, the sample and the population are identical.

The decennial (every 10 years) **census** is a survey that includes *all* the people in the United States. Every household is required to fill out a census form, but certain questions are asked only in a sample of 1 in 6 households.

A **random sample** is a sample that gives all members of the population the same chance of being selected. Random samples give more dependable information than those that are not random.

Did You Know?

The word *census* comes from the Latin word *censere*, which means "to tax." Censuses were taken by the ancient Babylonians, Egyptians, Greeks, and Romans. Their major purpose was to find the number and location of households that could be taxed.

Example   Suppose you want to estimate what percentage of the population will vote for Mr. Jenkins.

- If you use a sample of Mr. Jenkins's 100 best friends, the sample is *not* a random sample. People who do not know Mr. Jenkins have no chance of being selected.

- A sample of best friends will not fairly represent the entire population. It will not furnish a dependable estimate of how the entire population will vote.

Organizing Data

Once the data have been collected, it helps to organize them in order to make them easier to understand. **Line plots** and **tally charts** are two methods of organizing data.

Example Ms. Halko's class got the following scores on a 20-word vocabulary test. Make a line plot and a tally chart to show the data below.

20 15 18 17 20 12 15 17 19 18 20 16 16
17 14 15 19 18 18 15 10 20 19 18 15 18

Scores on a 20-Word Vocabulary Test

Number of Students

```
                                    X
                      X             X
                      X             X           X
                      X         X   X   X   X
                      X   X   X   X   X   X
 X          X         X   X   X   X   X   X   X
 +---+---+---+---+---+---+---+---+---+---+
 10  11  12  13  14  15  16  17  18  19  20
```
Number Correct

In the line plot, there are 5 Xs above 15.

In the tally chart, there are 5 tallies to the right of 15.

The 5 Xs and the 5 tallies each show that a score of 15 appeared 5 times in the class list of test scores.

Scores on a 20-Word Vocabulary Test

Number Correct	Number of Students
10	/
11	
12	/
13	
14	/
15	ЖТ
16	//
17	///
18	ЖТ /
19	///
20	////

Both the line plot and the tally chart help to organize the data.

They make it easier to describe the data. For example,

♦ 4 students had 20 correct (a perfect score).

♦ 18 correct is the score that came up most often.

♦ 10, 12, and 14 correct are scores that came up least often.

♦ 0 to 9, 11, and 13 correct are scores that did not occur at all.

Check Your Understanding

Here are the numbers of hits made by 18 players in a baseball game:
3 1 0 4 0 2 1 0 0 2 3 0 2 2 1 2 0 0
Organize the data.

1. Make a tally chart.   **2.** Make a line plot.   Check your answers on page 418.

Sometimes the data are spread over a wide range of numbers. This makes a tally chart or a line plot difficult to draw. In such cases, you can make a tally chart in which the results are **grouped.** Or, you may organize the data by making a **stem-and-leaf plot.**

For a health project, the students in Mr. Preston's class took each other's pulse rates. (A *pulse rate* is the number of heartbeats per minute.) These were the results:

92 72 90 86 102 78 88 75 72 82

90 94 70 94 78 75 90 102 65 94

70 94 85 88 105 86 78 75 86 108

94 75 88 86 99 78 86

Tally Chart of Grouped Data

The data have been sorted or grouped into intervals of 10. This is called a tally chart of **grouped data.**

The chart shows that most of the students had a pulse rate from 70 to 99. More students had a pulse rate in the 70s than in any other interval.

Stem-and-Leaf Plot

In a stem-and-leaf plot, the digit or digits in the left column (the **stem**) are combined with a single digit in the right column (the **leaf**) to form a numeral.

Each row has as many entries as there are digits in the right column. For example, the row with 8 in the left column has 10 entries: 86, 88, 82, 85, 88, 86, 86, 88, 86, and 86.

Pulse Rates of Students

Number of Heartbeats	Number of Students
60–69	/
70–79	⊬⊬⊬ ⊬⊬⊬ //
80–89	⊬⊬⊬ ⊬⊬⊬
90–99	⊬⊬⊬ ⊬⊬⊬
100–109	////

Pulse Rates of Students

Stems (10s)	Leaves (1s)
6	5
7	2 8 5 2 0 8 5 0 8 5 5 8
8	6 8 2 5 8 6 6 8 6 6
9	2 0 0 4 4 0 4 4 4 9
10	2 2 5 8

Check Your Understanding

The first manned spaceflight occurred during the early 1960s. The ages of the first 10 space travelers were:

27   25   43   37   32   31   39   36   28   26

Organize the data.
1. Make a tally chart of grouped data.   2. Make a stem-and-leaf plot.

Check your answers on page 419.

Statistical Landmarks

The **landmarks** for a set of data are used to describe the data.

◆ The **minimum** is the smallest value.
◆ The **maximum** is the largest value.
◆ The **range** is the difference between the maximum and the minimum.
◆ The **mode** is the value (or values) that occurs most often.
◆ The **median** is the middle value.

Example   Here is a record of children's absences for one week at Henry Esmond School. Find the landmarks for the data.

Day	Number Absent
Monday	24
Tuesday	21
Wednesday	9
Thursday	13
Friday	13

Minimum (lowest) number: 9
Maximum (highest) number: 24
Range of numbers: 24 − 9 = 15
Mode (most frequent number): 13

To find the median (middle value), list the numbers in order. Then, find the middle number.

9   13   13   21   24
↑
median

Example   The **line plot** shows students' scores on a 20-word vocabulary test. Find the landmarks for the data.

Minimum: 10
Maximum: 20
Range: 20 − 10 = 10
Mode: 18

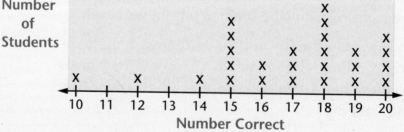

Scores on a 20-Word Vocabulary Test

Number of Students

Number Correct

10 12 14 15 15 15 15 15 16 16 17 17 [17 18] 18 18 18 18 18 19 19 19 20 20 20 20

middle scores

There are two middle scores, 17 and 18. The median is 17.5, which is the number halfway between 17 and 18

Check Your Understanding

Here are the math quiz scores (number correct): 4 2 1 2 0 2 1 0 0 2 1 3
Find the minimum, maximum, range, mode, and median for this set of data.

Check your answers on page 419.

The Mean (or Average)

The **mean** of a set of numbers is often called the *average*. To find the mean, do the following:

Step 1: Add the numbers.

Step 2: Divide the sum by the number of addends.

Example On a 4-day trip, Kenny's family drove 200, 120, 160, and 260 miles. What is the mean number of miles they drove per day?

Step 1: Add the numbers: 200 + 120 + 160 + 260 = 740.
Step 2: Divide by the number of addends: 740 ÷ 4 = 185.

The mean is 185.
They drove an average of 185 miles per day.

If you use a calculator key in:
200 [+] 120 [+] 160 [+] 260 [=]

Divide this sum by 4. 740 [÷] 4 [=] Answer: 185

When calculating the mean of a large set of numbers, it often helps to organize the data in a tally chart or line plot first.

Note

The mean can also be thought of as a landmark.

The mean and the median are often the same or almost the same.

Both the mean and the median can be thought of as a "typical" number for the data set.

Did You Know?

In Olympic diving events, seven judges score each dive. Then the highest score and lowest score are thrown out before the mean score is calculated. This is called a *trimmed mean*.

Example Students measured each other's height in inches. What is the mean height?

The students organized the measurements in a tally chart (the first two columns).

On calculators, they multiplied each height by the number of students of that height and kept a running total. Then, they divided the final total by the number of students.

The mean, rounded to the nearest inch, is 54 inches.

Height (inches)	Number of Students	Key in:	Calculator Answer
49	//	2 [×] 49 [=]	98
50	//	[+] 2 [×] 50 [=]	198
51	/	[+] 51 [=]	249
52	////	[+] 4 [×] 52 [=]	457
53	/	[+] 53 [=]	510
54	//	[+] 2 [×] 54 [=]	618
55	///	[+] 3 [×] 55 [=]	783
56	~~////~~	[+] 5 [×] 56 [=]	1063
57			
58	//	[+] 2 [×] 58 [=]	1179
Total	22	[÷] 22 [=]	53.590909

Check Your Understanding

Megan received these scores on math tests: 80 75 80 75 80 90 80 85 80 90 75
Use your calculator to find Megan's mean score.

Check your answers on page 419.

Bar Graphs

A **bar graph** is a drawing that uses bars to represent numbers. Bar graphs display information in a way that makes it easy to show comparisons.

The title of a bar graph describes the information in the graph. Each bar has a label. Units are given to show how something was counted or measured. When possible, the graph gives the source of the information.

Did You Know?

The first bar graph was drawn by William Playfair in 1786. It showed the cash value of Scotland's trade with 18 other countries.

Example   This is a **vertical bar graph.**

- Each bar represents the number of pet cats in the country that is named below the bar.

- It is easy to compare cat populations by comparing the bars. China has about 4 or 5 times as many cats as Russia, Brazil, and France. The ratio of cats in the U.S. to cats in China is about 75 to 50, or about 3 to 2.

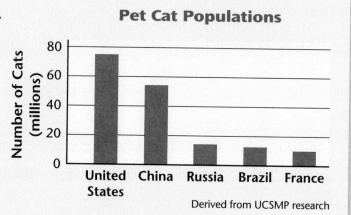

Pet Cat Populations

Derived from UCSMP research

Example   This is a **horizontal bar graph.**

- Each bar represents the percent of persons in an age group who live alone.

- The bars show a trend. As the age for a group increases, the percent of persons in that group who live alone increases. Only 5% of 5- up to 25-year-olds live alone. But 34% of 45- up to 65-year-olds live alone.

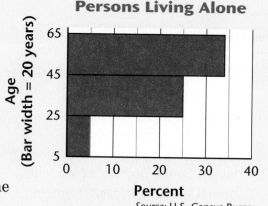

Persons Living Alone

Source: U.S. Census Bureau

This type of graph is also called a **histogram.** The vertical axis is a number line. The bars split the number line into equal width intervals of 20 years. The bars share sides at ages 25 and 45. An age on a bar side is counted in the bar above the side.

Check Your Understanding

Refer to the "Persons Living Alone" bar graph above. According to the graph:

1. What is the width of each bar?
2. What percent of people aged 25 up to 45 years old live alone?

Check your answers on page 419.

Side-by-Side and Stacked Bar Graphs

Sometimes two or more bar graphs are related to the same situation. Related bar graphs are often combined into a single graph. The combined graph saves space and makes it easier to compare the data. The examples below show two different ways to draw combined bar graphs.

Example   The first bar graph shows road miles from Boston to different cities. The second bar graph shows air miles.

The graphs are combined into a **side-by-side bar graph** by drawing the related bars side by side in different colors. It is easy to compare road miles and air miles on the side-by-side graph.

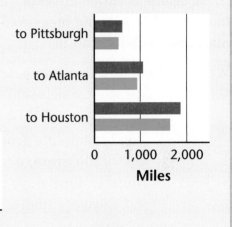

Distances from Boston

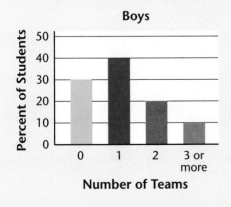

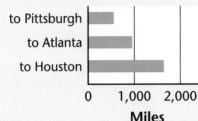

Source: The World Almanac and National Geodetic Survey

Key   Road Miles
       Air Miles

Example   The bar graphs below show the number of sports teams that boys and girls joined during a 1-year period.

The bars within each graph can be stacked on top of one another. The **stacked bar graph** includes each of the stacked bars.

Number of Sports Teams

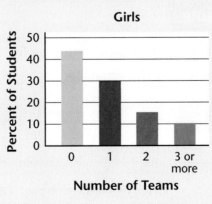

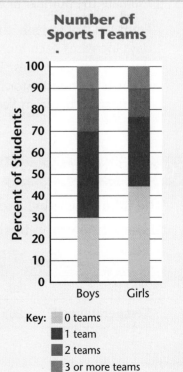

Number of Sports Teams

Key: 0 teams
       1 team
       2 teams
       3 or more teams

Line Graphs

Line graphs are used to display information that shows trends. They often show how something has changed over a period of time.

Line graphs are often called **broken-line graphs.** Line segments connect the points on the graph. The segments joined end to end look like a broken line.

Line graphs have a horizontal and a vertical scale. Each of these scales is called an **axis** (plural: **axes**). Each axis is labeled to show what is being measured or counted and what the unit of measure or count unit is.

When looking at a line graph, try to determine the purpose of the graph. See what conclusions you can draw from it.

Broken-Line Graph

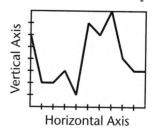

Joined end to end, the segments look like a broken line.

Example The broken-line graph at the right shows the average number of thunderstorm days for each month in Chicago, Illinois.

The horizontal axis shows each month of the year. The average number of thunderstorm days for a month is shown with a dot above the label for that month. The labels on the vertical axis are used to estimate the number of days represented by that dot.

From January to June, the number of thunderstorm days increases each month. From June to January, the number decreases. The greatest change in number of thunderstorm days from one month to the next occurs from September to October.

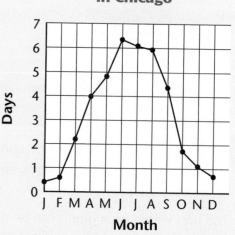

Average Number of Thunderstorm Days in Chicago

Check Your Understanding

The following table shows average temperatures for Boston, Massachusetts. Make a line graph to show this information.

Average Temperatures for Boston, Massachusetts												
Month	Jan	Feb	Mar	Apr	May	Jun	Jul	Aug	Sep	Oct	Nov	Dec
Temperature (°F)	29	32	39	48	59	68	74	72	65	54	45	35

Check your answer on page 419.

Step Graphs

Step graphs are used to describe situations in which changes in values are not gradual but occur in jumps.

Example The step graph at the right shows the cost of renting a bicycle from B & H Rentals.

According to the graph, it costs $10 to rent a bike for 1 hour or less and $2.50 for each additional half-hour or fraction of a half-hour. For example, it costs $10 whether you rent a bike for $\frac{1}{2}$ hour, 45 minutes, or 1 hour. It costs $12.50 whether you rent a bike for $1\frac{1}{2}$ hours or just 1 hour and 1 minute.

Note the dot at the end of the segment for the first hour. It indicates that the cost is $10 for 1 hour. There is no dot at the beginning of the segment for the second hour. This indicates that $12.50 is not the cost for 1 hour. The other dots in the graph are interpreted in the same way.

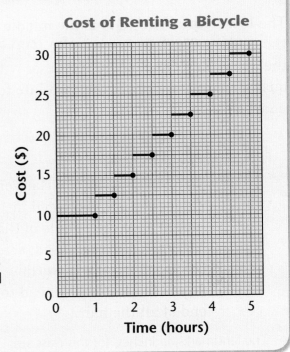

Cost of Renting a Bicycle

Telephone and parking lot rates are other examples of situations that can be shown by step graphs. Such rates do not change gradually over time but change at the end of intervals of time, such as minutes or hours.

Check Your Understanding

The step graph at the right shows the cost of taking a cab various distances.

1. Find the cost of taking a cab for each distance.
   a. 1 mile   b. 1.2 miles   c. 2 miles
   d. 4 miles   e. 3.5 miles

2. a. What is the cost of taking a cab for a distance of 1 mile or less?
   b. What is the cost for each additional mile or fraction of a mile?

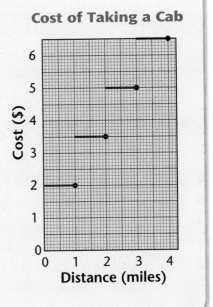

Cost of Taking a Cab

Check your answers on page 419.

Box-and-Whiskers Plots

Box-and-whiskers plots are used to display the spread of a data set with 5 landmarks. These are the minimum, the lower quartile, the median, the upper quartile, and the maximum.

The **lower quartile (Q1)** is the middle value of the data below the median. The **upper quartile (Q3)** is the middle value of the data above the median. Together with the median, these quartile landmarks split the data into four quarters, or quartiles.

> **Note**
>
> Box-and-whiskers plots are commonly called box plots.

> **Note**
>
> The median is sometimes called the *middle quartile (Q2)*.

Example The hair lengths of a class of students were measured in inches from the middle of the hairline on the forehead, across the top along the centerline, and down the back of the head to the end of all the hair.

Start at hairline, center front.

Measure to hairline, center back.

If hair is longer, continue measuring from back hairline to tips of hair.

The landmarks, in inches, for this class are:

Minimum	Lower quartile	Median	Upper quartile	Maximum
14	16	20	25	32

The box plot below shows the spread of the hair length data around these landmarks.

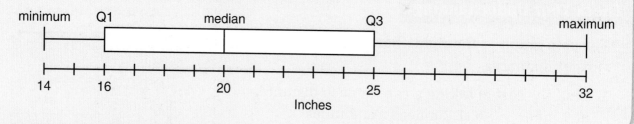

According to the plot, of the students in this class:
- About one quarter have hair lengths from 14 to 16 inches.
- About one quarter have hair lengths from 16 to 20 inches.
- About one quarter have hair lengths from 20 to 25 inches.
- About one quarter have hair lengths from 25 to 32 inches.
- The middle half have hair lengths from 16 to 25 inches.

The length of the interval between the lower and upper quartiles is called the **interquartile range (IQR).** You can think of the IQR as the distance between the upper and lower quartiles. For the hair length data, the IQR = Q3 – Q1 = 25 – 16 = 9 inches. So, overall, the range of class hair lengths is 32 – 14 = 18 inches. But for the middle half of students, the range is only 9 inches.

> **Note**
>
> We also say that the middle half of the data is *in* the interquartile range.

How to Draw a Box-and-Whiskers Plot

The calories in some popular chicken snacks are given in the table below.

Company	Chicken Snack	Total Calories
Brand A	Chicken Nuggets	250
Brand A	Chicken Strips	630
Brand B	Chicken Tenders	250
Brand B	Chicken Fries	260
Brand C	Chicken Nuggets	230
Brand D	Chicken Tenders	630
Brand D	Popcorn Chicken	531
Brand E	Chicken Strips	630
Brand F	Chicken Strips	1,270
Brand G	Popcorn Chicken	550
Brand H	Chicken Strips	710
Brand I	Chicken Rings	340

Source: www.acaloriecounter.com

Follow these steps to make a box-and-whiskers plot of the calorie data.

Step 1: Order the data and find the minimum, maximum, and median numbers of calories.

230, 250, 250, 260, 340, 531, 550, 630, 630, 630, 710, 1,270

minimum   median = 540.5   maximum

> **Note**
>
> If there are one or more data values at the median, do not include them when finding the upper and lower quartiles.

Step 2: Find the lower quartile (Q1) of the data. It is the middle value of the data *below* the median. Then find the upper quartile (Q3), the middle value of the data *above* the median.

230, 250, 250, 260, 340, 531, 550, 630, 630, 630, 710, 1,270

minimum   Q1 = 255   median = 540.5   Q3 = 630   maximum

Step 3: Draw a number line large enough to include the 5 landmarks you found in Steps 1 and 2. Choose a convenient scale. Here, the range of data from the 230 minimum to the 1,270 maximum is 1,270 − 230 = 1,040, or about 1,000. So we chose a scale of 100.

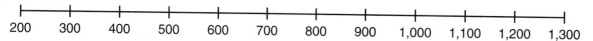

Step 4: Draw tick marks above the number line to mark the 5 landmarks.

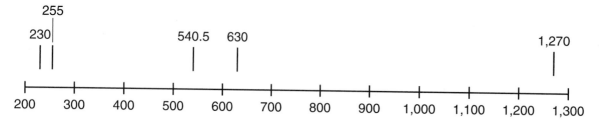

Step 5: Draw the box and whiskers. Connect the tick marks at Q1 and Q3 with horizontal lines to form the box. Draw one whisker from the minimum tick mark to the Q1 box end. Draw the other whisker from the Q3 box end to the maximum tick mark. Finish the plot with a title.

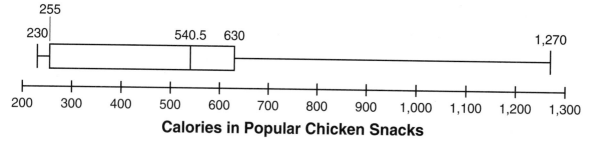

Calories in Popular Chicken Snacks

Comparing Box Plots

Box plots can be useful to compare different groups of data that measure the same thing with the same units. The calories in popular beef burgers are shown in the table below and plotted on the next page.

Company	Beef Burger	Total Calories
Brand A	Regular Burger	290
Brand B	Large Burger	470
Brand C	Large Burger	350
Brand D	Regular Burger	310
Brand E	Regular Burger	390
Brand F	Regular Burger	310
Brand G	Regular Burger	250
Brand H	Junior Burger	310
Brand I	Junior Burger	230
Brand J	Regular Burger	140

Below, a box plot for the beef burger calorie data is drawn above the chicken snack box plot. By using the same scale on the number line, you can compare the landmarks and spreads of the different types of foods.

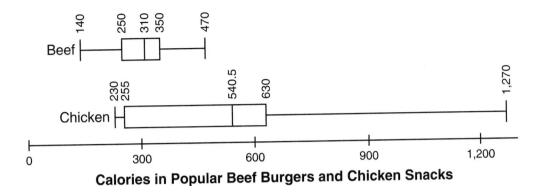

Calories in Popular Beef Burgers and Chicken Snacks

Mean Absolute Deviation

The *range* of data in a set is a rough measure of the set's variability based only on the minimum and maximum values. The *interquartile range* is a more detailed measure based on the data in quartile groupings. An even more detailed statistic for variability is calculated using *each* data value in the set. The **mean absolute deviation (m.a.d.)** in a data set is the average (mean) distance between individual data values and the mean of those values.

There are 3 steps to calculating the mean absolute deviation.

Step 1: Calculate the mean of the data.

Step 2: Find the distance of each data value from the mean found in Step 1.

Step 3: Calculate the mean of the distances found in Step 2.

> **Note**
>
> You can also calculate the *median absolute deviation* by substituting the median of the data for the mean in the steps at left.

Example   Use a spreadsheet to calculate the mean absolute deviation of the beef burger data on page 141C.

Step 1: Cell B12 shows the average of the data in column B.

Step 2: Column D shows the deviations (differences) of the column B calorie data from the mean entered in column C. Column E shows the absolute values of the column D deviations.

Step 3: Cell E12 shows the average of the column E absolute deviations.

◇	A	B Total Calories	C Mean Calories	D Deviation	E Absolute Deviation
1					
2		290	305	-15	15
3		470	305	165	165
4		350	305	45	45
5		310	305	5	5
6		390	305	85	85
7		310	305	5	5
8		250	305	-55	55
9		310	305	5	5
10		230	305	-75	75
11		140	305	-165	165
12	Mean	305		0	62

The m.a.d. of the beef burger data is 62 calories.

History and Uses of Spreadsheets

Everyday Rentals–Debit Statement for May, 1964

Company	Type	Invoice #	Invoice Amount	Amount Paid	Balance Due
Electric	Utility	2704–3364	342.12	100.00	242.12
Gas	Utility	44506–309	129.43	50.00	79.43
Phone	Utility	989–2209	78.56	78.56	0.00
Water	Utility	554–2–1018	13.12		13.12
NW Bank	Mortgage	May 1964	1,264.00	1,264.00	0.00
Waste Removal	Garbage	387–219	23.00		23.00
NW Lumber	Supplies	e–318	239.47	50.00	189.47
Total			2,089.70	1,542.56	547.14

Above is a copy of a financial record for Everyday Rentals Corporation for May, 1964. A financial record often had more columns of figures than would fit on one sheet of paper, so accountants taped several sheets together. They folded the sheets for storage and spread them out to read or make entries. Such sheets came to be called **spreadsheets.**

Note that the "Balance Due" **column** and the "Total" **row** are calculated from other numbers in the spreadsheet. Before they had computers, accountants wrote spreadsheets by hand. If an accountant changed a number in one row or column, several other numbers would have to be erased, recalculated, and reentered.

For example, when Everyday Rentals Corporation pays the $23 owed to Waste Removal, the accountant must enter that amount in the "Amount Paid" column. That means the total of the "Amount Paid" column must be changed as well. That's not all–making a payment changes the amount in the "Balance Due" column and the total of the "Balance Due" column. One entry requires three other changes to **update** (revise) the spreadsheet.

When personal computers were developed, spreadsheet programs were among the first applications. Spreadsheet programs save time by making changes automatically. Suppose the record at the top of this page is on a computer spreadsheet. When the accountant enters the payment of $23, the computer automatically recalculates all of the numbers that are affected by that payment.

Did You Know?

In mathematics and science, computer spreadsheets are used to store large amounts of data and to perform complicated calculations. People use spreadsheets at home to keep track of budgets, payments, and taxes.

Spreadsheets and Computers

A **spreadsheet program** enables you to use a computer to evaluate formulas quickly and efficiently. On a computer screen, a spreadsheet looks like a table. Each **cell** in the table has an **address** made up of a letter and a number. The letter identifies the column, and the number identifies the row in which the cell is found. For example, cell B3 is in column B, row 3.

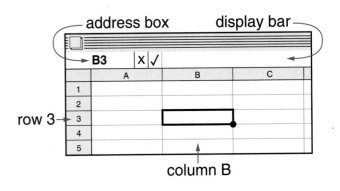

To enter information in a cell, you can use a computer mouse to click on the cell; the address of the cell will appear in the **address box.** Then, you type the information you want to enter in the cell; the information will appear in the **display bar.**

There are three kinds of information that may be entered in a spreadsheet.

♦ **Labels** (may consist of words, numbers, or both) These are used to display information about the spreadsheet, such as the headings for the columns and rows. Numbers in labels are never used in calculations. When a label is entered from the keyboard, it is stored in its address and shown in its cell on the screen.

♦ **Numbers** (those not included in labels) These are used in calculations. When a number is entered from the keyboard, it is stored in its address and appears in its cell on the screen.

♦ **Formulas** These tell the computer what calculations to make on the numbers in other cells. When a formula is entered from the keyboard, it is stored in its address but is *not* shown in its cell on the screen. Instead, a number is shown in the cell. This number is the result of applying the formula to numbers in other cells.

Example   Study the spreadsheet at the right.

	A	B	C	D
	D3	$\times$ $\checkmark$ =B3*C3		
	A	B	C	D
1	item name	unit price	quantity	totals
2				
3	pencils	0.29	6	1.74
4	graph paper	1.19	2	2.38
5	ruler	0.50	1	0.50
6	book	5.95	1	5.95
7				
8	Subtotal			10.57
9	tax 7%			0.74
10	Total			11.31

All the entries in row 1 and column A are *labels*.

The entries in cells B3 through B6 and cells C3 through C6 are *numbers* that are not labels. They have been entered from the keyboard and are used in calculations.

Cells D3 through D10 also display numbers, but they have not been entered from the keyboard. Instead, a *formula* was entered in each of these cells. The computer program used these formulas to calculate the numbers that appear in column D.

The numbers in column D are the results of calculations.

Example Study the spreadsheet at the right.

The *address box* shows that cell D3 has been selected; the *display bar* shows "= B3 * C3." This stands for the formula D3 = B3 * C3. It is not necessary to enter D3, since D3 is already identified as the address of the cell. This formula is stored in the computer; it is not shown in cell D3.

D3	X ✓	=B3*C3	
	A	B	C	D
1	item name	unit price	quantity	totals
2				
3	pencils	0.29	6	1.74
4	graph paper	1.19	2	2.38
5	ruler	0.50	1	0.50
6	book	5.95	1	5.95
7				
8	Subtotal			10.57
9	tax 7%			0.74
10	Total			11.31

When the formula is entered, the program multiplies the number in cell B3 (0.29) by the number in cell C3 (6) and displays the product (1.74) in cell D3.

Suppose that you clicked on cell C3 and changed the 6 to an 8. The entry in cell D3 would change automatically to 2.32 (= 0.29 * 8). At the same time, the entries in cells D8, D9, and D10 would also change automatically. The entry in cell D8 is the result of a calculation involving the entry in cell D3. The entry in cell D8 is used to calculate the entry in cell D9. And the entries in cells D8 and D9 are used to calculate the entry in cell D10.

Check Your Understanding

The following spreadsheet gives budget information for a class picnic.

	A	B	C	D
		Class Picnic ($$)		
1		budget for class picnic		
2				
3	quantity	food items	unit price	cost
4	6	packages of hamburgers	2.79	16.74
5	5	packages of hamburger buns	1.29	6.45
6	3	bags of potato chips	3.12	9.36
7	3	quarts of macaroni salad	4.50	13.50
8	4	bottles of soft drinks	1.69	6.76
9			subtotal	52.81
10			8% tax	4.23
11			total	57.04

Use the spreadsheet to answer the following questions.

1. What kind of information is shown in column B?

2. What information is shown in cell C7?

3. Which cell shows the title of the spreadsheet?

4. What information is shown in cell A5?

5. Which occupied cells do not hold labels or formulas?

6. Which column holds formulas?

Check your answers on page 419.

How to Use the Percent Circle

A **compass** is a device for drawing circles. You can also use your **Geometry Template** to draw circles.

An **arc** is a piece of a circle. If you mark two points on a circle, these points and the part of the circle between them form an arc.

The region inside a circle is called its **interior.**

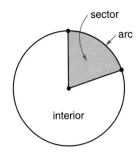

A **sector** is a wedge-shaped piece of a circle and its interior. A sector consists of two **radii** (singular: **radius**), one of the arcs determined by their endpoints, and the part of the interior of the circle bounded by the radii and the arc.

A **circle graph** is sometimes called a **pie graph** because it looks like a pie that has been cut into several pieces. Each "piece" is a sector of the circle.

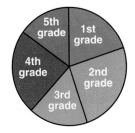

You can use the **Percent Circle** on your Geometry Template to find what percent of the circle graph each sector represents. Here are two methods for using the Percent Circle.

The circle graph shows the distribution of students in grades 1 to 5 at Elm School.

Method 1: Direct Measure

- ♦ Place the center of the Percent Circle over the center of the circle graph.
- ♦ Rotate the template so that the 0% mark is aligned with one side (line segment) of the sector you are measuring.
- ♦ Read the percent at the mark on the Percent Circle located over the other side of the sector. This tells what percent the sector represents.

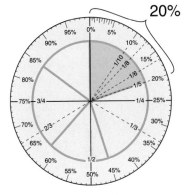

For example, the sector for first grade represents 20%.

Method 2: Difference Comparison

- ♦ Place the center of the Percent Circle over the center of the circle graph.
- ♦ Note the percent reading for one side of the sector you are measuring.
- ♦ Find the percent reading for the other side of the sector.
- ♦ Find the difference between these readings.

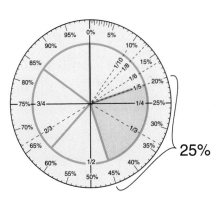

The sector for second grade represents 45% − 20%, or 25%.

Check Your Understanding

What percents are represented by the other three sectors in the above circle graph?

Check your answers on page 419.

How to Draw a Circle Graph Using a Percent Circle

Example Draw a circle graph to show the following information. The students in Mr. Zajac's class were asked to name their favorite colors: 9 students chose blue, 7 students chose green, 4 students chose yellow, and 5 chose red.

Step 1: Find what percent of the total each part represents. The total number of students who voted was $9 + 7 + 4 + 5 = 25$.

- 9 out of 25 chose blue.

  $\frac{9}{25} = \frac{36}{100} = 36\%$, so 36% chose blue.

- 7 out of 25 chose green.

  $\frac{7}{25} = \frac{28}{100} = 28\%$, so 28% chose green.

- 4 out of 25 chose yellow.

  $\frac{4}{25} = \frac{16}{100} = 16\%$, so 16% chose yellow.

- 5 out of 25 chose red.

  $\frac{5}{25} = \frac{20}{100} = 20\%$, so 20% chose red.

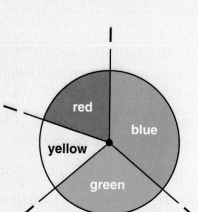

Step 3

Step 2: Check that the sum of the percents is 100%.
$36\% + 28\% + 16\% + 20\% = 100\%$

Step 3: Draw a circle. Then use the Percent Circle on the Geometry Template to mark off the sectors.

- To mark off 36%, place the center of the Percent Circle over the center of the circle graph. Make a mark at 0% and 36%.

- To mark off 28%, make a mark at 64% ($36\% + 28\% = 64\%$), without moving the Percent Circle.

- To mark off 16%, make a mark at 80% ($64\% + 16\% = 80\%$).

- Check that the final sector represents 20%.

Step 4

Step 4: Draw the sector lines (radii). Label each sector. Color the sectors.

Check Your Understanding

Draw a circle graph to display the following information:

- The basketball team scored 30 points in one game.
- Frank scored 3 points.
- Leah scored 6 points.
- Jill scored 6 points.
- Dave scored 15 points.

Check your answer on page 419.

How to Draw a Circle Graph Using a Protractor

Example Draw a circle graph to show the following information: In the month of April, there were 5 cloudy days, 6 sunny days, and 19 partly cloudy days.

Step 1: Find out what fraction or percent of the total each part represents. April has 30 days.
- 5 out of 30 were cloudy days.
  $\frac{5}{30} = \frac{1}{6}$, so $\frac{1}{6}$ of the days were cloudy.
- 6 out of 30 were sunny days.
  $\frac{6}{30} = \frac{1}{5}$, so $\frac{1}{5}$ of the days were sunny.
- 19 out of 30 were partly cloudy days.
  $\frac{19}{30} = 0.633 \ldots \approx 63.3\%$, so about 63.3% of the days were partly cloudy.

Step 2: Calculate the degree measure of the sector for each piece of data.

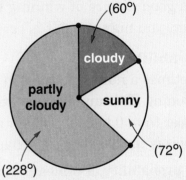

- The number of cloudy days in April was $\frac{1}{6}$ of the total number of days. Therefore, the degree measure of the sector for cloudy days is $\frac{1}{6}$ of 360°. $\frac{1}{6}$ of 360° = $\frac{1}{6} * 360° = 60°$.
- The number of sunny days in April was $\frac{1}{5}$ of the total number of days. Therefore, the degree measure of the sector for sunny days is $\frac{1}{5}$ of 360°. $\frac{1}{5}$ of 360° = $\frac{1}{5} * 360° = 72°$.
- The number of partly cloudy days in April was 63.3% of the total number of days. Therefore, the degree measure of the sector for partly cloudy days is 63.3% of 360°. $0.633 * 360° = 228°$, rounded to the nearest degree.

Step 3: Check that the sum of the degree measures of the sectors is 360°.
$60° + 72° + 228° = 360°$

Step 4: Draw a circle. Use a protractor to draw 3 sectors with degree measures of 60°, 72°, and 228°. Label each sector.

Check Your Understanding

Use your protractor to make a circle graph to display the following information.

What is the degree measure of each sector, rounded to the nearest degree?

Favorite Subjects

Subject	Number of Students
Reading	3
Art	1
Math	6
Music	3
Social Studies	2
Science	5

Check your answers on page 419.

Chance and Probability

Chance

Things that happen are called **events.** There are many events that you can be sure about:

♦ You are **certain** that the sun will set today.

♦ It is **impossible** for you to grow to be 10 feet tall.

There are also many events that you *cannot* be sure about.

♦ You cannot be sure that you will get a letter tomorrow.

♦ You cannot be sure whether it will be sunny next Friday.

You sometimes talk about the **chance** that something will happen. If Paul is a good chess player, you may say, "Paul has a *good chance* of winning the game." If Paul is a poor player, you may say, "It is *very unlikely* that Paul will win."

Probability

Sometimes a number is used to tell the chance of something happening. This number is called a **probability.** It is a number from 0 to 1. The closer a probability is to 1, the more likely it is that an event will happen.

♦ A probability of 0 means the event is *impossible*. The probability is 0 that you will live to the age of 150.

♦ A probability of 1 means that the event is *certain*. The probability is 1 that the sun will rise tomorrow.

♦ A probability of $\frac{1}{2}$ means that, in the long run, an event will happen about 1 in 2 times (half of the time, or 50% of the time). The probability that a tossed coin will land heads up is $\frac{1}{2}$. We often say that the coin has a "50-50 chance" of landing heads up.

A probability can be written as a fraction, a decimal, or a percent. The **Probability Meter** is often used to record probabilities. It is marked to show fractions, decimals, and percents between 0 (or 0%) and 1 (or 100%).

The phrases printed on the bar of the Probability Meter may be used to describe probabilities in words. For example, suppose that the probability of snow tomorrow is 70%. The 70% mark falls within that part of the bar where "LIKELY" is printed. So you can say that "Snow is *likely* tomorrow," instead of stating the probability as 70%.

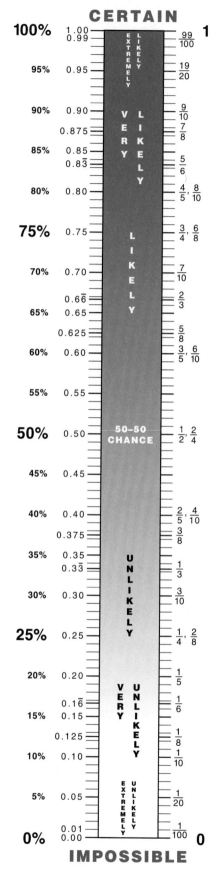

Calculating a Probability

Four common ways for finding probabilities are shown below.

Make a Guess

Vince guesses that he has a 40% chance (a 4 in 10 chance) of returning home by 9 o'clock.

Conduct an Experiment

Elizabeth dropped 67 tacks: 48 landed point up and 19 landed point down. The fraction of tacks that landed point up is $\frac{48}{67}$.

Elizabeth estimates the probability that the next tack she drops will land point up is $\frac{48}{67}$, or about 72%.

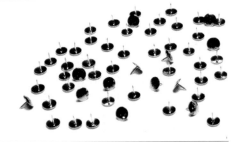

Use a Data Table

Kenny got 32 hits in his last 100 times at bat. He estimates the probability that he will get a hit the next time at bat is $\frac{32}{100}$, or 32%.

Hits	32
Walks	14
Outs	54
Total	**100**

Assume that All Possible Results Have the Same Chance

A standard die has 6 faces and is shaped like a cube. You can assume that each face has the same $\frac{1}{6}$ chance of landing up.

A 20-sided die has 20 faces and is shaped like a regular icosahedron. You can assume that each face has the same $\frac{1}{20}$ chance of landing up.

A standard deck of playing cards has 52 cards. Suppose the cards are shuffled and one card is drawn. You can assume that each card has the same $\frac{1}{52}$ chance of being drawn.

Suppose that a spinner is divided into 12 equal sections. When you spin the spinner, you can assume that each section has the same $\frac{1}{12}$ chance of being landed on.

Naming a Probability

A spinner is divided into 10 equal sections, numbered 1 through 10. Each of the following statements has the same meaning:

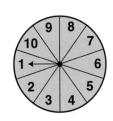

♦ The probability (chance) of landing on 4 is $\frac{1}{10}$.

♦ The probability (chance) of landing on 4 is **0.1.**

♦ The probability (chance) of landing on 4 is **10%.**

♦ The probability (chance) of landing on 4 is **1 out of 10.**

♦ There is a **1 in 10** probability (chance) of landing on 4.

♦ If you spin many times, you can expect to land on 4 **about $\frac{1}{10}$ of the time.**

Equally Likely Outcomes

In solving probability problems, it is often useful to list all of the possible results for a situation. Each possible result is called an **outcome.** If all of the possible outcomes have the same probability, they are called **equally likely outcomes.**

Example If you roll a 6-sided die, the number of dots on the face that lands up may be 1, 2, 3, 4, 5, or 6. There are six possible outcomes. You can assume that each face of the die has the same $\frac{1}{6}$ chance of landing up. So, the outcomes are equally likely.

Example The blue spinner is divided into 10 equal parts. If you spin the spinner, it will land on one of the numbers 1 through 10. There are ten possible outcomes. You can assume that each number has a $\frac{1}{10}$ chance of being landed on because the sections are equal parts of the spinner. So, the outcomes are equally likely.

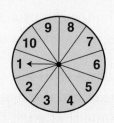

Example The red and blue spinner is divided into two sections. If you spin the spinner, it will land either on red or on blue. So red and blue are the two possible outcomes. The outcomes are *not* equally likely because there is a greater chance of landing on red than on blue

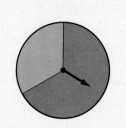

A Probability Formula

Finding the probability of an event is easy *if all of the outcomes are equally likely.* Follow these steps:

1. List all of the possible outcomes.

2. Look for any outcomes that will make the event happen. These outcomes are called **favorable outcomes.** Circle each favorable outcome.

3. Count the number of possible outcomes. Count the number of favorable outcomes. The probability of the event is:

$$\frac{\text{number of favorable outcomes}}{\text{number of possible outcomes}}$$

Example | Amy, Beth, Carol, Dave, Edgar, Frank, George, and Hank are on a camping trip. They decide to choose a leader. Each child writes his or her name on an index card. The cards are put into a paper bag and mixed. One card will be drawn, and the child whose name is drawn will become leader. Find the probability that a girl will be selected.

What is the *event* you want to find the probability of? Draw a girl's name.

How many *possible outcomes* are there? Eight. Any of 8 names might be drawn.

Are the outcomes *equally likely?* Yes
Names were written on identical cards and the cards were mixed in the bag. Each name has the same $\frac{1}{8}$ chance of being drawn.

Which of the possible outcomes are *favorable outcomes?* Amy, Beth, and Carol
Drawing any one of these 3 names will make the event happen.

List the possible outcomes and circle the favorable outcomes.

(Amy)   (Beth)   (Carol)   Dave   Edgar   Frank   George   Hank

The probability of drawing a girl's name equals $\dfrac{\text{number of favorable outcomes}}{\text{number of possible outcomes}} = \frac{3}{8}$.

Example | Use 1 each of the number cards 1, 4, 6, 8, 10, 14, and 20. Draw 1 card without looking. What is the probability that the number is less than 10?

Event: Get a number less than 10.

Possible outcomes: 1, 4, 6, 8, 10, 14, and 20

The card is drawn without looking. So the outcomes are equally likely.

Favorable outcomes: 1, 4, 6, and 8   ① ④ ⑥ ⑧ 10   14   20

The probability of getting a number less than 10 equals

$$\dfrac{\text{number of favorable outcomes}}{\text{number of possible outcomes}} = \frac{4}{7}, \text{ or } 0.571, \text{ or } 57.1\%.$$

Listing all of the possible outcomes is sometimes confusing. Study the example below.

Example   What are the possible outcomes for the spinner shown here?

The spinner is divided into 10 equal sections. When you spin the spinner, it may land on any one of those 10 sections. So, there are 10 possible outcomes. But how do you list the 10 sections?

If you include both the number and the color in your list it will look like this:

1 blue   2 red   3 yellow   4 blue   5 orange   6 yellow   7 red   8 yellow   9 blue   10 orange

If you list only the number for each section, it will look like this:

1   2   3   4   5   6   7   8   9   10

The short list of numbers is good enough. If you know the number, you can always look at the spinner to find the color that goes with that number.

Example   What is the probability that the spinner shown above will land on a prime number?

Event: Land on a prime number.

Possible outcomes: 1, 2, 3, 4, 5, 6, 7, 8, 9, 10

The sections are equal parts of the spinner, so the outcomes are equally likely.

Favorable outcomes: the prime numbers 2, 3, 5, and 7

1   ② ③ 4   ⑤ 6   ⑦ 8   9   10

The probability of landing on a prime number equals

$$\frac{\text{number of favorable outcomes}}{\text{number of possible outcomes}} = \frac{4}{10}, \text{ or } 0.4, \text{ or } 40\%.$$

Example   What is the probability that the spinner shown above will land on a blue or yellow section that has an even number?

The possible outcomes are 1, 2, 3, 4, 5, 6, 7, 8, 9, and 10.

The favorable outcomes are the sections that have an even number *and* have the color blue or yellow. Only three sections meet these conditions: the sections numbered 4, 6, and 8.

1   2   3   ④ 5   ⑥ 7   ⑧ 9   10

The probability of landing on a section that has an even number and is blue or yellow equals

$$\frac{\text{number of favorable outcomes}}{\text{number of possible outcomes}} = \frac{3}{10}, \text{ or } 0.3, \text{ or } 30\%.$$

In some problems, there may be several outcomes that look exactly the same.

Example   Two red blocks and 3 blue blocks are placed in a bag. All of the blocks are cubes of the same size. One block is drawn without looking. What is the probability of drawing a red block?

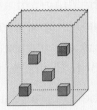

What is the *event* you want to find the probability of? Draw a red block.

How many *possible outcomes* are there? Five

It is true that the red blocks look the same and the blue blocks look the same. But there are 5 *different blocks* in the bag. So when you draw a block from the bag, there are 5 *possible results* or outcomes.

When you list the possible outcomes, be sure to include each of the 5 blocks in your list.

red   red   blue   blue   blue

Are the outcomes *equally likely?* Yes. All 5 blocks are cubes of the same size. We can assume that each block has the same $\frac{1}{5}$ chance of being drawn.

Which of the possible outcomes in your list are *favorable outcomes?* The 2 red outcomes. Drawing a red block will make the event happen.

Circle the favorable outcomes.   ⬭red⬭ ⬭red⬭ blue   blue   blue

The probability of drawing a red block equals $\dfrac{\text{number of favorable outcomes}}{\text{number of possible outcomes}} = \frac{2}{5}$.

Example   A bag contains 1 green, 2 blue, and 3 red counters. The counters are the same, except for color. One counter is drawn. What is the probability of drawing a blue counter?

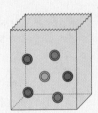

Event: Draw a blue counter.

Possible outcomes: green, blue, blue, red, red, red
The counters are the same, except for color.
So, the 6 outcomes are equally likely.

Favorable outcomes: blue, blue   green ⬭blue⬭ ⬭blue⬭ red   red   red

The probability of drawing a blue counter equals $\dfrac{\text{number of favorable outcomes}}{\text{number of possible outcomes}} = \frac{2}{6}$, or $\frac{1}{3}$.

Check Your Understanding

Six red, 4 green, and 3 blue blocks are placed in a bag. The blocks are the same, except for color. One block is drawn without looking.

Find the probability of each event.

1. Draw a green block.   **2.** Draw a block that is *not* green.   **3.** Draw a blue block.

4. Draw a red block.   **5.** Draw a block that is *not* red.   **6.** Draw any block.

Check your answers on page 419.

Tree Diagrams

The diagram at the right represents a maze. Without retracing their steps, people walk through the maze, taking paths at random without any pattern or preference. Some people end up in Room A; some end up in Room B.

The diagrams in the example below represent this maze. Notice that they look like upside-down trees.

Tree diagrams like those below can help in analyzing probability situations. Tree diagrams are especially useful in probability situations that consist of two or more choices or stages. In a tree diagram, the branches represent different paths, possibilities, or cases.

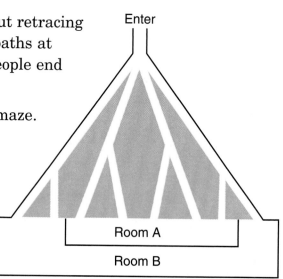

Tree diagrams can help answer questions about a maze.

Example Suppose 40 people walk through the maze at the top of the page. How many can be expected to end up in Room A? In Room B?

Step 1: In the top box of the tree diagram, record the number of people (40) who reach the first intersection where the path divides.

Step 2: The path divides. Since there is an equal chance of selecting any one of the next paths, an equal number of people should select each path. Write 10 in each box in the second row of boxes.

$\frac{1}{4}$ of 40 is 10.

Step 3: Each path divides at the next intersection. Again, an equal number of people should select each of the next paths. Write 5 in the next set of boxes.

$\frac{1}{2}$ of 10 is 5.

Step 4: Add to find how many people reach each room. Five boxes are labeled A, representing exits into Room A. 25 people end up in Room A. Three boxes are labeled B, representing exits into Room B. 15 people end up in Room B.

So, 25 people can be expected to end up in Room A and 15 people can be expected to end up in Room B.

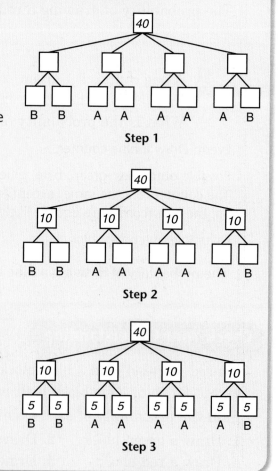

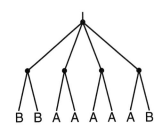

A different kind of tree diagram can help find probabilities.

The tree diagram at the right represents the maze on page 154.

The branching paths are shown, but there are no boxes. You can use this kind of tree diagram to calculate probabilities.

Example   Find the probability of ending up in Room A and the probability of ending up in Room B.

Step 1: At the first place where the path divides, there is an equal chance of selecting each path leading from it. Write this probability next to each path. There are four paths, so the probability of taking any one path is $\frac{1}{4}$.

Step 2: Each of these paths divides at the next intersection. Again, an equal number of people should select each of the next paths. Each path divides into two paths, so the probability of taking any one path is $\frac{1}{2}$. Write the appropriate probability next to each path.

Step 3: There are 8 exits. To find the probability of reaching a particular exit, multiply the probabilities of taking the paths leading to that exit.

For any exit, the probability of taking the first path to it is $\frac{1}{4}$. The probability of the next path is $\frac{1}{2}$. So, the probability of reaching a particular exit is $\frac{1}{2}$ of $\frac{1}{4} = \frac{1}{2} * \frac{1}{4} = \frac{1}{8}$.

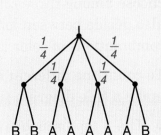

Step 1

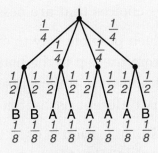

Steps 2 and 3

Step 4: Add to find the probability of entering each room.
There are 5 exits to Room A, each with a probability of $\frac{1}{8}$.
There are 3 exits to Room B, each with a probability of $\frac{1}{8}$.

So, the probability of entering Room A is $\frac{1}{8} + \frac{1}{8} + \frac{1}{8} + \frac{1}{8} + \frac{1}{8}$, or $\frac{5}{8}$.

The probability of entering Room B is $\frac{1}{8} + \frac{1}{8} + \frac{1}{8}$, or $\frac{3}{8}$.

Note that the sum of all the final probabilities is 1 because $\frac{5}{8} + \frac{3}{8} = 1$.

Check Your Understanding

The map shows the roads from the town Alpha. Suppose you start in Alpha and drive south. When the road divides, you choose the next road at random. Draw a tree diagram to help you find the probability of getting from Alpha:

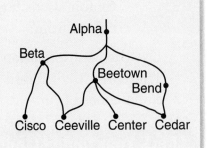

1. to Cisco.   **2.** to Ceeville.   **3.** to Center.   **4.** to Cedar.

Check your answers on page 420.

Tree Diagrams and the Multiplication Counting Principle

Many situations require two or more choices. **Tree diagrams** can be used to count the number of different ways to make those choices.

For example, suppose Vince is buying a new shirt. He must choose among three colors—white, blue, and green. He must also decide between long or short sleeves. How many different combinations of color and sleeve length are there?

To count the different combinations and see what they are, make a tree diagram like the one at the right.

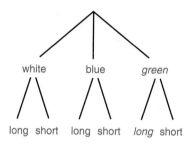

♦ The 3 top branches of the diagram are labeled white, blue, and green to show the color choices.

♦ The 2 branches below each color show the sleeve-length choices that are possible.

Each possible way to choose a shirt is found by following a path from the top to the bottom of the diagram. One possible choice is shown in italics: *green-long*. Counting shows that there are six different paths. Six different shirt choices are possible.

Multiplication is used to solve many types of counting problems that involve two or more choices.

Multiplication Counting Principle

Suppose you can make a first choice in m ways and a second choice in n ways. There are $m * n$ ways of making the first choice followed by the second choice.

Note

Cases with 3 or more choices are counted in the same way.

Suppose you can make the first choice in x ways, the second choice in y ways, and the third choice in z ways. Then there are $x * y * z$ different ways to make the 3 choices.

Example

Vince has pants in 4 different colors and shirts in 8 different colors. How many different color combinations for pants and shirts can Vince choose from?

Use the Multiplication Counting Principle: $4 * 8 = 32$. There are 32 different color combinations that Vince could choose from.

Check Your Understanding

Draw a tree diagram that shows all 32 combinations for the above example.

Check your answer on page 420.

Geometry and Constructions

What Is Geometry?

Geometry is the mathematical study of space and objects in space.

Plane geometry concerns 2-dimensional objects (figures) on a flat surface (a plane). Figures studied in plane geometry include lines, line segments, angles, polygons, and circles.

Solid geometry is the study of 3-dimensional objects. These objects are often simplified versions of familiar everyday things. For example, *cylinders* are suggested by food cans, *spheres* by balls, and *prisms* by boxes.

Graphs, coordinates, and coordinate grids have also been part of *Everyday Mathematics*. The branch of geometry dealing with figures on a coordinate grid is called **coordinate geometry,** or **analytic geometry.** Analytic geometry combines algebra with geometry.

Transformation geometry is the study of geometric properties under motion. **Translations** (slides), **reflections** (flips), and **rotations** (turns) are familiar operations in transformation geometry.

A world globe is 3-dimensional and shows Earth correctly. Flat maps are 2-dimensional; all flat views of Earth have some distortions.

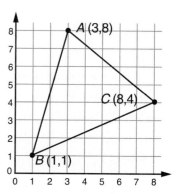

These windmills illustrate rotational symmetry. Each 90° (quarter-turn) rotation of the windmill blades will leave the picture unchanged.

Geometry originated in ancient Egypt and Mesopotamia as a practical tool for surveying land and constructing buildings. (The word *geometry* comes from the Greek words *ge,* meaning *the Earth,* and *metron,* meaning *to measure.*) About 300 B.C., the Greek mathematician Euclid gathered the geometric knowledge of his time into a book known as the *Elements.* Euclid's *Elements* is one of the great achievements of human thought. It begins with ten unproved statements, called postulates and common notions. In modern wording, one of the postulates reads, "Through a given point not on a given line, there is exactly one line parallel to the given line."

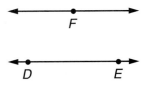

There is exactly one line through point *F* parallel to $\overleftrightarrow{DE}$.

Euclid used logic to deduce several hundred propositions (theorems) from the postulates and common notions—for example, "In any triangle, the side opposite the greater angle is greater."

Mathematicians began to develop other forms of geometry in the seventeenth century, beginning with René Descartes' analytic geometry (1637). The problem of perspective in drawings and paintings led to **projective geometry.**

In the nineteenth century, mathematicians explored the results of changing Euclid's postulate about parallel lines quoted above. In **non-Euclidean geometry,** there are either no lines or at least two lines parallel to a given line through a given point.

Topology, a modern branch of geometry, deals with properties of geometric objects that do not change when their shapes are changed. You can read more about how to decide if two objects are topologically equivalent on pages 184 and 185.

Even though the rails in this track are parallel and always the same distance apart, they appear to meet at a common point on the horizon. This is one of the laws of perspective.

A coffee mug and a doughnut are topologically equivalent because they both have one hole.

A lidless teapot with a handle and a trophy cup with two handles are topologically equivalent because they both have two holes.

Angles

An angle is formed by 2 rays or 2 line segments that share the same endpoint. The symbol for an angle is ∠.

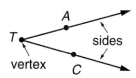

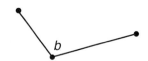

angle formed by 2 rays
names: ∠T, or ∠ATC, or ∠CTA

angle formed by 2 segments
name: ∠b

Measuring Angles

The **protractor** is a tool used to measure angles. Angles are measured in **degrees.** A degree is the unit of measure for the size of an angle.

The **degree symbol** (°) is often used in place of the word *degrees.* The measure of ∠T above is 30 degrees, or 30°.

Sometimes there is confusion about which angle should be measured. The small curved arrow in each picture below shows which angle opening should be measured.

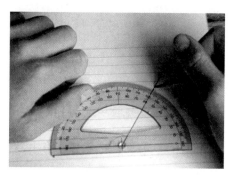

This half-circle protractor may be used to measure and to draw angles of any size.

 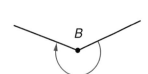 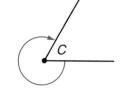

Measure of ∠A is 60°.   Measure of ∠B is 225°.   Measure of ∠C is 300°.

Classifying Angles

Angles may be classified according to their size.

   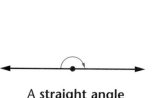 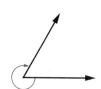

A **right angle** measures 90°.

An **acute angle** measures between 0° and 90°.

An **obtuse angle** measures between 90° and 180°.

A **straight angle** measures 180°.

A **reflex angle** measures between 180° and 360°.

Parallel Lines and Segments

Parallel lines are lines on a flat surface that never meet or cross. Think of a railroad track that goes on forever. The two rails are parallel lines. The rails never meet or cross and are always the same distance apart.

Parallel line segments are segments that are parts of lines that are parallel. The top and bottom edges of this page are parallel. If each edge were extended forever in both directions, the lines would be parallel.

The symbol for *parallel* is a pair of vertical lines ‖. If $\overline{BF}$ and $\overline{TG}$ are parallel, write $\overline{BF} \parallel \overline{TG}$.

If lines or segments cross or meet each other, they **intersect.** Lines or segments that intersect and form right angles are called **perpendicular** lines or segments.

The symbol for *perpendicular* is ⊥, which looks like an upside-down letter *T*. If $\overleftrightarrow{SU}$ and $\overleftrightarrow{XY}$ are perpendicular, write $\overleftrightarrow{SU} \perp \overleftrightarrow{XY}$.

Examples

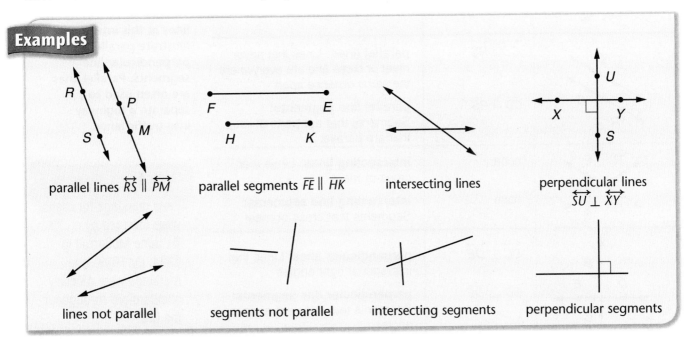

parallel lines $\overleftrightarrow{RS} \parallel \overleftrightarrow{PM}$ parallel segments $\overline{FE} \parallel \overline{HK}$ intersecting lines   perpendicular lines $\overleftrightarrow{SU} \perp \overleftrightarrow{XY}$

lines not parallel   segments not parallel   intersecting segments   perpendicular segments

Check Your Understanding

Draw and label the following.

1. Parallel line segments *AB* and *JK*

2. A line segment that is perpendicular to both $\overline{EF}$ and $\overline{CD}$

Check your answers on page 420.

Line Segments, Rays, Lines, and Angles

Figure	Symbol	Name and Description
• A	point A	**point:** A location in space
F, C endpoints	$\overline{CF}$, or $\overline{FC}$	**line segment:** A straight path between two points, called its endpoints
K, L endpoint	$\overrightarrow{LK}$	**ray:** A straight path that goes on forever in one direction from an endpoint
P, R	$\overleftrightarrow{PR}$, or $\overleftrightarrow{RP}$	**line:** A straight path that goes on forever in both directions
vertex, S, R, P	$\angle R$, or $\angle SRP$, or $\angle PRS$	**angle:** Two rays or line segments with a common endpoint, called the vertex
C, D, S, R	$\overleftrightarrow{CD} \parallel \overleftrightarrow{RS}$; $\overline{CD} \parallel \overline{RS}$	**parallel lines:** Lines that never meet or cross and are everywhere the same distance apart; **parallel line segments:** Segments that are parts of lines that are parallel
R, E, D, S	none ; none	**intersecting lines:** Lines that cross or meet; **intersecting line segments:** Segments that cross or meet
B, D, E, C	$\overleftrightarrow{BC} \perp \overleftrightarrow{DE}$; $\overline{BC} \perp \overline{DE}$	**perpendicular lines:** Lines that intersect at right angles; **perpendicular line segments:** Segments that intersect at right angles

The painted dividing lines at this intersection illustrate parallel and perpendicular line segments. Parallel lines are often used to separate a highway into traffic lanes.

Did You Know?

Lane markings for roads were invented by Dr. June McCarroll in 1924. By 1939, lane markings were officially standardized throughout the U.S.

Check Your Understanding

Draw and label each of the following.

1. point X   **2.** $\overleftrightarrow{PQ}$ **3.** $\angle DEF$ **4.** $\overline{LM}$ **5.** $\overleftrightarrow{ST} \parallel \overleftrightarrow{JK}$ **6.** $\overrightarrow{TU}$

Check your answers on page 420.

Parallel Lines and Angle Relationships

Two angles that have a common vertex and common side and whose interiors do not overlap are called **adjacent angles.** Angles *a* and *b* in Figure 1 are adjacent angles.

When two lines intersect, the angles opposite each other are called **vertical angles,** or **opposite angles.** The measures of any pair of vertical angles are equal.

The sum of the measures of adjacent angles, formed by two intersecting lines, is 180°. Two angles whose measures total 180° are called **supplementary angles.**

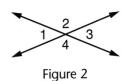

Figure 1

> **Examples** The intersecting lines in Figure 2 form four angles, numbered 1, 2, 3, and 4.
>
> ∠1 and ∠3 are vertical angles and have the same measure. ∠1 and ∠2 are supplementary angles. The sum of their measures is 180°.

Figure 2

A line that crosses two lines is called a **transversal.** Any two angles formed by one of the lines and the transversal are either vertical or supplementary angles.

> **Examples** ∠*b* and ∠*d* in Figure 3 are vertical angles. They have the same measure.
>
> ∠*a* and ∠*b* are supplementary angles. The sum of their measures is 180°.

Any pair of angles *between* two parallel lines that are *on the same side* of the transversal are supplementary angles.

> **Example** ∠*c* and ∠*f* in Figure 3 are supplementary angles. The sum of their measures is 180°.

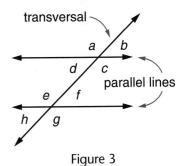

Figure 3

Check Your Understanding

1. In Figure 3 above, which of the following pairs of angles have the same measure? Which are supplementary angles?
   a. ∠*e* and ∠*g*   **b.** ∠*d* and ∠*e*   **c.** ∠*b* and ∠*g*
   d. ∠*d* and ∠*c*   **e.** ∠*c* and ∠*e*   **f.** ∠*h* and ∠*b*

2. In the parallelogram at the right, what is the measure of
   a. ∠1?   **b.** ∠2?   **c.** ∠3?

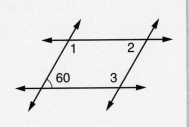

Check your answers on page 420.

Polygons

A **polygon** is a flat, **2-dimensional** figure made up of line segments called **sides.** A polygon can have any number of sides, as long as it has at least three. The **interior** (inside) of the polygon is not a part of the polygon. A polygon can enclose only one interior.

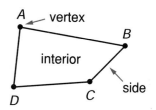

- The sides of a polygon are connected end to end and make one closed path.

- The sides of a polygon do not cross.

Each endpoint where two sides of a polygon meet is called a **vertex.** The plural of vertex is **vertices.**

Figures That Are Polygons

4 sides, 4 vertices   3 sides, 3 vertices   7 sides, 7 vertices

Figures That Are NOT Polygons

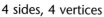 

All sides of a polygon must be line segments. Curved lines are not line segments.

The sides of a polygon must form a closed path.

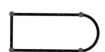

A polygon must have at least 3 sides.

The sides of a polygon must not cross.

Prefixes	
tri-	3
quad-	4
penta-	5
hexa-	6
hepta-	7
octa-	8
nona-	9
deca-	10
dodeca-	12

Polygons are named after the number of their sides. The prefix for a polygon's name tells the number of sides it has.

Convex Polygons

A **convex** polygon is a polygon in which all the sides are pushed outward. The polygons below are all convex.

triangle

quadrangle
(or quadrilateral)

pentagon

hexagon

heptagon

octagon

nonagon

decagon

Nonconvex (Concave) Polygons

A **nonconvex,** or **concave,** polygon is a polygon in which at least two sides are pushed in. The four polygons at the right are nonconvex.

quadrangle
(or quadrilateral)

pentagon

hexagon

octagon

Regular Polygons

A polygon is a **regular polygon** if (1) the sides all have the same length and (2) the angles inside the figure all have the same measure. A regular polygon is always convex. The polygons below are all regular.

equilateral triangle

square

regular pentagon

regular hexagon

regular octagon

regular nonagon

Check Your Understanding

1. What is the name of a polygon that has
   a. 8 sides?   b. 4 sides?   c. 6 sides?

2. a. Draw a convex heptagon.   b. Draw a concave decagon.

3. Explain why a journal page is not a regular polygon.

Check your answers on page 420.

Triangles

Triangles are the simplest type of polygon. The prefix *tri-* means *three*. All triangles have 3 vertices, 3 sides, and 3 angles.

For the triangle shown here:

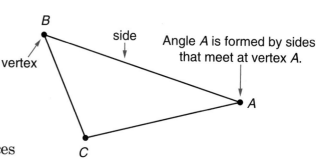

- The vertices are the points *B*, *C*, and *A*.
- The sides are $\overline{BC}$, $\overline{BA}$, and $\overline{CA}$.
- The angles are $\angle B$, $\angle C$, and $\angle A$.

Triangles have 3-letter names. You name a triangle by listing the letter names for the vertices *in order*. The triangle above has 6 possible names: triangle *BCA*, *BAC*, *CAB*, *CBA*, *ABC*, and *ACB*.

Triangles may be classified according to the length of their sides.

A **scalene triangle** is a triangle whose sides all have different lengths.

An **isosceles triangle** is a triangle that has two sides of the same length.

An **equilateral triangle** is a triangle whose sides all have the same length.

Triangles may also be classified according to the size of their angles.

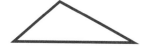 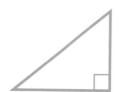 

An **acute triangle** is a triangle whose angles all are acute.

A **right triangle** is a triangle with one right angle.

An **obtuse triangle** is a triangle with one obtuse angle.

Check Your Understanding

1. Draw and label an equilateral triangle named *JKL*. Write the five other possible names for this triangle.
2. Draw an isosceles triangle.
3. Draw a right scalene triangle.

Check your answers on page 420.

The Theorem of Pythagoras

A right triangle is a triangle that has a right angle (90°). In a right triangle, the side opposite the right angle is called the **hypotenuse.** The other two sides are called **legs.**

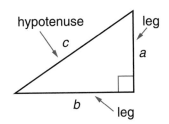

In the diagram at the right, a and b represent the lengths of the legs, and c represents the length of the hypotenuse.

There is a surprising connection between the lengths of the three sides of any right triangle. It is probably the most useful property in all of geometry and is known as the Pythagorean theorem. It is stated in the box at the right.

Nobody knows when this relationship was first discovered. The Babylonians, Egyptians, and Chinese knew of it before the Greeks. But Pythagoras, a Greek philosopher born about 572 B.C., was the first person to prove that the relationship is true for any right triangle. It is called a **theorem** because it is a statement that has been proved.

> **Pythagorean Theorem**
>
> If the legs of a right triangle have lengths a and b, and the hypotenuse has length c, then $a^2 + b^2 = c^2$.

A Chinese proof of the Pythagorean theorem (written about A.D. 40) is shown below. Two identical squares, each with sides of length $a + b$, are partitioned in different ways.

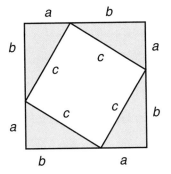

This square contains four identical right triangles and one square whose area is $c * c = c^2$.

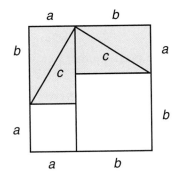

This square contains four identical right triangles and two squares whose areas are a^2 and b^2.

> **Did You Know?**
>
> By analyzing the writing on ancient clay tablets, historians have determined that the Babylonians knew of the Pythagorean theorem.

The four right triangles inside each square all have the same area. Therefore, the area of the large square (c^2) inside the first square must be equal to the total area of the two smaller squares ($a^2 + b^2$) that are inside the second square; that is, c^2 must equal $a^2 + b^2$.

Quadrangles

A **quadrangle** is a polygon that has 4 sides. Another name for quadrangle is **quadrilateral.** The prefix *quad-* means *four.* All quadrangles have 4 vertices, 4 sides, and 4 angles.

For the quadrangle shown here:

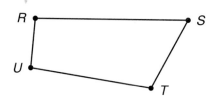

♦ The sides are $\overline{RS}$, $\overline{ST}$, $\overline{TU}$, and $\overline{UR}$.

♦ The vertices are *R*, *S*, *T*, and *U*.

♦ The angles are ∠*R*, ∠*S*, ∠*T*, and ∠*U*.

A quadrangle is named by listing the letter names for the vertices in order. The quadrangle above has 8 possible names:

RSTU, RUTS, STUR, SRUT, TURS, TSRU, URST, and *UTSR*

Some quadrangles have two pairs of parallel sides. These quadrangles are called **parallelograms.**

Reminder: Two sides are parallel if they never cross or meet, no matter how far they are extended.

Figures That Are Parallelograms

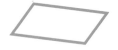

Opposite sides are parallel in each figure.

Figures That Are NOT Parallelograms

No parallel sides   Only 1 pair of parallel sides   3 pairs of parallel sides

A parallelogram must have exactly 2 pairs of parallel sides.

Check Your Understanding

1. Draw and label a quadrangle named *EFGH* that has exactly one pair of parallel sides.

2. Is *EFGH* a parallelogram?

3. Write the seven other possible names for this quadrangle.

Check your answers on page 421.

Special types of quadrangles have been given names. Some of these are parallelograms; others are not.

The tree diagram below shows how the different types of quadrangles are related. For example, quadrangles are divided into two major groups— "parallelograms" and "not parallelograms." The special types of parallelograms include rectangles, rhombuses, and squares.

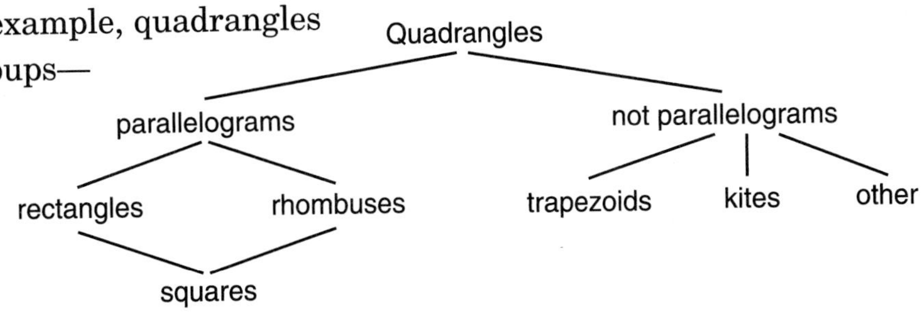

Quadrangles

parallelograms → rectangles, rhombuses → squares

not parallelograms → trapezoids, kites, other

Quadrangles That Are Parallelograms		
rectangle	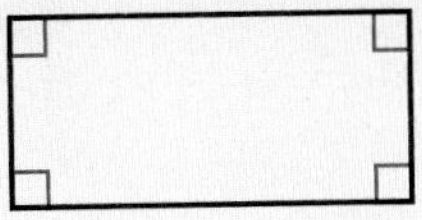	**Rectangles** are parallelograms. A rectangle has 4 right angles (square corners). The sides do not all have to be the same length.
rhombus	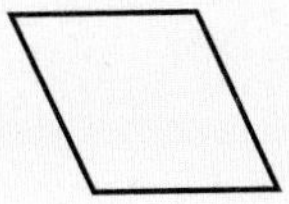	**Rhombuses** are parallelograms. A rhombus has 4 sides that are all the same length. The angles of a rhombus are usually not right angles, but they may be.
square	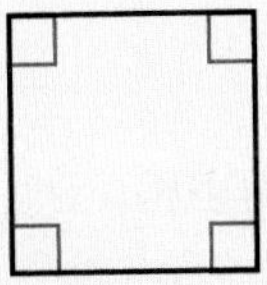	**Squares** are parallelograms. A square has 4 right angles (square corners). Its 4 sides are all the same length. *All* squares are rectangles. *All* squares are also rhombuses.

Quadrangles That Are NOT Parallelograms		
trapezoid	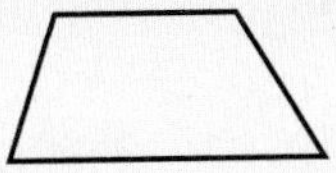	**Trapezoids** have exactly 1 pair of parallel sides. The 4 sides of a trapezoid can all have different lengths.
kite	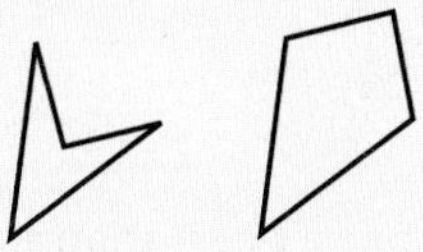	**Kites** are quadrangles with 2 pairs of equal sides. The equal sides are next to each other. The 4 sides cannot all have the same length. (A rhombus is not a kite.)
other	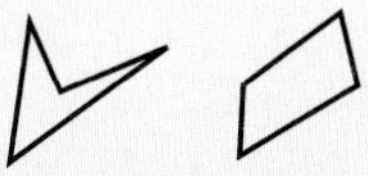	Any polygon with 4 sides that is not a parallelogram, a trapezoid, or a kite.

Check Your Understanding

What is the difference between the quadrangles in each pair below?

1. a rhombus and a rectangle
2. a trapezoid and a square
3. a kite and a parallelogram

Check your answers on page 421.

Geometry and Constructions

Geometric Solids

Polygons and circles are flat, **2-dimensional** figures. The surfaces they enclose take up a certain amount of area, but they do not have any thickness and do not take up any volume.

Three-dimensional shapes have length, width, *and* thickness. They take up volume. Boxes, chairs, books, cans, and balls are all examples of 3-dimensional shapes.

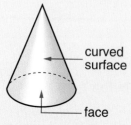

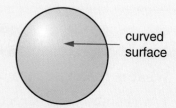

A **geometric solid** is the surface or surfaces that surround a 3-dimensional shape. The surfaces of a geometric solid may be flat, curved, or both. Despite its name, a geometric solid is hollow. It does not include the points within its interior.

♦ A **flat surface** of a solid is called a **face.**

♦ A **curved surface** of a solid does not have any special name.

Examples Describe the surfaces of each geometric solid.

A cube has 6 square faces that are the same size.

A rectangular pyramid has 4 triangular faces and 1 rectangular face.

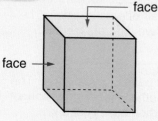

A cone has 1 circular face and 1 curved surface. The circular face is called its **base.**

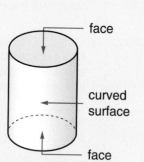

A cylinder has 1 curved surface. It has 2 circular faces that are the same size and are parallel. The two faces are called its **bases.**

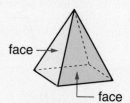

A sphere has 1 curved surface.

 170 one hundred seventy

The **edges** of a geometric solid are the line segments or curves where surfaces meet.

A corner of a geometric solid is called a **vertex** (plural, *vertices*).

A vertex is a point at which edges meet. The vertex of a cone is an isolated corner completely separated from the edge of the cone.

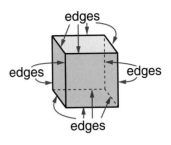

A cone has 1 edge and 1 vertex. This vertex is opposite the circular base and is called the **apex.**

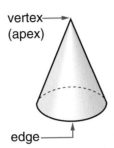

vertex (apex)

edge

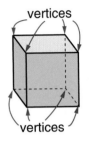

A cube has 12 edges and 8 vertices.

The pyramid shown below has 8 edges and 5 vertices. The vertex opposite the rectangular base is called the **apex.**

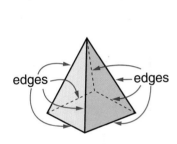

edges — edges

vertices

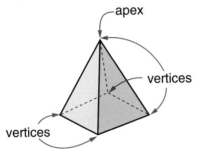

apex

vertices

vertices

edge

A cylinder has 2 edges. It has no vertices.

edge

A sphere has no edges and no vertices.

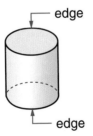

Check Your Understanding

1. **a.** How are cylinders and spheres alike? **b.** How are they different?
2. **a.** How are pyramids and cones alike? **b.** How are they different?

Check your answers on page 421.

Polyhedrons

A **polyhedron** is a geometric solid whose surfaces are all formed by polygons. These surfaces are the faces of the polyhedron. A polyhedron does not have any curved surfaces.

Two important groups of polyhedrons are shown below. These are **pyramids** and **prisms.**

Pyramids

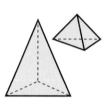
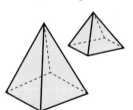

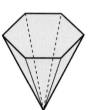

| triangular pyramids | rectangular pyramids | pentagonal pyramid | hexagonal pyramid |

Prisms

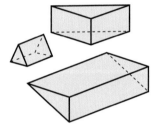

| triangular prisms | rectangular prisms | hexagonal prism |

Many polyhedrons are neither pyramids nor prisms. Some examples are shown below.

Polyhedrons That Are NOT Pyramids or Prisms

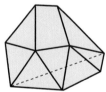
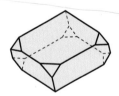

To find out why these are neither pyramids nor prisms, read pages 173 and 174.

To find out why these are neither pyramids nor prisms, read pages 173 and 174.

Did You Know?

A **geodesic dome** is a polyhedron whose faces are all formed by triangles. The vertices of all of the triangles lie on the surface of a sphere. From a distance, the dome looks like a sphere.

The main building of the Amundsen-Scott South Pole Station is the top half of a geodesic dome.

Check Your Understanding

1. **a.** How many faces does a rectangular pyramid have?
   b. How many faces have a rectangular shape?

2. **a.** How many faces does a triangular prism have?
   b. How many faces have a triangular shape?

Check your answers on page 421.

Prisms

All of the geometric solids below are **prisms.**

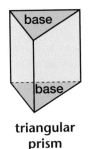

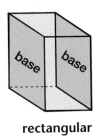

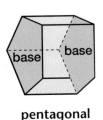

triangular
prism

rectangular
prism

pentagonal
prism

hexagonal
prism

The two shaded faces of each prism are called **bases.**

♦ The bases have the same size and shape.

♦ The bases are parallel. This means that the bases will never meet, no matter how far they are extended.

♦ All other faces connect the bases and are parallelograms.

The shape of its bases is used to name a prism. If the bases are triangular shapes, it is called a **triangular prism.** If the bases are rectangular shapes, it is called a **rectangular prism.** Rectangular prisms have three possible pairs of bases.

The number of faces, edges, and vertices that a prism has depends on the shape of the base.

Note

Notice that the edges connecting the bases of a prism are parallel to each other.

Did You Know?

A triangular prism made of glass may be used to separate light into colors. The band of separated colors is called the *spectrum.*

Example The triangular prism shown here has 5 faces—3 rectangular faces and 2 triangular bases. It has 9 edges and 6 vertices.

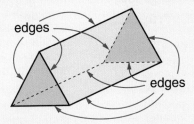

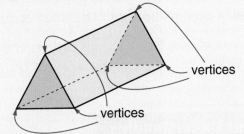

Check Your Understanding

1. **a.** How many faces does a pentagonal prism have?
   b. How many edges?
   c. How many vertices?
2. What is the name of a prism that has 10 faces?

Check your answers on page 421.

Pyramids

All of the geometric solids below are **pyramids.**

triangular pyramid

square pyramid

pentagonal pyramid

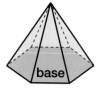
hexagonal pyramid

The shaded face of each pyramid is called the **base** of the pyramid.

♦ The polygon that forms the base can have any number of sides.

♦ The faces that are not a base all have a triangular shape.

♦ The faces that are not a base all meet at the same vertex.

The shape of its base is used to name a pyramid. If the base is a triangle shape, it is called a **triangular pyramid.** If the base is a square shape, it is called a **square pyramid.**

The great pyramids of Giza were built near Cairo, Egypt, around 2600 B.C. They have square bases and are square pyramids.

The number of faces, edges, and vertices that a pyramid has depends on the shape of the base.

The largest of the Giza pyramids covers about 55,000 square meters at its base and is 137 meters high.

Example The hexagonal pyramid shown here has 7 faces— 6 triangular faces and a hexagonal base.

It has 12 edges. Six edges surround the hexagonal base. The other 6 edges meet at the apex (tip) of the pyramid.

It has 7 vertices. Six vertices are on the hexagonal base. The remaining vertex is the apex of the pyramid.

apex

The apex is the vertex opposite the base.

Check Your Understanding

1. **a.** How many faces does a triangular pyramid have?
   b. How many edges?
   c. How many vertices?

2. What is the name of a pyramid that has 10 edges?

3. **a.** How are prisms and pyramids alike?
   b. How are they different?

Check your answers on page 421.

Regular Polyhedrons

A polyhedron is **regular** if:

♦ Each face is formed by a regular polygon.

♦ The faces all have the same size and shape.

♦ Every vertex looks exactly the same as every other vertex.

There are only five kinds of regular polyhedrons:

Regular Polyhedrons

regular tetrahedron
(pyramid)
(4 faces)

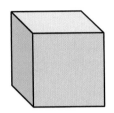
cube
(prism)
(6 faces)

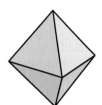
regular
octahedron
(8 faces)

regular
dodecahedron
(12 faces)

regular
icosahedron
(20 faces)

The pictures below show each regular polyhedron with its faces unfolded. There is more than one way to unfold each polyhedron.

regular tetrahedron
(4 equilateral triangles)

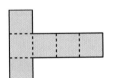
cube
(6 squares)

regular octahedron
(8 equilateral triangles)

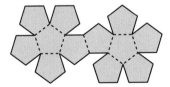
regular dodecahedron
(12 regular pentagons)

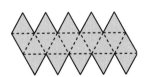
regular icosahedron
(20 equilateral triangles)

Check Your Understanding

1. a. How many edges does a tetrahedron have? **b.** How many vertices?

2. a. How are regular tetrahedrons and regular icosahedrons alike? **b.** How are they different?

Check your answers on page 421.

Circles

A **circle** is a curved line that forms a closed path on a flat surface. All of the points on a circle are the same distance from the **center of the circle.**

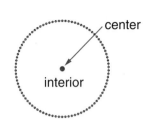

The center is not part of the circle. The interior is not part of the circle.

The **compass** is a tool used to draw circles.

◆ The point of a compass (called the **anchor**) is placed at the center of the circle.

◆ The pencil in a compass traces out a circle. Every point on the circle is the same distance from the anchor.

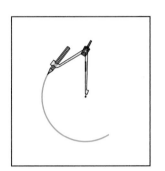

The **radius** (plural, *radii*) of a circle is any line segment that connects the center of the circle with any point on the circle. The word *radius* can also refer to the length of this segment.

The **diameter** of a circle is any line segment that passes through the center of the circle and has its endpoints on the circle. The word *diameter* can also refer to the length of this segment.

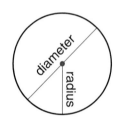

An **arc** is a part of a circle, from one point on the circle to another. For example, a **semicircle** is an arc. Its endpoints are the endpoints of a diameter of the circle.

All circles are similar because they have the same shape, but all circles do not have the same size.

arcs

Example   Many pizzas have a circular shape. You can order a pizza by saying the diameter that you want.

A "12-inch pizza" means a pizza with a 12-inch diameter.

A "16-inch pizza" means a pizza with a 16-inch diameter.

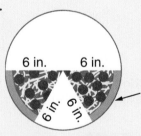

6 in.   6 in.

6 in.   6 in.

A 12-inch pizza

This pizza is 12 inches across. The diameter is 12 inches.

Each slice is a wedge that has 6-inch-long sides.

Spheres

A **sphere** is a geometric solid that has a single curved surface shaped like a ball, a marble, or a globe. All of the points on the surface of the sphere are the same distance from the **center of the sphere.**

All spheres have the same shape. But spheres do not all have the same size. The size of a sphere is the distance across its center.

♦ The line segment *RS* passes through the center of the sphere. This line segment is called a **diameter of the sphere.**

♦ The length of line segment *RS* is also called the diameter of the sphere.

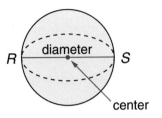

Globes and basketballs are examples of spheres that are hollow. The interior of each is empty. The hollow interior is not part of the sphere. The sphere includes only the points on the curved surface.

Marbles and baseballs are examples of spheres that have solid interiors. In cases like these, think of the solid interior as part of the sphere.

Example

Earth is shaped very nearly like a sphere.

The diameter of Earth is about 8,000 miles.

The distance from Earth's surface to the center of Earth is about 4,000 miles.

Every point on Earth's surface is about 4,000 miles from the center of Earth.

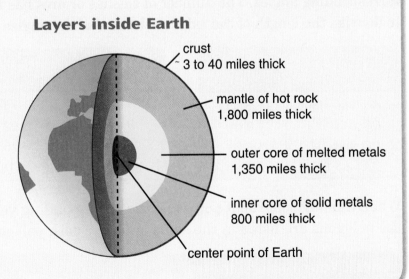

Layers inside Earth

- crust 3 to 40 miles thick
- mantle of hot rock 1,800 miles thick
- outer core of melted metals 1,350 miles thick
- inner core of solid metals 800 miles thick
- center point of Earth

Congruent Figures

Two figures that are the same size and have the same shape are **congruent.**

◆ All line segments have the same shape. Two line segments are congruent if they have the same length.

◆ All circles have the same shape. Two circles are congruent if their diameters have the same length.

◆ All squares have the same shape. Two squares are congruent if their sides have the same length.

◆ Two angles are congruent if they have the same degree measure.

Two figures are congruent if they match exactly when one figure is placed on top of the other. The matching sides of congruent polygons are called **corresponding sides,** and their matching angles are called **corresponding angles.** Each pair of corresponding sides of congruent polygons is the same length, and each pair of corresponding angles has the same degree measure.

Slash marks and arcs are used to identify pairs of corresponding sides and angles. Sides with the same number of slash marks are corresponding sides, and angles with the same number of arcs are corresponding angles. The number of slashes or arcs has nothing to do with the length of the sides or the size of the angles.

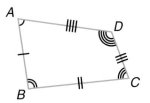

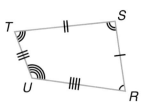

Quadrangles *ABCD* and *RSTU* are congruent.

Corresponding Sides	Length
$\overline{AB}$ and $\overline{RS}$	1.9 cm
$\overline{BC}$ and $\overline{ST}$	2.8 cm
$\overline{CD}$ and $\overline{TU}$	1.3 cm
$\overline{DA}$ and $\overline{UR}$	2.6 cm

Corresponding Angles	Degree Measure
$\angle A$ and $\angle R$	75°
$\angle B$ and $\angle S$	92°
$\angle C$ and $\angle T$	75°
$\angle D$ and $\angle U$	118°

When naming congruent polygons, the corresponding vertices of the polygons are listed in the same order for both polygons.

Check Your Understanding

Which of these triangles is not congruent to the other three?

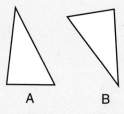

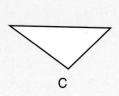

A   B   C   D

Check your answer on page 421.

Similar Figures

Figures that have exactly the same shape are called **similar figures.**
They may also be the same size, but do not have to be. If two figures
are similar, one figure is an enlargement of the other. The **size-
change factor** tells the amount of enlargement or reduction.

In any enlargement, the size of the angles does not change. But if one
polygon is an enlargement of another polygon, each of the sides of the
smaller polygon is enlarged by the same size-change factor. Each side
and its enlargement form a pair of sides called **corresponding sides.**

Example Triangles *BAT* and *HOG* are similar.

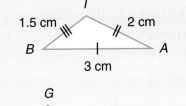

Length of Corresponding Sides	$\overline{HO}$: 6 cm $\overline{BA}$: 3 cm	$\overline{OG}$: 4 cm $\overline{AT}$: 2 cm	$\overline{GH}$: 3 cm $\overline{TB}$: 1.5 cm
Ratio of Lengths	$\frac{6}{3} = \frac{2}{1}$	$\frac{4}{2} = \frac{2}{1}$	$\frac{3}{1.5} = \frac{2}{1}$

The size-change factor is 2X. Each side in
the larger triangle is twice the size of the
corresponding side in the smaller triangle.

Example Quadrangles *ABCD* and *MNOP* are similar.
What is the length of $\overline{AB}$?

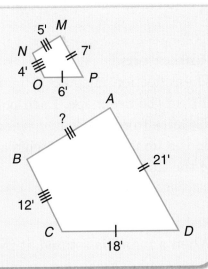

Since the quadrangles are similar, the ratios of the lengths
of any corresponding sides are equal. Find the ratio of the
length of a longer side to the length of the corresponding
shorter side. Choose any pair of corresponding sides.

$\frac{(\text{length of } \overline{BC})}{(\text{length of } \overline{NO})} = \frac{12}{4} = \frac{3}{1}$

To find the length of side *AB*, solve $\frac{3}{1} = \frac{x}{5}$ ← (length of $\overline{AB}$)
← (length of $\overline{MN}$)

$\frac{3}{1} = \frac{3 * 5}{1 * 5} = \frac{15}{5}$

The length of side *AB* is 15 feet.

Check Your Understanding

Polygons *BETH* and *MARY* are similar.

Find the length of these sides.

1. side *MA*   2. side *HT*

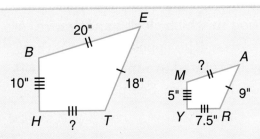

Check your answers on page 421.

SRB

Isometry Transformations

In geometry, a **transformation** is an operation on a figure that produces another figure. Transformations are sometimes thought of as motions that take a figure from one place to another. **Reflections** (flips), **rotations** (turns), and **translations** (slides) are familiar operations in **transformation geometry.**

Each of these transformations produces a new figure called the **image.** The image has the same size and shape as the original figure, called the **preimage.** The image and preimage are congruent figures.

Reflections, translations, and rotations are called **isometry transformations.** They do not change the distances between points. The term *isometry* comes from the Greek words *iso,* meaning *same,* and *metron,* meaning *measure.*

This photograph shows an approximate reflection. The line of reflection is the water's edge, along the bank.

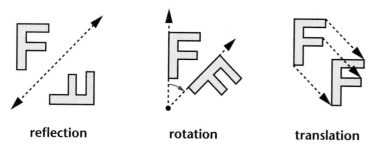

reflection   rotation   translation

Reflections

The reflection image of a figure appears to be a reversal, or flip, of the preimage. Each point on the preimage is the same distance from the *line of reflection* as the corresponding point on the image. The preimage and image are on opposite sides of the line of reflection.

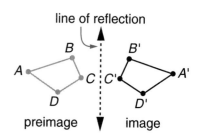

Rotations

When a figure is rotated, it is turned a specific number of degrees in a specific direction around a specific point. A figure can be rotated *clockwise* (the direction in which clock hands move) or *counterclockwise* (the opposite direction). A figure may be rotated around a point outside, on, or inside the figure. See page 183.

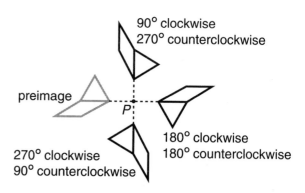

Rotation around the point *P*

Translations

When a figure is translated, each point on the preimage slides the same distance in the same direction to create the image.

Imagine a figure on a coordinate grid. If the same number (for example, 5) is added to the x-coordinates of all the points in the figure and the y-coordinates are not changed, the result is a **horizontal translation.**

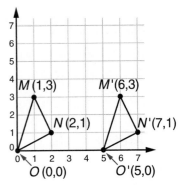

horizontal translation

If the same number (for example, 4) is added to all the y-coordinates and the x-coordinates are not changed, the result is a **vertical translation.**

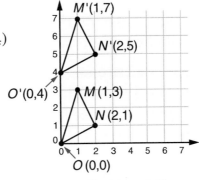

vertical translation

Suppose that the same number (for example, 5) is added to all the x-coordinates and another number (for example, 4) is added to all the y-coordinates. The result is a **diagonal translation.**

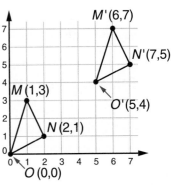

diagonal translation

Check Your Understanding

1. Name the word that results when this figure is reflected over $\overleftrightarrow{AB}$.

2. Triangle $C'A'T'$ is the image of triangle CAT after a slide. Point C is at $(1,1)$, point A at $(3,9)$, point T at $(8,5)$, and point C' at $(4,7)$. What are the coordinates of points A' and T'?

3. Which figure is a 270° clockwise rotation of the preimage?

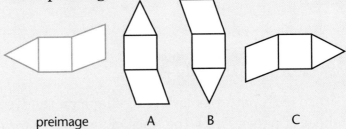

preimage   A   B   C

Check your answers on page 421.

Symmetry

Line Symmetry

A figure is **symmetric about a line** if the line divides the figure into two parts so that both parts look alike but are facing in opposite directions. In a symmetric figure, each point in one of the halves of the figure is the same distance from the **line of symmetry** as the corresponding point in the other half.

Example   The figure shown here is symmetric about the dashed line. The dashed line is its line of symmetry.

Points *A* and *A'* (read as "A prime") are corresponding points. The shortest distance from point *A* to the line of symmetry is equal to the shortest distance from point *A'* to the line of symmetry. The same is true of points *B* and *B'*, points *C* and *C'*, and any other pair of corresponding points.

The line of symmetry is the **perpendicular bisector** of line segments that connect corresponding points such as points *B* and *B'*. It bisects each line segment and is perpendicular to each line segment.

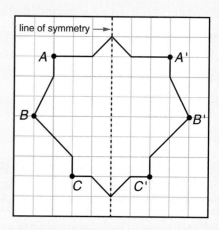

An easy way to check whether a figure has **line symmetry** is to fold it in half. If the two halves match exactly, the figure is symmetric about the fold.

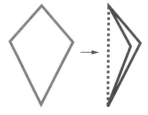

Some figures have more than one line of symmetry.   Some figures have no lines of symmetry.

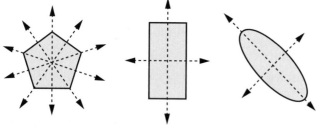 

Check Your Understanding

1. Trace each pattern-block (PB) shape on the Geometry Template onto a sheet of paper. Draw the lines of symmetry for each shape.

2. How many lines of symmetry does a circle have?

Check your answers on page 421.

Rotation Symmetry

A figure has **rotation symmetry** if it can be rotated around a point in such a way that the resulting figure exactly matches the original figure. The rotation must be more than 0 degrees, but less than 360 degrees. The figure may not be flipped over.

So, a figure has rotation symmetry if it can be rotated so the image and preimage match exactly without flipping the figure over.

If a figure has rotation symmetry, its **order of rotation symmetry** is the number of different ways it can be rotated to match itself exactly. "No rotation" is counted as one of the ways.

Example Consider square *ABCD*. The points *A*, *B*, *C*, and *D* are labeled to help identify the rotations. They are not part of the figure.

The square can be rotated in four different ways to match itself exactly (without flipping it over).

The order of rotation symmetry for square *ABCD* is 4.

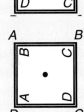

original position   90° rotation   180° rotation   270° rotation

Point Symmetry

A figure has **point symmetry** if it can be rotated 180° around a point to match the original figure exactly. Point symmetry is a special kind of rotation symmetry. A figure with point symmetry has rotation symmetry of at least order 2.

Examples This figure has point symmetry.   This figure does not have point symmetry.

original position   180° rotation

original position   180° rotation

Check Your Understanding

1. Draw a square. Color or shade the interior of the square so that the resulting figure still has rotation symmetry of order 4.

2. Draw a rectangle. Color or shade the interior of the rectangle so that the figure has point symmetry.

Check your answers on page 422.

Topology

Topology is a branch of mathematics that has many connections with geometry. In topology, two geometric shapes are equivalent if one shape can be stretched, squeezed, crumpled, twisted, or turned inside out until it looks like the other shape. These changes are called **topological transformations.** Tearing, breaking, and sticking together are not allowed. These changes are *not* called topological transformations.

Topology studies the properties of geometric shapes that are not changed by topological transformations. The number of holes in an object is one of these properties. For this reason, it is sometimes said that a topologist (a mathematician

who studies topology) can't tell the difference between a coffee mug and a doughnut. They are **topologically equivalent.** A coffee mug and a doughnut will always have one hole that passes through them, no matter how they are transformed.

Consider a juice glass and a doughnut. They are *not* topologically equivalent. No matter how much stretching, twisting, or squeezing you might do, a juice glass cannot be transformed into a doughnut, except by punching a hole in it. This is because a juice glass has no holes through it, whereas a doughnut has one.

In topology, geometric shapes are sorted by the number of holes they have. This property is called the **genus** (pronounced "GEE-nuss") of the shape. Objects with the same genus are topologically equivalent.

The objects below have genus 0.

The objects below have genus 1.

The objects below have genus 2.

  8  B

Topology is sometimes called **rubber sheet geometry.** If you think of a figure as if it were drawn on a rubber sheet or a shape as if it were made from rubber, it may help you to see which properties remain the same after a topological transformation.

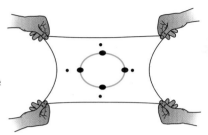

For example, the top diagram at the right shows a circle drawn on a rubber sheet. When the rubber sheet is stretched, the circle is transformed into other figures, as shown in the other two diagrams below.

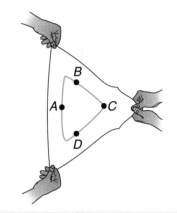

No matter how the circle is transformed, the resulting figure is still a closed curve. Points that were originally inside remain inside, and points that were originally outside remain outside. Points on the circle remain in the same position relative to each other—for example, point B remains between point A and point C.

Properties that do not change when a figure is distorted are called **topological properties.**

Check Your Understanding

1. Triangle XYZ is drawn on a rubber sheet. The sheet is stretched to represent a topological transformation. Which of the following statements are true?

   a. The distance from point X to point Y remains the same.

   b. The measure of angle X remains the same.

   c. The image of side XY might not be a line segment.

   d. The image of triangle XYZ is a triangle.

   e. Figure XYZ remains closed.

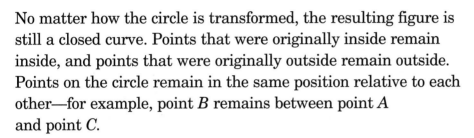

2. **Challenge** Imagine that this shape is made of clay. It may not seem possible, but the loops can be separated—without cutting or tearing them—by a series of topological transformations.

Draw a series of diagrams to show how this might be done.

From *You Are a Mathematician* by David Wells. Copyright (c) 1995 by David Wells. Reprinted by permission of John Wiley & Sons, Inc.

Check your answers on page 422.

The Geometry Template

The **Geometry Template** has many uses.

The template has two rulers. The inch scale measures in inches and fractions of an inch. The centimeter scale measures in centimeters or millimeters. Use either edge of the template as a straightedge for drawing line segments.

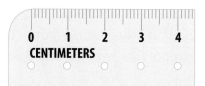

There are 17 different geometric figures on the template. The figures labeled "PB" are **pattern-block shapes.** These are half the size of real pattern blocks. There is a hexagon, a trapezoid, two different rhombuses, an equilateral triangle, and a square. These will come in handy for some of the activities you do this year.

Each triangle on the template is labeled with the letter T and a number. Triangle "T1" is an equilateral triangle whose sides all have the same length. Triangles "T2" and "T5" are right triangles. Triangle "T3" has sides that all have different lengths. Triangle "T4" has two sides of the same length.

The remaining shapes are circles, squares, a regular octagon, a regular pentagon, a kite, a rectangle, a parallelogram, and an ellipse.

The two circles near the inch scale can be used as ring-binder holes so you can store your template in your notebook.

Use the **half-circle** and **full-circle protractors** at the bottom of the template to measure and draw angles. Use the **Percent Circle** at the top of the template to construct and measure circle graphs. The Percent Circle is divided into 1% intervals, and some common fractions of the circle are marked.

Notice the tiny holes near the 0-, $\frac{1}{4}$-, $\frac{2}{4}$-, and $\frac{3}{4}$-inch marks of the inch scale and at each inch mark from 1 to 7. On the centimeter side, the holes are placed at each centimeter mark from 0 to 10. These holes can be used to draw circles.

Did You Know?

A transit is an instrument used by surveyors to measure both horizontal and vertical angles. Using a transit, you can measure an angle to within $\frac{1}{60}$ of a degree.

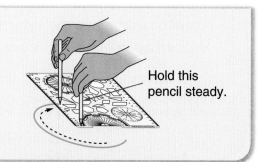

Example   Draw a circle with a 3-inch radius.

Place one pencil point in the hole at 0 inches. Place another pencil point in the hole at 3 inches. Hold the pencil at 0 inches steady while rotating the pencil at 3 inches (along with the template) to draw the circle.

Hold this pencil steady.

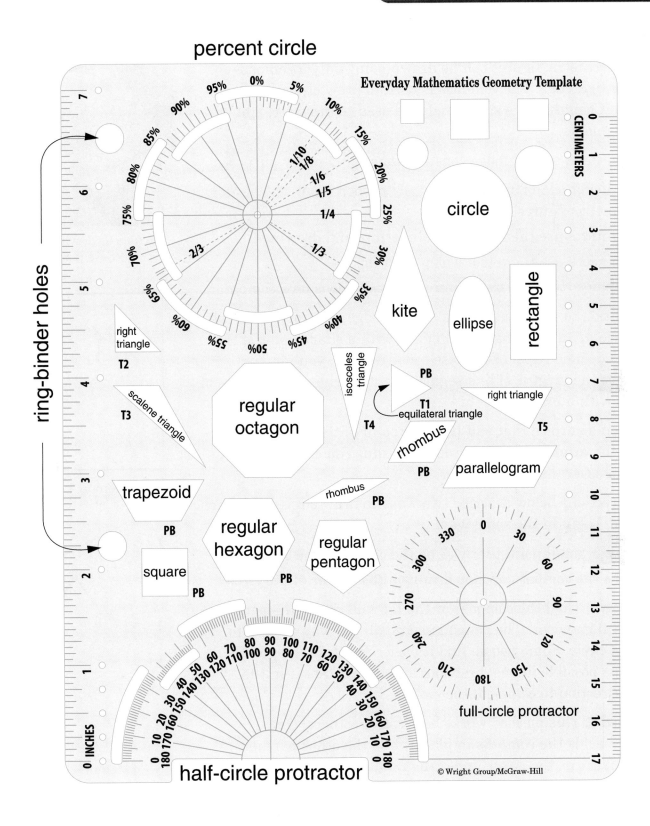

percent circle

ring-binder holes

Everyday Mathematics Geometry Template

circle

rectangle

kite

ellipse

right triangle
T2

scalene triangle
T3

regular octagon

isosceles triangle
T4

equilateral triangle
T1
PB

right triangle
T5

rhombus
PB

parallelogram

trapezoid
PB

rhombus
PB

regular hexagon
PB

regular pentagon

square
PB

full-circle protractor

half-circle protractor

© Wright Group/McGraw-Hill

Compass-and-Straightedge Constructions

Many geometric figures can be drawn using only a compass and straightedge. The compass is used to draw circles and mark off lengths. The straightedge is used to draw straight line segments.

Compass-and-straightedge constructions serve many purposes:

♦ Mathematicians use them to study properties of geometric figures.

♦ Architects use them to make blueprints and drawings.

♦ Engineers use them to develop their designs.

♦ Graphic artists use them to create illustrations on a computer.

Architect's drawing of a house plan

In addition to a compass and a straightedge, the only materials you need are a drawing tool (a pencil with a sharp point is the best) and some paper. For these constructions, you may not measure the lengths of line segments with a ruler or the sizes of angles with a protractor.

Draw on a surface that will hold the point of the compass (also called the **anchor**) so that it does not slip. You can draw on a stack of several sheets of paper.

Method 1

The directions below describe two ways to draw circles. For each method, begin in the same way:

♦ Draw a small point that will be the center of the circle.

♦ Press the compass anchor firmly on the center of the circle.

Method 1 Hold the compass at the top and rotate the pencil around the anchor. The pencil must go all the way around to make a circle. Some people find it easier to rotate the pencil as far as possible in one direction, and then rotate it in the other direction to complete the circle.

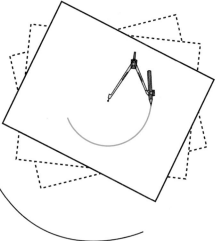

Method 2 This method works best with partners. One partner holds the compass in place while the other partner carefully turns the paper under the compass to form the circle.

Method 2

Check Your Understanding

Concentric circles are circles that have the same center.

Use a compass to draw 3 concentric circles.

concentric circles

Copying a Line Segment

Follow each step carefully. Use a clean sheet of paper.

Step 1: Draw line segment AB.

Step 2: Draw a second line segment that is longer than segment AB. Label one of its endpoints as A' (read as "A prime").

Step 3: Place the compass anchor at A and the pencil point at B.

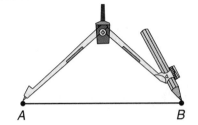

Step 4: Without changing your compass opening, place the compass anchor on A' and draw a small arc that crosses the line segment. Label the point where the arc crosses the line segment as B'.

The segments $A'B'$ and AB have the same length.

Line segment $A'B'$ is **congruent** to line segment AB.

Check Your Understanding

Draw a line segment. Using a compass and a straightedge only, copy the line segment.

After you make your copy, measure the segments with a ruler to see how accurately you copied the original line segment.

Copying a Triangle

Follow each step carefully. Use a clean sheet of paper.

Step 1: Draw triangle *ABC*. Draw a line segment that is longer than line segment *AB*. Copy line segment *AB* onto the segment you just drew. (See page 189.) Label the endpoints of the copy as *A'* and *B'* (read as "A prime" and "B prime").

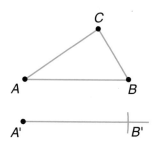

Step 2: Place the compass anchor at *A* and the pencil point at *C*. Without changing your compass opening, place the compass anchor on *A'* and draw an arc.

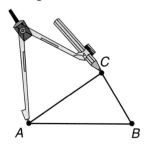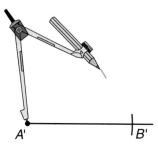

Step 3: Place the compass anchor at *B* and the pencil point at *C*. Without changing your compass opening, place the compass anchor on *B'* and draw another arc. Label the point where the arcs intersect as *C'*.

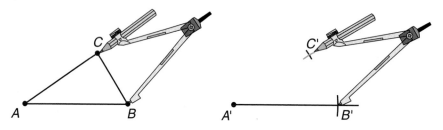

Step 4: Draw line segments *A'C'* and *B'C'*.

Triangles *ABC* and *A'B'C'* are congruent. That is, they are the same size and shape.

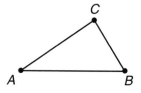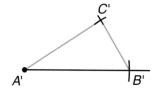

Check Your Understanding

Draw a triangle. Using a compass and a straightedge, copy the triangle.

Cut out the copy and place it on top of the original triangle to check that the triangles are congruent.

Constructing a Parallelogram

Follow each step carefully. Use a clean sheet of paper.

Step 1: Draw angle *ABC*.

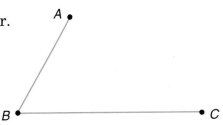

Step 2: Place the compass anchor at *B* and the pencil point at *C*. Without changing your compass opening, place the compass anchor on *A* and draw an arc.

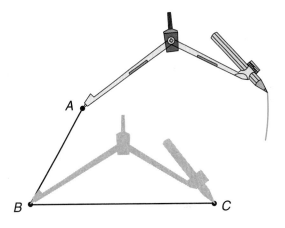

Step 3: Place the compass anchor at *B* and the pencil point at *A*. Without changing your compass opening, place the compass anchor on *C* and draw another arc that crosses the first arc. Label the point where the arcs cross as *D*.

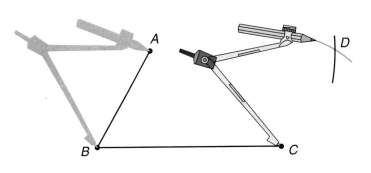

Step 4: Draw line segments *AD* and *CD*.

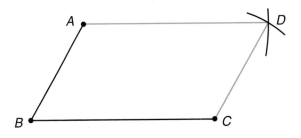

Check Your Understanding

Use a compass and a straightedge to construct a parallelogram.

Use a compass and a straightedge to construct a rhombus. (Hint: A rhombus is a parallelogram whose sides are all the same length.)

Constructing a Regular Inscribed Hexagon

Follow each step carefully. Use a clean sheet of paper.

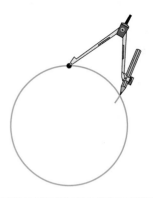

Step 1: Draw a circle and keep the same compass opening. Make a dot on the circle. Place the compass anchor on the dot and make a mark with the pencil point on the circle. Keep the same compass opening for Steps 2 and 3.

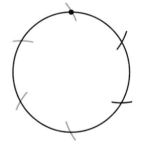

Step 2: Place the compass anchor on the mark you just made. Make another mark with the pencil point on the circle.

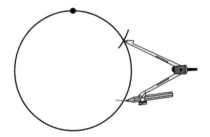

Step 3: Do this four more times to divide the circle into 6 equal parts. The sixth mark should be on the dot you started with or very close to it.

Step 4: With your straightedge, connect the 6 marks on the circle to form a regular hexagon inside the circle.

Use your compass to check that the sides of the hexagon are all the same length.

The hexagon is **inscribed** in the circle because each vertex of the hexagon is on the circle.

Check Your Understanding

Draw a circle. Using a compass and a straightedge, construct a regular hexagon that is inscribed in the circle.

Draw a line segment from the center of the circle to each vertex of the hexagon to form 6 triangles. Use your compass to check that the sides of each triangle are the same length.

Constructing an Inscribed Square

Follow each step carefully. Use a clean sheet of paper.

Step 1: Draw a circle with a compass.

Step 2: Draw a line segment through the center of the circle with endpoints on the circle. Label the endpoints as *A* and *B*.

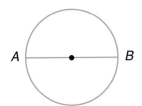

Step 3: Increase the compass opening. Place the compass anchor on *A*. Draw an arc below the center of the circle and another arc above the center of the circle.

Step 4: Without changing the compass opening, place the compass anchor on *B*. Draw arcs that cross the arcs you draw in Step 3. Label the points where the arcs intersect as *C* and *D*.

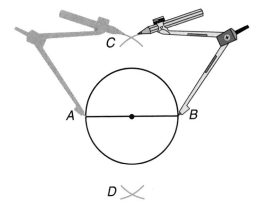

Step 5: Draw a line through points *C* and *D*.

Label the points where line *CD* intersects the circle as *E* and *F*.

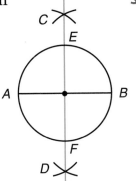

Step 6: Draw line segments *AE, EB, BF,* and *FA*.

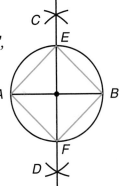

Use your compass to check that all four line segments are the same length. Use the corner of your straightedge or some other square corner to check that all four angles are right angles.

The square is **inscribed** in the circle because all vertices of the square are on the circle.

Check Your Understanding

Use a compass and a straightedge to construct an inscribed square.

Did You Know?

Using only a compass and a straightedge, the ancient Greeks knew how to construct inscribed regular polygons with 5, 10, and 15 sides. Around 1800, Karl Gauss constructed an inscribed regular polygon with 17 sides. These constructions are all complicated.

Constructing a Perpendicular Bisector of a Line Segment

Follow each step carefully. Use a clean sheet of paper.

Step 1: Draw line segment *AB*.

Step 2: Open your compass so that the compass opening is greater than half the distance between point *A* and point *B*. Place the anchor on *A*. Draw a small arc below $\overline{AB}$ and another small arc above $\overline{AB}$.

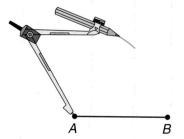

Step 3: Without changing the compass opening, place the anchor on *B*. Draw an arc below $\overline{AB}$ and another arc above $\overline{AB}$ so that the arcs cross the first arcs you drew. Where pairs of arcs intersect, label the points as *M* and *N*.

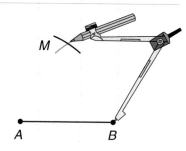

Step 4: Draw $\overline{MN}$. Label the point where $\overline{MN}$ intersects $\overline{AB}$ as *O*.

Line segment *MN* **bisects** line segment *AB* at point *O*. The distance from *A* to *O* is the same as the distance from *B* to *O*.

Line segments *MN* and *AB* are perpendicular because the angles formed where they intersect are right angles. Line segment *MN* is a **perpendicular bisector** of line segment *AB*.

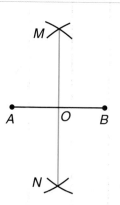

Check Your Understanding

Draw a line segment. Use a compass and a straightedge to bisect it. Then measure to check that the line segment has been divided into two equal parts. Use a protractor to check that the segments are perpendicular.

Constructing a Perpendicular Line Segment (Part 1)

Let *P* be a point *on* the line segment $\overline{AB}$. You can construct a line
segment that is perpendicular to $\overline{AB}$ at the point *P*.

Follow each step carefully. Use a clean sheet of paper.

Step 1: Draw line segment *AB*. Make a dot on $\overline{AB}$
and label it as *P*.

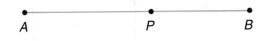

Step 2: Place the compass anchor on point *P*, and
draw an arc that crosses $\overline{AB}$. Label the point
where the arc crosses the segment as *C*.

Keeping the compass anchor on point *P* and
the same compass opening, draw another arc
that crosses $\overline{AB}$. Label the point where the
arc crosses the segment as *D*.

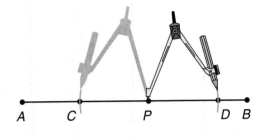

Step 3: Make sure the compass opening is greater
than the length of $\overline{CP}$. Place the compass
anchor on *C* and draw an arc above $\overline{AB}$.

Keeping the same compass opening, place the
compass anchor on *D* and draw another arc
above $\overline{AB}$ that crosses the first arc.

Label the point where the two arcs cross as *Q*.

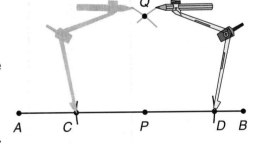

Step 4: Draw $\overline{QP}$.

$\overline{QP}$ is **perpendicular** to $\overline{AB}$.

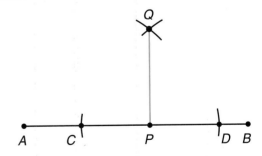

Check Your Understanding

Draw a line segment. Draw a point on the segment
and label it as *R*.

Use a compass and a straightedge. Construct a line segment through point *R* that
is perpendicular to the segment you drew.

Use a protractor to check that the segments are perpendicular.

Constructing a Perpendicular Line Segment (Part 2)

Let M be a point that is *not on* the line segment $\overline{PQ}$. You can construct a line segment with one endpoint at M that is perpendicular to $\overline{PQ}$.

Follow each step carefully. Use a clean sheet of paper.

Step 1: Draw line segment PQ. Draw a point M not on $\overline{PQ}$.

Step 2: Place the compass anchor on M and draw an arc that crosses $\overline{PQ}$ at two points.

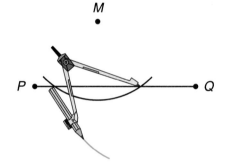

Step 3: Place the compass anchor on one of the points and draw an arc below $\overline{PQ}$

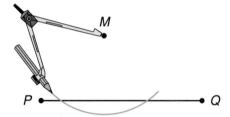

Step 4: Keeping the same compass opening, place the compass anchor on the other point and draw another arc that crosses the first arc.

Label the point where the two arcs cross as N. Then draw the line segment MN.

$\overline{MN}$ is **perpendicular** to $\overline{PQ}$.

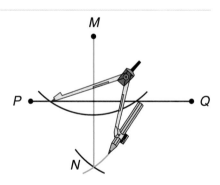

Check Your Understanding

Draw a line segment KL and a point J above the line segment. Using a compass and a straightedge, construct a line segment from point J that is perpendicular to $\overline{KL}$.

Use the Geometry Template to draw a parallelogram. Then construct a line segment to show the height of the parallelogram.

Bisecting an Angle

Follow each step carefully. Use a clean sheet of paper.

Step 1: Draw angle *ABC*.

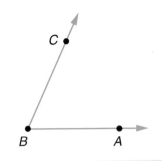

Step 2: Place the compass anchor on *B*, and draw an arc that intersects both $\overrightarrow{BA}$ and $\overrightarrow{BC}$. Label the points where the arcs cross $\overrightarrow{BA}$ and $\overrightarrow{BC}$ as *M* and *N*.

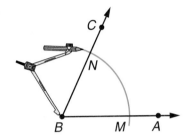

Step 3: Place the compass anchor on *M* and draw a small arc inside ∠*ABC*.

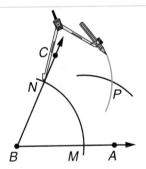

Step 4: Without changing the compass opening, place the compass anchor on *N* and draw a small arc that intersects the one you drew in Step 3. Label the point where the two arcs meet as *P*.

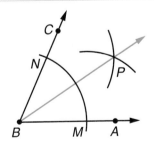

Step 5: Draw a ray from point *B* through point *P*.

Ray *BP* **bisects** ∠*ABC*. The measure of ∠*ABP* is equal to the measure of ∠*CBP*.

Check Your Understanding

Draw an obtuse angle. Use a compass and a straightedge to bisect it.

Copying an Angle

Follow each step carefully. Use a clean sheet of paper.

Step 1: Draw angle *B*.

Step 2: To start copying the angle, draw a ray. Label the endpoint of the ray as *B'*.

Step 3: Place the compass anchor on *B*. Draw an arc that crosses both sides of angle *B*. Label the point where the arc crosses one side as *A*. Label the point where the arc crosses the other side as *C*.

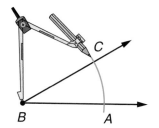

Step 4: Without changing the compass opening, place the compass anchor on *B'*. Draw an arc about the same size as the one you drew in Step 3. Label the point where the arc crosses the ray as *A'*.

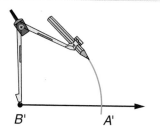

Step 5: Place the compass anchor on *A* and the pencil point on *C*.

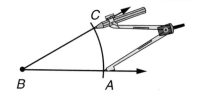

Step 6: Without changing the compass opening, place the compass anchor on *A'*. Draw a small arc where the pencil point crosses the larger arc and label the crossing point as *C'*.

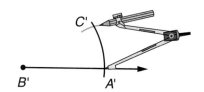

Step 7: Draw a ray from point *B'* through point *C'*.

∠*A'B'C'* is **congruent** to ∠*ABC*. That is, the two angles have the same degree measure.

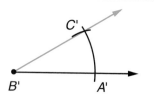

Check Your Understanding

Draw an angle. Use a compass and a straightedge to copy the angle. Then measure the two angles with a protractor to check that they are the same size.

Copying a Quadrangle

Before you can copy a quadrangle with a compass and a straightedge, you need to know how to copy line segments and angles. Those constructions are described on pages 189 and 198.

Follow each step carefully. Use a clean sheet of paper.

Step 1: Draw quadrangle *ABCD*. Copy ∠*BAD*. Label the vertex of the new angle as *A'*. The sides of your new angle should be longer than $\overline{AB}$ and $\overline{AD}$.

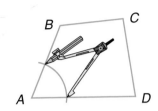

Step 2: Mark off the distance from *A* to *D* on the horizontal side of your new angle. Label the endpoint as *D'*.

Mark off the distance from *A* to *B* on the other side of your new angle. Label the endpoint as *B'*.

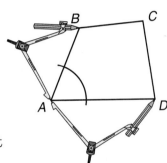
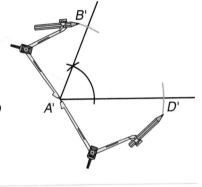

Step 3: Place the compass anchor on *B* and the pencil point on *C*. Without changing the compass opening, place the compass anchor on *B'* and make an arc.

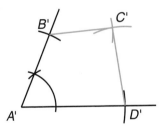

Step 4: Place the compass anchor on *D* and the pencil point on *C*. Without changing the compass opening, place the compass anchor on *D'* and make an arc that crosses the arc you made in Step 3. Label the point where the two arcs meet as *C'*.

Step 5: Draw $\overline{B'C'}$ and $\overline{D'C'}$.

Quadrangle *A'B'C'D'* is **congruent** to quadrangle *ABCD*. The two quadrangles are the same size and shape.

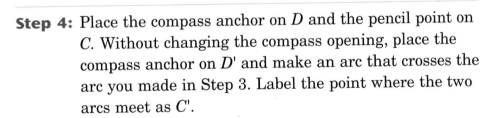

Check Your Understanding

Draw a quadrangle. Use a compass and a straightedge to copy the quadrangle.

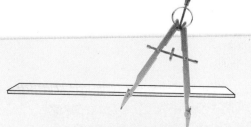

Constructing Parallel Lines

Follow each step carefully. Use a clean sheet of paper.

Step 1: Draw line *AB* and ray *AC*.

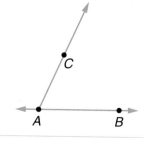

Step 2: Place the compass anchor on *A*. Draw an arc that crosses both $\overrightarrow{AB}$ and $\overrightarrow{AC}$. Label the point where the arc crosses $\overrightarrow{AB}$ as *D*. Label the point where the arc crosses $\overrightarrow{AC}$ as *E*.

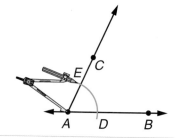

Step 3: Without changing the compass opening, place the compass anchor on *C*. Draw an arc the same size as the one you drew in Step 2. Label the point where the arc crosses $\overrightarrow{AC}$ as *F*.

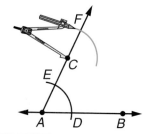

Step 4: Place the compass anchor on *E* and the pencil point on *D*.

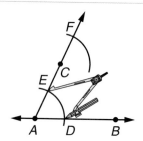

Step 5: Without changing the compass opening, place the compass anchor on *F*. Draw a small arc where the pencil point crosses the larger arc. Label the point where the small arc crosses the larger arc as *G*.

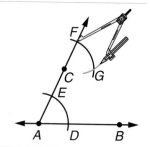

Step 6: Draw a line through points *C* and *G*.

Line *CG* is **parallel** to line *AB*. ∠*CAB* is **congruent** to ∠*FCG*.

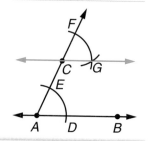

Check Your Understanding

Draw a line. Use a compass and a straightedge to draw a line that is parallel to it.

People around the world weave textiles, such as fabric, blankets, and carpets. Many modern textiles are woven by machine. Others, woven by hand, require a great deal of time, effort, and skill. Weavers often create complex designs that incorporate repeating patterns, geometric shapes, and symmetry.

Handweaving

There is often much creativity and cultural meaning expressed in the traditional weaving of textiles.

◄ *Kente* is a ceremonial cloth made by the people of Ghana. The bold, geometric patterns are often named for culturally-significant things, such as historical events, important chiefs, queen mothers, and plants.

◄ In Ghana, the men weave long strips of cloth, four to eight inches wide, which are cut into shorter lengths and sewn together to make a single *kente* cloth.

◄ Using wool from sheep they raise and dyes from local plants, Navajo weavers create textiles with elaborate repeating borders and intricate geometric designs. Because traditional artists clean, dye, and spin the wool for their yarn, it can take 350 hours to make a 3' x 5' rug.

Materials and Equipment

Weaving by hand is a laborious process that requires patience, time, and the right materials and equipment. Much time is spent gathering and preparing the plant or animal fibers that will be spun into thread or yarn.

◄ Wool must be sheared from sheep, as this man is doing, then cleaned and combed. Between 10 and 18 pounds of wool come from one merino sheep, enough to make four sweaters.

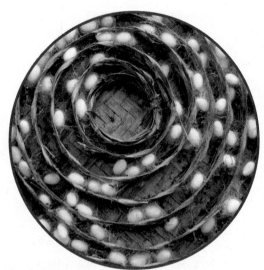

Silk, nature's strongest fiber, comes from the thread silkworms spin for their cocoons. The time and effort it takes to raise the silkworms and then unroll the super-fine filament from the cocoons make silk very expensive. It can take 1,000 cocoons to produce 100 grams of silk thread. ➤

◄ Short fibers, such as wool, cotton, and flax, are spun into thread or yarn by drawing out fibers and twisting them together into a continuous strand. These women from Nepal are spinning fibers to make yarn.

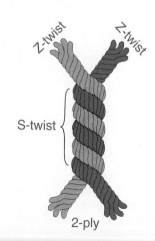

Thread or yarn is said to have a Z- or S-twist, depending on how it was spun. To make a stronger thread, the single threads are sometimes "plied," or twisted together. ➤

S-twist   Z-twist

Z-twist   Z-twist

S-twist

2-ply

Weaving is done by interlacing two sets of threads at right angles. The vertical threads are called the **warp** and the horizontal threads are the **weft.** Different designs are made by varying the way the weft threads cross the warp threads. ➤

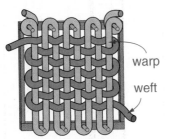

warp

weft

◄ The warp threads are held in place parallel to each other on a loom, or frame, and the weft threads are woven over and under the warp threads, back and forth, one row at a time. This Navajo weaver teaches her granddaughter how to weave on a traditional loom.

This Mayan woman works at a portable backstrap loom. The loom's top bar is attached to an immovable object, such as a pole or tree. The bottom bar is attached to a strap that goes around the weaver's back. By leaning forward or backward, the weaver can adjust the tension on the warp threads. ▼

▲ These Turkish women work at a frame loom. A frame loom is stationary. The warp threads are wound tightly around beams at the top and the bottom of the loom.

Oriental Carpet Weaving

Oriental carpets are textiles created within what is traditionally called the Orient—a swath of land that stretches from Istanbul, Turkey to Western China. Oriental carpets are some of the oldest, sturdiest, and most sought after carpets in the world. The intricate designs have been passed down from generation to generation.

Two basic types of carpet designs are **rectilinear** and **curvilinear.** Rectilinear designs, as in this carpet, have angular patterns and shapes. ▼

Curvilinear designs often have floral patterns with curved outlines. ▼

▲

Weavers often tell stories by including repeating symbols, or **motifs** in their designs. Animals are common motifs. This ram symbol represents heroism and power.

▲

This symbol, a person with her hands on her hips, signifies pride. A weaver might include it to show the pride felt after the birth of a child.

In an Oriental carpet, small loops or knots of yarn are added between the rows of weft threads to create the design. These knots form the **pile,** the velvety surface of the rug. There are two ways of tying these knots.

The **Turkish Knot**, or double knot, results in a strong, durable carpet. It is a strand of yarn that circles two warp threads. The loose ends are drawn tightly between the two warp threads. ▼

The **Persian Knot**, or single knot, is less durable. It is a strand of yarn that circles one warp thread and winds loosely around another warp thread. One loose end is pulled through the two warps, while the other end goes to the outside of the paired warps. ▼

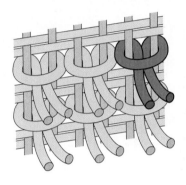

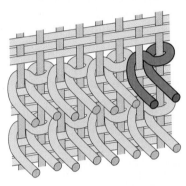

A carpet's **knot density,** the number of knots per square inch, is often a factor in determining a rug's value. "The Ardebil," a magnificent rug made for a Persian King in 1540, is approximately 17' by 34', with over 32 million tightly-woven knots. ▼

Machineweaving

Although handweaving is still done by people throughout the world, most of the everyday, affordable carpets, fabrics, and other textiles used in industrial countries are woven by machines.

The Jacquard loom, invented in the early 1800s, revoluntionized the textile industry. It was the first machine able to create intricate patterns by using punch cards that controlled the raising and lowering of the warp threads, a job that had previously been done by workers. Charles Babbage later adapted the punch card idea in his design for the first automated computer. ➤

◄ A worker in a 1920s textile factory repairs a Jaquard-style loom. The panels at the top of the photo are the punch cards that store information for the machine to follow, like an algorithm.

A modern loom, like this one from a textile factory in France, has features similar to the Jacquard loom and handlooms. With this loom, however, the weaving is controlled by computer software. ➤

What textiles do you see around you? What patterns do you see in those textiles?

Measurement

Natural Measures and Standard Units

Systems of weights and measures have been used in many parts of the world since ancient times. People measured lengths and weights for centuries before there were rulers and scales.

Ancient Measures of Weight

Shells and grains such as wheat or rice were often used as units of weight. For example, a small item might be said to weigh 300 grains of rice. Large weights were often compared to the load that could be carried by a man or a pack animal.

Ancient Measures of Length

People used **natural measures** based on the human body to measure length and distance. Some of these units are shown below.

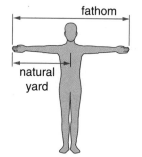
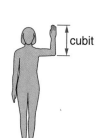
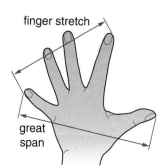
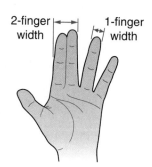

Standard Units of Length and Weight

Using shells and grains to measure weight is not exact. Even if the shells and grains are of the same type, they vary in size and weight.

Using body lengths to measure length is not exact. Body measures depend upon the person who is doing the measuring. The problem is that different persons have hands and arms of different lengths.

One way to solve this problem is to make **standard units** of length and weight. Most rulers are marked off using inches and centimeters as standard units. Bath scales are marked off using pounds and kilograms as standard units. Standard units never change and are the same for everyone. If two people measure the same object using standard units, their measurements will be the same or almost the same.

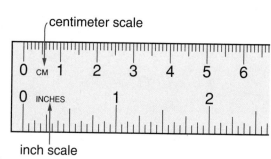

The Metric System and the U.S. Customary System

About 200 years ago, a system of weights and measures called the **metric system** was developed. It uses standard units for length, weight, and temperature. In the metric system:

♦ The **meter** is the standard unit for length. The symbol for a meter is **m.** A meter is about the width of a front door.

♦ The **gram** is the standard unit for weight. The symbol for a gram is **g.** A paper clip weighs about $\frac{1}{2}$ gram.

♦ The **Celsius degree,** or **°C,** is the standard unit for temperature. Water freezes at 0°C and boils at 100°C. Normal room temperature is about 20°C.

Two paper clips weigh about 1 gram.

Scientists almost always measure with metric units. The metric system is easy to use because it is a base-10 system. Larger and smaller units are defined by multiplying or dividing the units named above by powers of 10: 10, 100, 1,000, and so on.

Examples All metric units of length are based on the meter. Each unit is defined by multiplying or dividing the meter by a power of 10.

Units of Length Based on the Meter	Prefix	Meaning
1 millimeter (mm) = $\frac{1}{1,000}$ meter	milli-	$\frac{1}{1,000}$
1 centimeter (cm) = $\frac{1}{100}$ meter	centi-	$\frac{1}{100}$
1 decimeter (dm) = $\frac{1}{10}$ meter	deci-	$\frac{1}{10}$
1 kilometer (km) = 1,000 meters	kilo-	1,000

Note

The U.S. customary system is not based on powers of 10. This makes it more difficult to use than the metric system. For example, in order to change inches to yards, you must know that 36 inches equals 1 yard.

The metric system is used in most countries around the world. In the United States, the **U.S. customary system** is used for everyday purposes. The U.S. customary system uses standard units like the **inch, foot, yard, mile, ounce, pound,** and **ton.**

Check Your Understanding

1. Which of these units are in the metric system?
   ton   millimeter   pound   mile   gram   kilometer   decimeter   ounce

2. **a.** What does the prefix "kilo"- mean?   **b.** 2 grams = ? kilograms

Check your answers on page 422.

Converting Units of Length

This table shows how different units of length in the metric system compare. You can use this table to rewrite a length using a different unit.

Comparing Metric Units of Length				Symbols for Units of Length	
1 cm = 10 mm	1 m = 1,000 mm	1 m = 100 cm	1 km = 1,000 m	mm = millimeter	cm = centimeter
1 mm = $\frac{1}{10}$ cm	1 mm = $\frac{1}{1,000}$ m	1 cm = $\frac{1}{100}$ m	1 m = $\frac{1}{1,000}$ km	m = meter	km = kilometer

Examples Use the table to rewrite each length using a different unit. Replace the unit given first with an equal length that uses the new unit.

Problem	Solution
56 centimeters = ? millimeters	56 cm = 56 * 10 mm = 560 mm
56 centimeters = ? meters	56 cm = 56 * $\frac{1}{100}$ m = $\frac{56}{100}$ m = 0.56 m
9.3 kilometers = ? meters	9.3 km = 9.3 * 1,000 m = 9,300 m
6.9 meters = ? centimeters	6.9 m = 6.9 * 100 cm = 690 cm

The table below shows how different units of length in the U.S. customary system compare. You can use this table to rewrite a length using a different unit.

Comparing U.S. Customary Units of Length				Symbols for Units of Length	
1 ft = 12 in.	1 yd = 36 in.	1 yd = 3 ft	1 mi = 5,280 ft	in. = inch	ft = foot
1 in. = $\frac{1}{12}$ ft	1 in. = $\frac{1}{36}$ yd	1 ft = $\frac{1}{3}$ yd	1 ft = $\frac{1}{5,280}$ mi	yd = yard	mi = mile

Examples Use the table to rewrite each length using a different unit. Replace the unit given first with an equal length that uses the new unit.

Problem	Solution
14 ft = ? inches	14 ft = 14 * 12 in. = 168 in.
21 ft = ? yards	21 ft = 21 * $\frac{1}{3}$ yd = $\frac{21}{3}$ yd = 7 yd
7 miles = ? feet	7 mi = 7 * 5,280 ft = 36,960 ft
180 inches = ? yards	180 in. = 180 * $\frac{1}{36}$ yd = $\frac{180}{36}$ yd = 5 yd

Personal References for Units of Length

Sometimes it is difficult to remember exactly how long a centimeter or a yard is or how a kilometer and a mile compare. You may not have a ruler, yardstick, or tape measure handy. When this happens, you can estimate lengths by using the lengths of common objects and distances that you know well.

Some examples of **personal references** for length are given below. A good personal reference is something that you see or use often, so you don't forget it. A good personal reference also doesn't change size. For example, a wooden pencil is not a good personal reference for length because it gets shorter as it is sharpened.

Personal References for Metric Units of Length

About 1 millimeter	About 1 centimeter
Thickness of a dime	Thickness of a crayon
Thickness of the point of a thumbtack	Width of the head of a thumbtack
Thickness of the thin edge of a paper match	Thickness of a pattern block

About 1 meter	About 1 kilometer
One big step (for an adult)	1,000 big steps (for an adult)
Width of a front door	Length of 10 football fields (including the end zones)
Tip of the nose to tip of the thumb, with arm extended (for an adult)	

Note

The personal references for 1 meter can also be used for 1 yard. 1 yard equals 36 inches, while 1 meter is about 39.37 inches. One meter is often called a "fat yard," which means 1 yard plus 1 hand width.

Personal References for U.S. Customary Units of Length

About 1 inch	About 1 foot
Length of a paper clip	A man's shoe length
Width (diameter) of a quarter	Length of a license plate
Width of a man's thumb	Length of this book

About 1 yard	About 1 mile
One big step (for an adult)	2,000 average-size steps (for an adult)
Width of a front door	Length of 15 football fields (including the end zones)
Tip of the nose to tip of the thumb, with arm extended (for an adult)	

Did You Know?

The U.S.–Mexican border is 1,950 miles long, the length of 29,000 football fields. It would take about 3,500,000 big steps for an adult to walk its length.

Perimeter

The **distance around** a polygon is called its **perimeter.** To find the perimeter of any polygon, add the lengths of all its sides.

Example   Find the perimeter of polygon *ABCDE*.

2 cm + 2 cm + 1.5 cm + 2 cm + 2.5 cm = 10 cm

The perimeter is 10 centimeters.

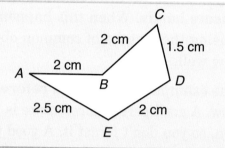

Perimeter Formulas		
Rectangles	**Squares**	**Regular Polygons**
$p = 2 * (l + w)$	$p = 4 * s$	$p = n * s$
p is the perimeter	*p* is the perimeter	*p* is the perimeter
l is the length of the rectangle	*s* is the length of one side of the square	*n* is the number of sides
w is the width of the rectangle		*s* is the length of a side

Examples   Find the perimeter of each polygon.

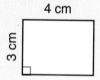

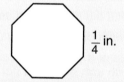

Rectangle

Use the formula $p = 2 * (l + w)$.
- length (*l*) = 4 cm
- width (*w*) = 3 cm
- perimeter (*p*) = 2 * (4 cm + 3 cm)
  $$= 2 * 7 \text{ cm} = 14 \text{ cm}$$

The perimeter is 14 centimeters.

Square

Use the formula $p = 4 * s$.
- length of side (*s*) = 9 ft
- perimeter (*p*) = 4 * 9 ft
  $$= 36 \text{ ft}$$

The perimeter is 36 feet.

Regular Octagon

Use the formula $p = n * s$.
- number of sides (*n*) = 8
- length of side (*s*) = $\frac{1}{4}$ in.
- perimeter (*p*) = 8 * $\frac{1}{4}$ in.
  $$= \frac{8}{4} \text{ in.} = 2 \text{ in.}$$

The perimeter is 2 inches.

Check Your Understanding

Solve. Include the unit in each answer.

1. Find the perimeter of a rectangle whose dimensions are 9 feet and 3 feet.

2. Find the perimeter of a regular hexagon whose sides are 15 yards long.

Check your answers on page 422.

Circumference

The perimeter of a circle is the **distance around** the circle. The perimeter of a circle has a special name. It is called the **circumference** of the circle.

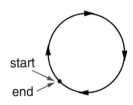

start
end

Example Most food cans are cylinders. Their tops and bottoms have circular shapes. The circumference of a circular can top is how far the can turns when opened by a can opener.

The **diameter** of a circle is any line segment that passes through the center of the circle and has both endpoints on the circle. The length of a diameter segment is also called the diameter.

If you know the diameter of a circle, there is a simple formula for finding its circumference.

Formula for the Circumference of a Circle
circumference = pi * diameter   or $c = \pi * d$
c is the circumference, and d is the diameter of the circle.

Note

The Greek letter π is called **pi**. It is approximately equal to 3.14. In your work with the number π, you can use 3.14 or $3\frac{1}{7}$ as the approximate value for π. You can also use a calculator with a π key.

Example Find the circumference of the circle.

Use the formula $c = \pi * d$.
• diameter (d) = 6 cm
• circumference (c) = $\pi * 6$ cm

Use either the π key on a calculator, or use 3.14 as an approximate value for π.

circumference (c) = 18.8 cm, rounded to the nearest tenth of a centimeter

The circumference of the circle is 18.8 cm.

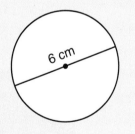

6 cm

Check Your Understanding

1. Measure the diameter of the dollar coin in millimeters.
2. Find the circumference of the dollar coin in millimeters.
3. What is the circumference of a pizza whose diameter is 14 inches?

Check your answers on page 422.

Area

Area is a measure of the amount of surface inside a closed boundary. You can find the area by counting the number of squares of a certain size that cover the region inside the boundary. The squares must cover the entire region. They must not overlap, have any gaps, or cover any surface outside the boundary.

Sometimes a region cannot be covered by an exact number of squares. In that case, first count the number of whole squares, then the fractions of squares that cover the region.

Area is reported in square units. Units of area for small regions are square inches (in.^{2}), square feet (ft^2), square yards (yd^2), square centimeters (cm^2), and square meters (m^2). For large areas, square miles (mi^2) are used in the United States, while square kilometers (km^2) are used in most other countries.

You may report area using any of the square units, but you should choose a square unit that makes sense for the region being measured.

1 square centimeter
(actual size)

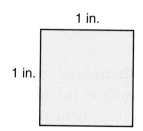

1 square inch
(actual size)

Examples The area of a field-hockey field is reported below in three different ways.

Area of the field is 6,000 square yards.

Area = 6,000 yd^2

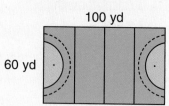

Area of the field is 54,000 square feet.

Area = 54,000 ft^2

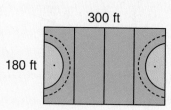

Area of the field is 7,776,000 square inches.

Area = 7,776,000 in.^{2}

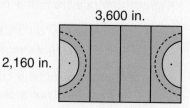

Although each of these measurements is correct, reporting the area in square inches really doesn't give us a good idea about the size of the field. It is hard to imagine 7,776,000 of anything!

Did You Know?

Tropical rain forests, where more than half the plant and animal species in the world live, once covered more than 8,000,000 square miles of Earth's surface. By 2004, because of destruction by people, fewer than 3,400,000 square miles remain.

Area of a Rectangle

When you cover a rectangular shape with unit squares, the squares can be arranged into rows. Each row will contain the same number of squares and fractions of squares.

Examples Find the area of the rectangle.

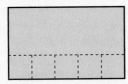 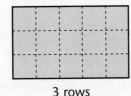

5 squares in a row   3 rows

3 rows with 5 squares in each row for a total of 15 squares

Area = 15 square units

To find the area of a rectangle, use either of these formulas:

Area = (the number of squares in 1 row) * (the number of rows)
Area = length of a base * height

height

base

Area Formulas	
Rectangles	**Squares**
$A = b * h$	$A = s^2$
A is the area, b is the length of a base, h is the height of the rectangle.	A is the area, s is the length of a side of the square.

Either pair of parallel sides in a rectangle can be chosen as its **bases.** The **height** of a rectangle is the shortest distance between its bases.

Examples Find the area of the rectangle.

Use the formula $A = b * h$.
• length of base (b) = 4 in.
• height (h) = 3 in.
• area (A) = 4 in. * 3 in.
          = 12 in.^{2}

3 in.

4 in.

The area of the rectangle is 12 in.^{2}.

Find the area of the square.

Use the formula $A = s^2$.
• length of a side (s) = 9 ft
• area (A) = 9 ft * 9 ft
          = 81 ft^2

9 ft

The area of the square is 81 ft^2.

Check Your Understanding

Find the area of these figures. Include the unit in each answer.

1.

3 units

2 units

2.

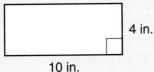

4 in.

10 in.

3.

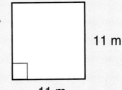

11 m

11 m

Check your answers on page 422.

Area of a Parallelogram

In a parallelogram, either pair of opposite sides can be chosen as its **bases.** The **height** of the parallelogram is the shortest distance between the two bases.

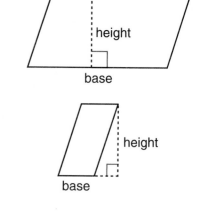

In the parallelograms at the right, the height is shown by a dashed line that is **perpendicular** (at a right angle) to the base. In the second parallelogram, the base has been extended, and the dashed height line falls outside the parallelogram.

Any parallelogram can be cut into two pieces and the pieces rearranged to form a rectangle whose base length and height are the same as the base length and height of the parallelogram. The rectangle has the same area as the parallelogram. So, you can find the area of the parallelogram in the same way you find the area of the rectangle—by multiplying the length of the base by the height.

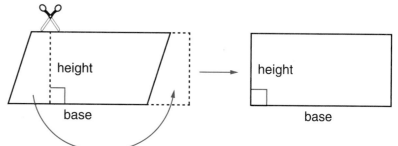

Formula for the Area of a Parallelogram

$$A = b * h$$

A is the area, *b* is the length of the base, *h* is the height of the parallelogram.

Example Find the area of the parallelogram.

Use the formula $A = b * h$.

- length of base (*b*) = 6 cm
- height (*h*) = 3.8 cm
- area (*A*) = 6 cm * 3.8 cm = 22.8 cm²

The area of the parallelogram is 22.8 cm².

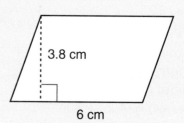

3.8 cm

6 cm

Check Your Understanding

Find the area of each parallelogram. Include the unit in each answer.

1.
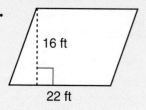
16 ft

22 ft

2.

12 in.

9 in.

3.
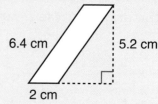
6.4 cm   5.2 cm

2 cm

Check your answers on page 422.

Area of a Triangle

Any of the sides of a triangle can be chosen as its **base.** The **height** of the triangle (for that base) is the shortest distance between the base and the **vertex** opposite the base.

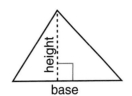
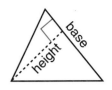

In the triangles at the right, the height is shown by a dashed line that is **perpendicular** (at a right angle) to the base. In one of the triangles, the base has been extended and the dashed height line falls outside the triangle. In the right triangle shown, the height line is one of the sides of the triangle.

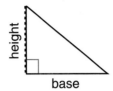
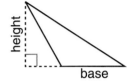

Any triangle can be combined with a second triangle of the same size and shape to form a parallelogram. Each triangle at the right has the same size base and height as the parallelogram. The area of each triangle is half the area of the parallelogram. Therefore, the area of a triangle is half the product of the base length multiplied by the height.

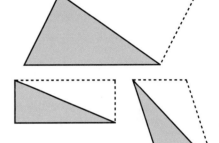

Area Formulas	
Parallelograms	**Triangles**
$A = b * h$	$A = \frac{1}{2} * (b * h)$
A is the area, *b* is the length of a base, *h* is the height.	*A* is the area, *b* is the length of the base, *h* is the height.

Example Find the area of the triangle.

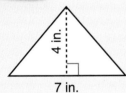

Use the formula $A = \frac{1}{2} * (b * h)$.
- length of base (*b*) = 7 in.
- height (*h*) = 4 in.
- area (*A*) = $\frac{1}{2} * (7$ in. $* 4$ in.$) = \frac{1}{2} * 28$ in.$^2 = \frac{28}{2}$ in.$^2 = 14$ in.^{2}

The area of the triangle is 14 in.^{2}.

Check Your Understanding

Find the area of each triangle. Include the unit in each answer.

1.
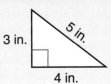
3 in.
5 in.
4 in.

2.
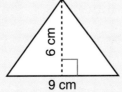
6 cm
9 cm

3.
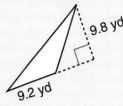
9.8 yd
9.2 yd

Check your answers on page 422.

Area of a Circle

The **radius** of a circle is any line segment that connects the center of the circle with any point on the circle. The length of a radius segment is also called the radius.

The **diameter** of a circle is any segment that passes through the center of the circle and has both endpoints on the circle. The length of a diameter segment is also called the diameter.

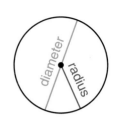

If you know either the radius or the diameter of a circle, you can find the other length using these formulas:

$$\text{diameter} = 2 * \text{radius} \qquad \text{radius} = \frac{1}{2} * \text{diameter}$$

If you know the radius, there is a simple formula for finding the area of a circle.

> **Note**
>
> If d is the diameter, r is the radius, and c is the circumference of a circle, then $c = \pi * d$. Since $d = 2 * r$, another formula for circumference is $c = 2 * \pi * r$.

Formula for the Area of a Circle

Area = pi * (radius squared)   or $A = \pi * r^2$

A is the area, and *r* is the radius of the circle.

> **Note**
>
> The Greek letter π is called **pi**, and it is approximately equal to 3.14. When you work with the number π, use 3.14 or $3\frac{1}{7}$ as the approximate value for π, or use a calculator with a π key.

Example   Find the area of the circle.

Use the formula $A = \pi * r^2$.

• radius (*r*) = 5 in.
• area (*A*) = π * 5 in. * 5 in.

Use either the π key on a calculator, or use 3.14 as an approximate value for π.

• area (*A*) = 78.5 in.², rounded to the nearest tenth of a square inch.

The area of the circle is 78.5 in.².

Check Your Understanding

1. Measure the diameter of the nickel in millimeters.
2. What is the radius of the nickel in millimeters?
3. Find the area of the nickel in square millimeters.

Check your answers on page 422.

Volume and Capacity

Volume

The **volume** of a solid object such as a brick or a ball is a measure of *how much space the object takes up.* The volume of a container such as a freezer is a measure of *how much the container will hold.*

Volume is measured in **cubic units,** such as cubic inches (in.^{3}), cubic feet (ft^3), and cubic centimeters (cm^3). It is easy to find the volume of an object that is shaped like a cube or other rectangular prism. For example, picture a container in the shape of a 10-centimeter cube (that is, a cube that is 10 cm by 10 cm by 10 cm). It can be filled with exactly 1,000 centimeter cubes. Therefore, the volume of a 10-centimeter cube is 1,000 cubic centimeters (1,000 cm^3).

All you need to know to find the volume of a rectangular prism are the length and width of its base and its height. The length, width, and height are called the **dimensions** of the prism.

You can also find the volume of other solids (such as triangular prisms, pyramids, cones, and spheres) by measuring their dimensions. It is even possible to find the volume of irregular objects, such as rocks or your own body.

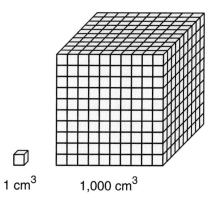

1 cm^3 1,000 cm^3

The Dimensions of a Rectangular Prism

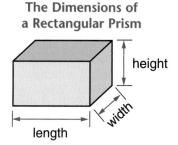

height
length
width

Capacity

We often measure things that can be poured into or out of containers, such as liquids, grains, salt, and so on. The volume of a container that is filled with a liquid or a solid that can be poured is often called its **capacity.**

Capacity is usually measured in units such as **gallons, quarts, pints, cups, fluid ounces, liters,** and **milliliters.**

The tables at the right compare different units of capacity. These units of capacity are not cubic units, but liters and milliliters are easily converted to cubic units.

$$1 \text{ milliliter} = 1 \text{ cm}^3 \qquad 1 \text{ liter} = 1,000 \text{ cm}^3$$

U.S. Customary Units

1 gallon (gal) = 4 quarts (qt)
1 gallon = 2 half-gallons
1 half-gallon = 2 quarts
1 quart = 2 pints (pt)
1 pint = 2 cups (c)
1 cup = 8 fluid ounces (fl oz)
1 pint = 16 fluid ounces
1 quart = 32 fluid ounces
1 half-gallon = 64 fluid ounces
1 gallon = 128 fluid ounces

Metric Units

1 liter (L) = 1,000 (mL)
1 milliliter = $\frac{1}{1,000}$ liter
1 liter = 1,000 cubic centimeters
1 milliliter = 1 cubic centimeter

two hundred nineteen   **219**

Volume of a Geometric Solid

You can think of the volume of a geometric solid as the total number of whole unit cubes and fractions of unit cubes that are needed to fill the interior of the solid without gaps or overlaps.

Prisms and Cylinders

In a prism or cylinder, the cubes can be arranged in layers, each containing the same number of cubes or fractions of cubes.

> **Example** Find the volume of the prism.
>
> 8 cubes in 1 layer   3 layers
>
> 3 layers with 8 cubes in each layer makes a total of 24 cubes.
>
> Volume = 24 cubic units

Note

For a *right rectangular prism* with side lengths of *l*, *w*, and *h* units, the volume *V* can be found using the formula
$$V = l * w * h.$$

The **height** of a prism or a cylinder is the shortest distance between its **bases.** The volume of a prism or a cylinder is the product of the area of the base (the number of cubes in one layer) multiplied by its height (the number of layers).

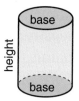

Pyramids and Cones

The height of a pyramid or a cone is the shortest distance between its base and the vertex opposite its base.

If a prism and a pyramid have the same size base and height, then the volume of the pyramid is one-third the volume of the prism. If a cylinder and a cone have the same size base and height, then the volume of the cone is one-third the volume of the cylinder.

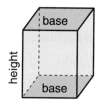

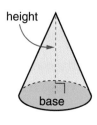

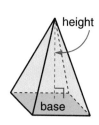

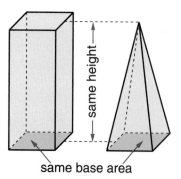

same height

same base area

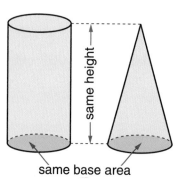

same height

same base area

Volume of a Rectangular or Triangular Prism

Volume of a Prism	Area of a Rectangle	Area of a Triangle
$V = B * h$	$A = b * h$	$A = \frac{1}{2} * (b * h)$
V is the volume, B is the area of the base, h is the height of the prism.	A is the area, b is the length of the base, h is the height of the rectangle.	A is the area, b is the length of the base, h is the height of the triangle.

Example Find the volume of the rectangular prism.

Step 1: Find the area of the base (B). Use the formula $A = b * h$.
- length of the rectangular base (b) = 8 cm
- height of the rectangular base (h) = 5 cm
- area of base (B) = 8 cm * 5 cm = 40 cm^2

Step 2: Multiply the area of the base by the height of the rectangular prism. Use the formula $V = B * h$.
- area of base (B) = 40 cm^2
- height of prism (h) = 6 cm
- volume (V) = 40 cm^2 * 6 cm = 240 cm^3

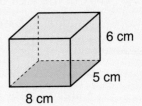

The volume of the rectangular prism is 240 cm^3.

Example Find the volume of the triangular prism.

Step 1: Find the area of the base (B). Use the formula $A = \frac{1}{2} * (b * h)$.
- length of the triangular base (b) = 5 in.
- height of the triangular base (h) = 4 in.
- area of base (B) = $\frac{1}{2}$ * (5 in. * 4 in.) = 10 in.^{2}

Step 2: Multiply the area of the base by the height of the triangular prism. Use the formula $V = B * h$.
- area of base (B) = 10 in.^{2}
- height of prism (h) = 6 in.
- volume (V) = 10 in.^{2} * 6 in. = 60 in.^{3}

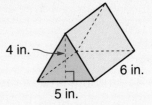

The volume of the triangular prism is 60 in.^{3}.

Check Your Understanding

Find the volume of each prism. Include the unit in each answer.

1.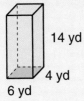
14 yd
4 yd
6 yd

2.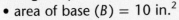
12 cm
12 cm
12 cm

3.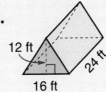
12 ft
24 ft
16 ft

Check your answers on page 422.

Volume of a Cylinder or Cone

Volume of a Cylinder	Volume of a Cone	Area of a Circle
$V = B * h$	$V = \frac{1}{3} * (B * h)$	$A = \pi * r^2$
V is the volume, B is the area of the base, h is the height of the cylinder.	V is the volume, B is the area of the base, h is the height of the cone.	A is the area, r is the radius of the circle.

Example   Find the volume of the cylinder.

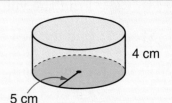

4 cm

5 cm

Step 1: Find the area of the base (B). Use the formula $A = \pi * r^2$.

- radius of base (r) = 5 cm
- area of base (B) = $\pi * 5$ cm $* 5$ cm

Use the π key on a calculator or 3.14 as an approximate value for π.

- area of base (B) = 78.5 cm^2, rounded to the nearest tenth of a square centimeter.

Step 2: Multiply the area of the base by the height of the cylinder.

Use the formula $V = B * h$.

- area of base (B) = 78.5 cm^2
- height of cylinder (h) = 4 cm
- volume (V) = 78.5 cm^2 $* 4$ cm = 314.0 cm^3

The volume of the cylinder is 314.0 cm^3.

Example   Find the volume of the cone.

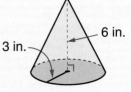

6 in.

3 in.

Step 1: Find the area of the base (B). Use the formula $A = \pi * r^2$.

- radius of base (r) = 3 in.
- area of base (B) = $\pi * 3$ in. $* 3$ in.

Use the π key on a calculator or 3.14 as an approximate value for π.

- area of base (B) = 28.3 in.^{2}, rounded to the nearest tenth of a square inch.

Step 2: Find $\frac{1}{3}$ of the product of the area of the base multiplied by the height of the cone. Use the formula $V = \frac{1}{3} * (B * h)$.

- area of base (B) = 28.3 in.^{2}
- height of cone (h) = 6 in.
- volume (V) = $\frac{1}{3} * (28.3$ in.^{2} $* 6$ in.$) = 56.6$ in.^{3}

The volume of the cone is 56.6 in.^{3}.

Volume of a Rectangular or Triangular Pyramid

Volume of a Pyramid	Area of a Rectangle	Area of a Triangle
$V = \frac{1}{3} * (B * h)$	$A = b * h$	$A = \frac{1}{2} * (b * h)$
V is the volume, B is the area of the base, h is the height of the pyramid.	A is the area, b is the length of the base, h is the height of the rectangle.	A is the area, b is the length of the base, h is the height of the triangle.

Example Find the volume of the rectangular pyramid.

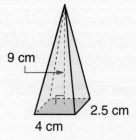

Step 1: Find the area of the base (B). Use the formula $A = b * h$.
- length of the rectangular base (b) = 4 cm
- height of the rectangular base (h) = 2.5 cm
- area of base (B) = 4 cm * 2.5 cm = 10 cm²

Step 2: Find $\frac{1}{3}$ of the product of the area of the base multiplied by the height of the pyramid. Use the formula $V = \frac{1}{3} * (B * h)$.
- area of base (B) = 10 cm²
- height of pyramid (h) = 9 cm
- volume (V) = $\frac{1}{3} * (10$ cm² $* 9$ cm$) = 30$ cm³

The volume of the rectangular pyramid is 30 cm³.

Example Find the volume of the triangular pyramid.

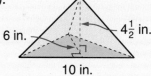

Step 1: Find the area of the base (B). Use the formula $A = \frac{1}{2} * (b * h)$.
- length of the triangular base (b) = 10 in.
- height of the triangular base (h) = 6 in.
- area of base (B) = $\frac{1}{2} * (10$ in. $* 6$ in.$) = 30$ in.²

Step 2: Find $\frac{1}{3}$ of the product of the area of the base multiplied by the height of the pyramid. Use the formula $V = \frac{1}{3} * (B * h)$.
- area of base (B) = 30 in.²
- height of pyramid (h) = $4\frac{1}{2}$ in.
- volume (V) = $\frac{1}{3} * (30$ in.² $* 4\frac{1}{2}$ in.$) = 45$ in.³

The volume of the triangular pyramid is 45 in.³.

Check Your Understanding

Find the volume of each pyramid. Include the unit in each answer.

1.
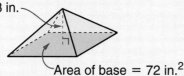

2.
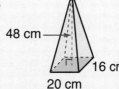

3.
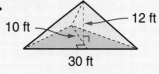

Check your answers on page 422.

Volume of a Sphere

Suppose you drew point A on a sheet of paper. Now imagine that you could draw every point on the sheet of paper that is 2 inches from point A. You would get a circle whose **center** is point A and whose **radius** is 2 inches long. Any points inside the circle are not a part of the circle; they form the **interior** of the circle.

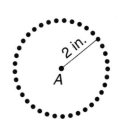

Now imagine all of the points in space that are 2 inches from point A in every direction. You would get a figure that looks like the surface of a ball. This figure is called a **sphere.** Point A is the **center of the sphere.** The distance from point A to any point on the sphere is the **radius of the sphere.**

Just as with the circle, the points inside the sphere are not a part of the sphere. A good way to think of a sphere is to picture a soap bubble. Another way is to imagine a circle with a rod passing through its center. If the circle is rotated around the rod, the path of the circle will form a sphere.

If a sphere is cut in half, each half is a figure called a **half-sphere.** The rim of the half-sphere is a circle whose center is the center of the sphere.

Formula for the Volume of a Sphere

$$V = \tfrac{4}{3} * \pi * r^3$$

V is the volume and r is the radius of the sphere.

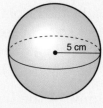

Example   Find the volume of the sphere.

Use the formula $V = \tfrac{4}{3} * \pi * r^3$.
- radius (r) = 5 cm
- volume (V) = $\tfrac{4}{3} * \pi * 5$ cm $* 5$ cm $* 5$ cm

Use the π key on a calculator or 3.14 as an approximate value for π.
- volume = 523.6 cm³, rounded to the nearest tenth of a cubic centimeter

The volume of the sphere is 523.6 cm³.

Check Your Understanding

Find the volume of each sphere to the nearest tenth of a cubic unit. Include the unit in each answer.

1. radius of sphere = 2 inches   **2.** diameter of sphere = 8 centimeters

Check your answers on page 422.

Surface Area of a Rectangular Prism

A rectangular prism has six flat surfaces called **faces.** The **surface area** of a rectangular prism is the sum of the areas of all six of its faces. Think of the six faces as three pairs of opposite, parallel faces. Since opposite faces have the same area, you find the area of one face in each pair of opposite faces. Then find the sum of these three areas and double the result.

The simplest rectangular prisms have all six of their faces shaped like rectangles. These prisms look like boxes. You can find the surface area of a box-like prism if you know its dimensions: length (l), width (w), and height (h).

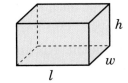

A box-like prism has all 6 faces shaped like rectangles.

Step 1: Find the area of one face in each pair of opposite faces.

area of	area of front	area of side
base = $l * w$	face = $l * h$	face = $w * h$

Step 2: Find the sum of the areas of the three faces.
  sum of areas = $(l * w) + (l * h) + (w * h)$

Step 3: Multiply the sum of the three areas by 2.
  surface area of prism = $2 * ((l * w) + (l * h) + (w * h))$

Surface Area of a Box–Like Rectangular Prism

$$S = 2 * ((l * w) + (l * h) + (w * h))$$

S is the surface area, l the length of the base, w the width of the base, h the height of the prism.

Example   Find the surface area of the box-like rectangular prism.

Use the formula $S = 2 * ((l * w) + (l * h) + (w * h))$.
- length (l) = 4 in.   width (w) = 3 in.   height (h) = 2 in.
- surface area (S) = $2 * ((4 \text{ in.} * 3 \text{ in.}) + (4 \text{ in.} * 2 \text{ in.}) + (3 \text{ in.} * 2 \text{ in.}))$
  $= 2 * (12 \text{ in.}^2 + 8 \text{ in.}^2 + 6 \text{ in.}^2)$
  $= 2 * 26 \text{ in.}^2 = 52 \text{ in.}^2$

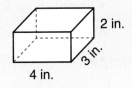

The surface area of the rectangular prism is 52 in.^{2}.

Check Your Understanding

Find the surface area of the box-like prism. Include the unit in your answer.

Check your answer on page 422.

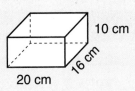

Surface Area of a Cylinder

The simplest cylinders look like food cans and are called **right cylinders.** Their bases are perpendicular to the line joining the centers of the bases.

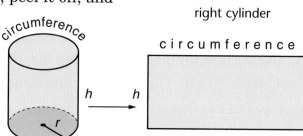

right cylinder

To find the area of the curved surface of a right cylinder, imagine a soup can with a label. If you cut the label perpendicular to the top and bottom of the can, peel it off, and lay it flat on a surface, you will have a rectangle. The length of the rectangle is the same as the circumference of a base of the cylinder. The width of the rectangle is the same as the height of the can. Therefore, the area of the curved surface is the product of the circumference of the base and the height of the can.

circumference of base $= 2 * \pi * r$

area of curved surface $= (2 * \pi * r) * h$

The surface area of a cylinder is the sum of the areas of the two bases $(2 * \pi * r^2)$ and the curved surface.

Surface Area of a Right Cylinder

$$S = (2 * \pi * r^2) + ((2 * \pi * r) * h)$$

S is the surface area, r is the radius of the base, h is the height of the cylinder.

Example Find the surface area of the right cylinder.

Use the formula $S = (2 * \pi * r^2) + ((2 * \pi * r) * h)$.

- radius of base $(r) = 3$ cm
- height $(h) = 5$ cm

Use the π key on a calculator or 3.14 as an approximate value for π.

- surface area $(S) = (2 * \pi * 3 \text{ cm} * 3 \text{ cm}) + ((2 * \pi * 3 \text{ cm}) * 5 \text{ cm})$
  $= (\pi * 18 \text{ cm}^2) + (\pi * 30 \text{ cm}^2)$
  $= 150.8 \text{ cm}^2$, rounded to the nearest tenth of a square centimeter

The surface area of the cylinder is 150.8 cm^2.

Check Your Understanding

Find the surface area of the right cylinder to the nearest tenth of a square inch. Include the unit in your answer.

Check your answer on page 422.

Temperature

Temperature is a measure of the hotness or coldness of something. To read a temperature in degrees, you need a reference frame that begins with a **zero point** and has a number-line **scale.** The two most commonly used temperature scales, Fahrenheit and Celsius, have different zero points.

Fahrenheit

This scale was invented in the early 1700s by the German physicist G.D. Fahrenheit. On the Fahrenheit scale, pure water freezes at 32°F and boils at 212°F. A saltwater solution freezes at 0°F (the zero point) at sea level. The normal temperature for the human body is 98.6°F. The Fahrenheit scale is used primarily in the United States.

Celsius

This scale was developed in 1742 by the Swedish astronomer Anders Celsius. On the Celsius scale, the zero point (0°C) is the freezing point of pure water. Pure water boils at 100°C. The Celsius scale divides the interval between these two points into 100 equal parts. For this reason, it is sometimes called the *centigrade* scale. The normal temperature for the human body is 37°C. The Celsius scale is the standard for most people outside of the United States and for scientists everywhere.

A **thermometer** measures temperature. The common thermometer is a glass tube that contains a liquid. When the temperature goes up, the liquid expands and moves up the tube. When the temperature goes down, the liquid shrinks and moves down the tube.

To convert between degrees Fahrenheit (°F) and degrees Celsius (°C), use these formulas:

$$F = \frac{9}{5} * C + 32 \quad \text{and} \quad C = \frac{5}{9} * (F - 32)$$

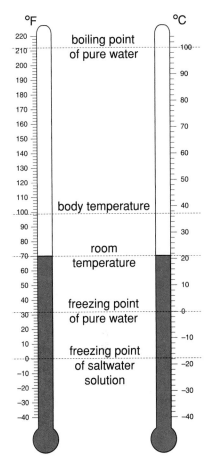

These thermometers show both the Fahrenheit and Celsius scales. Key reference temperatures, such as the boiling and freezing points of water, are indicated. A thermometer reading of 70°F (or about 21°C) is normal room temperature.

Example Find the Celsius equivalent of 82°F.

Use the formula $C = \frac{5}{9} * (F - 32)$, and replace F with 82:

$C = \frac{5}{9} * (82 - 32) = \frac{5}{9} * (50) = 27.77$

The Celsius equivalent of 82°F is about 28°C.

Note

Lava that erupts through the crater of a volcano usually has a temperature between 1,300°F and 2,300°F.

Weight

Today, in the United States, two different sets of standard units are used to measure weight.

♦ The standard unit for weight in the metric system is the **gram.** A small, plastic base-10 cube weighs about 1 gram. Heavier weights are measured in **kilograms.** One kilogram equals 1,000 grams.

♦ Two standard units for weight in the U.S. customary system are the **ounce** and the **pound.** Heavier weights are measured in pounds. One pound equals 16 ounces. Some weights are reported in both pounds and ounces. For example, we might say that "the suitcase weighs 14 pounds, 6 ounces."

Metric Units	U.S. Customary Units
1 gram (g) = 1,000 milligrams (mg)	1 pound (lb) = 16 ounces (oz)
1 milligram = $\frac{1}{1,000}$ gram	1 ounce = $\frac{1}{16}$ pound
1 kilogram (kg) = 1,000 grams	1 ton (t) = 2,000 pounds
1 gram = $\frac{1}{1,000}$ kilogram	1 pound = $\frac{1}{2,000}$ ton
1 metric ton (t) = 1,000 kilograms	
1 kilogram = $\frac{1}{1,000}$ metric ton	

Rules of Thumb	Exact Equivalents
1 ounce equals about 30 grams. 1 kilogram equals about 2 pounds.	1 ounce = 28.35 grams 1 kilogram = 2.205 pounds

Example A bicycle weighs 13 kilograms. How many pounds is that?

Rough Solution: Use the Rule of Thumb. Since 1 kg equals about 2 lb, 13 kg equals about 13 * 2 lb = 26 lb.

Exact Solution: Use the exact equivalent.
Since 1 kg = 2.205 lb, 13 kg = 13 * 2.205 lb = 28.665 lb.

Did You Know?

The Environmental Protection Agency estimates that about 2 million tons of old and broken computers, TVs, cell phones, and other electronic trash are dumped into U.S. landfills each year. That's more than 13 pounds per person each year.

Note

The Rules of Thumb Table shows how units of weight in the metric system compare to units in the U.S. customary system. You can use this table to convert between ounces and grams, and between pounds and kilograms. You need only remember the simple Rules of Thumb for most everyday purposes.

Check Your Understanding Solve each problem. Include the unit in each answer.

1. A softball weighs 6 ounces. How many grams is that? Use both a Rule of Thumb and an exact equivalent.

2. Ashley's sister weighs 36 pounds, 12 ounces. How many ounces is that?

Check your answers on page 422.

Capacity and Precision

Different kinds of scales are shown here. The **capacity** and **precision** are shown for each type of scale.

The **capacity** of a scale is the greatest weight that the scale can hold. For example, most infant scales have a capacity of about 25 pounds. Bath scales are used to weigh older children and adults and usually have a capacity of about 300 pounds.

The **precision** of a scale is its accuracy. If you can read a weight on a bath scale to the nearest pound, then the precision for that scale is 1 pound. On most infant scales, you can read a weight to the nearest ounce, so the precision is 1 ounce.

With a balance scale, you can measure weight to the nearest gram. A balance scale is much more precise than an infant scale because a gram is lighter than an ounce.

infant scale
capacity: 25 lb
precision: 1 oz

bath scale
capacity: 300 lb or 135 kg
precision: 0.1 lb or 50 g

balance scale
capacity: 2 kg
precision: 1 g

1g   1g   2g   2g   5g   10g   20g   20g   50g   100g   100g   200g   500g   1000g

weight set for balance scale

Some scales are extremely precise. They can weigh things that cannot be seen with the naked eye. Other scales are very large. They can be used to weigh objects that weigh as much as 1,000 tons (2,000,000 pounds). Many scales display weight in both metric and U.S. customary units.

platform scale
capacity: 1 T to 1,000 T
precision: $\frac{1}{4}$ lb to 1 T

food scale
capacity: 12 lb
precision: 1oz

market scale
capacity: 30 lb or 15 kg
precision: 0.01 lb or 5 g

spring scale
capacity: 500 g
precision: 20 g

Measuring and Drawing Angles

Angles are measured in **degrees.** When writing the measure of an angle, a small raised circle (°) is used as a symbol for the word *degree*.

Angles are measured with a tool called a **protractor.** You will find both a full-circle and a half-circle protractor on your Geometry Template. Since there are 360 degrees in a circle, a 1° angle marks off $\frac{1}{360}$ of a circle.

The **full-circle protractor** on the Geometry Template is marked off in 5° intervals from 0° to 360°. Although it can be used to measure angles, it cannot be used to draw angles of a given measure.

Sometimes you will use a full-circle protractor that is a paper cutout. This *can* be used to draw angles.

The **half-circle protractor** on the Geometry Template is marked off in 1° intervals from 0° to 180°.

It has two scales. Each scale starts at 0°. One scale is read clockwise, the other is read counterclockwise.

The half-circle protractor can be used both to measure angles and to draw angles of a given measure.

Two rays starting from the same endpoint form two angles. The smaller angle measures between 0° and 180°. The larger angle measures between 180° and 360°. The larger angle is called a **reflex angle.** The sum of the measures of the smaller angle and the reflex angle is 360°.

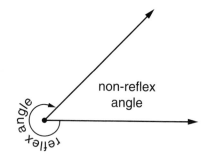

non-reflex angle

reflex angle

Measuring an Angle with a Full-Circle Protractor

Think of the angle as a rotation of the minute hand on a clock. One side of the angle represents the minute hand at the beginning of a time interval. The other side of the angle represents the minute hand some time later.

Example   Measure angle *ABC* with a full-circle protractor.

Step 1: Place the center of the protractor over the vertex of the angle, point *B*.

Step 2: Line up the 0° mark on the protractor with $\overrightarrow{BA}$.

Step 3: Read the degree measure where $\overrightarrow{BC}$ crosses the edge of the protractor.

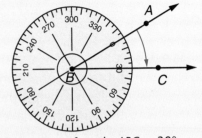

The measure of angle *ABC* = 30°.

Example   Measure reflex angle *EFG*.

Step 1: Place the center of the protractor over point *F*.

Step 2: Line up the 0° mark on the protractor with $\overrightarrow{FG}$.

Step 3: Read the degree measure where $\overrightarrow{FE}$ crosses the edge of the protractor.

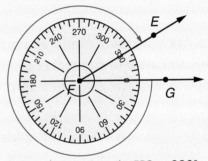

The measure of angle *EFG* = 330°.

Measuring an Angle with a Half-Circle Protractor

Example   Measure angle *PQR* with a half-circle protractor.

Step 1: Lay the baseline of the protractor on $\overrightarrow{QR}$.

Step 2: Slide the protractor so that the center of the baseline is over the vertex of the angle, point *Q*.

Step 3: Read the degree measure where $\overrightarrow{QP}$ crosses the edge of the protractor. There are two scales on the protractor. Use the scale that makes sense for the size of the angle that you are measuring.

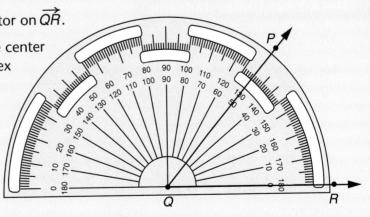

The measure of angle *PQR* = 50°.

Drawing an Angle with a Half-Circle Protractor

Example Draw a 40° angle.

Step 1: Draw a ray from point *A*.

Step 2: Lay the baseline of the protractor on the ray.

Step 3: Slide the protractor so that the center of the baseline is over point *A*.

Step 4: Make a mark at 40° near the protractor. There are two scales on the protractor. Use the scale that makes sense for the size of the angle that you are drawing.

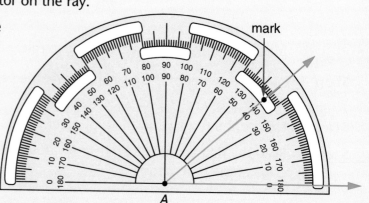

Step 5: Draw a ray from point *A* through the mark.

To draw a reflex angle using the half-circle protractor, subtract the measure of the reflex angle from 360°. Use this as the measure of the smaller angle.

Example Draw a 240° angle.

Step 1: Subtract: 360° − 240° = 120°.

Step 2: Draw a 120° angle.

The larger angle is a 240° reflex angle.

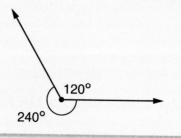

Check Your Understanding

Measure each angle to the nearest degree.

1.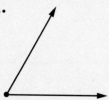

2.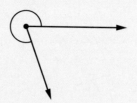

3.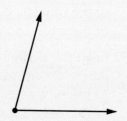

Draw each angle.

4. a 70° angle

5. a 280° angle

6. a 55° angle

Check your answers on page 422.

The Measures of the Angles of Polygons

Any polygon can be divided into triangles.

♦ The measures of the three angles of each triangle add up to 180°.

♦ To find the sum of the measures of all the angles inside a polygon, multiply the number of triangles in the polygon by 180°.

Example What is the sum of the measures of the angles of a hexagon?

Step 1: Draw any hexagon; then divide it into triangles. The hexagon can be divided into four triangles.

Step 2: Multiply the number of triangles by 180°.
4 * 180° = 720°

The sum of the measures of all the angles inside a hexagon equals 720°.

hexagon

Finding the Measure of an Angle of a Regular Polygon

All the angles of a regular polygon have the same measure. So the measure of one angle is equal to the sum of the measures of the angles of the polygon divided by the number of angles.

Example What is the measure of one angle of a regular hexagon?

The sum of the measures of the angles of *any* hexagon is 720°.
A regular hexagon has six congruent angles.

The measure of one angle of a regular hexagon is 720°/6 = 120°.

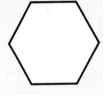

regular hexagon
(6 congruent sides and 6 congruent angles)

Check Your Understanding

1. Into how many triangles can you divide each polygon?
   a. a quadrilateral   **b.** a pentagon   **c.** an octagon   **d.** a 12-sided polygon

2. What is the sum of the measures of the angles of a pentagon?

3. What is the measure of an angle of a regular octagon?

4. Suppose that you know the number of sides of a polygon. How can you calculate the number of triangles into which it can be divided without drawing a picture?

Check your answers on page 422.

Plotting Ordered Number Pairs

A **rectangular coordinate grid** is used to name points in a plane. It is made up of two number lines called **axes** that meet at right angles at their zero points. The point where the two lines meet is called the **origin.**

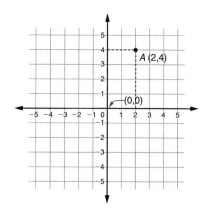

The ordered pair (0,0) names the origin.

Every point on a rectangular coordinate grid can be named by an **ordered number pair.** The two numbers that make up an ordered pair are called the **coordinates** of the point. The first coordinate is always the *horizontal* distance of the point from the vertical axis. The second coordinate is always the *vertical* distance of the point from the horizontal axis. For example, the ordered pair (2,4) names point A on the grid at the right. The numbers 2 and 4 are the coordinates of point A.

Example   Plot the ordered pair (4,2).

Locate 4 on the horizontal axis and draw a vertical line.

Locate 2 on the vertical axis and draw a horizontal line.

The point (4,2) is located at the intersection of the two lines.

The order of the numbers in an ordered pair is important. The ordered pair (4,2) does not name the same point as the ordered pair (2,4).

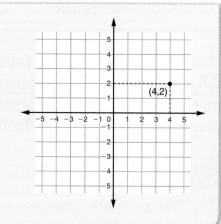

Example   Locate (−4,3), (−4,−3), and (2,0).

For each ordered pair:

Locate the first coordinate on the horizontal axis and draw a vertical line.

Locate the second coordinate on the vertical axis and draw a horizontal line.

The two lines intersect at the point named by the ordered pair.

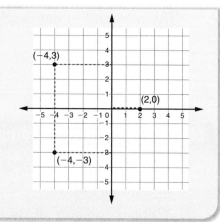

Check Your Understanding

Draw a coordinate grid on graph paper. Plot the following points:

1. (1,5)   **2.** (−4,−4)   **3.** (0,−3)   **4.** $(3\frac{1}{2},−2)$

Check your answers on page 423.

Latitude and Longitude

Earth is almost a perfect **sphere.** All points on Earth are about the same distance from its center. Earth rotates on an **axis,** which is an imaginary line that connects the **North Pole** and the **South Pole.**

Reference lines are drawn on globes and maps to make places easier to locate. Lines that go east and west around Earth are called **lines of latitude.** The lines of latitude are often called **parallels** because each one is a circle that is parallel to the equator. The **equator** is a special line of latitude. Every point on the equator is the same distance from both the North Pole and the South Pole.

The **latitude** of a place is measured in **degrees.** The symbol for degrees is (°). Lines north of the equator are labeled °N (degrees north); lines south of the equator are labeled °S (degrees south). The number of degrees tells how far north or south of the equator a place is. The area north of the equator is called the **Northern Hemisphere.** The area south of the equator is called the **Southern Hemisphere.**

Examples

The latitude of the North Pole is 90°N. The latitude of the South Pole is 90°S. The poles are the points farthest north and farthest south on Earth.

The latitude of Cairo, Egypt, is 30°N. We say that Cairo is 30 degrees north of the equator.

The latitude of Durban, South Africa, is 30°S. Durban is in the Southern Hemisphere.

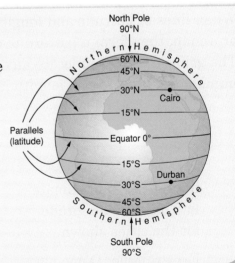

A second set of lines runs north to south. These are semicircles (half circles) that connect the two Poles. They are called **lines of longitude** or **meridians.**

Note

Meridians are not parallel since they meet at the Poles.

The **prime meridian** is a special meridian that is labeled 0°. The prime meridian passes through Greenwich, near London, England. Another special meridian is the **International Date Line.** This meridian is labeled 180° and is exactly opposite the prime meridian on the other side of the world.

The **longitude** of a place is measured in degrees. Lines west of the prime meridian are labeled °W. Lines east of the prime meridian are labeled °E. The number of degrees tells how far west or east of the prime meridian a place is. The area west of the prime meridian is called the **Western Hemisphere.** The area east of the prime meridian is called the **Eastern Hemisphere.**

Examples

The longitude of Greenwich, England, is 0° because Greenwich lies on the prime meridian.

The longitude of Durban, South Africa, is 30°E. Durban is in the Eastern Hemisphere.

The longitude of Gambia (a small country in Africa) is about 15°W. We say that Gambia is 15 degrees west of the prime meridian.

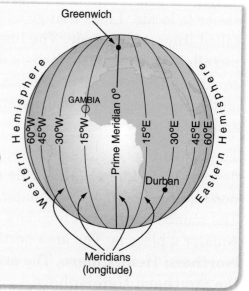

When lines of latitude and longitude are both shown on a globe or a map, they form a pattern of crossing lines called a **grid.** The grid can help you locate any place on the map by simply naming its latitude and longitude.

Example

This map may be used to find the approximate latitude and longitude for the cities shown.

For example, Denver, Colorado, is about 40° North and 105° West.

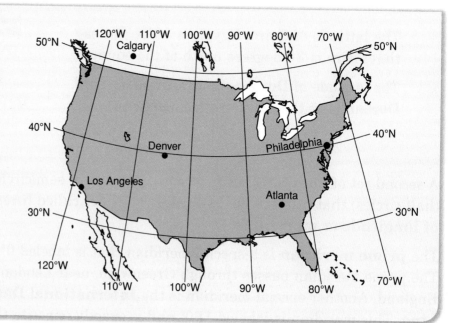

Algebra

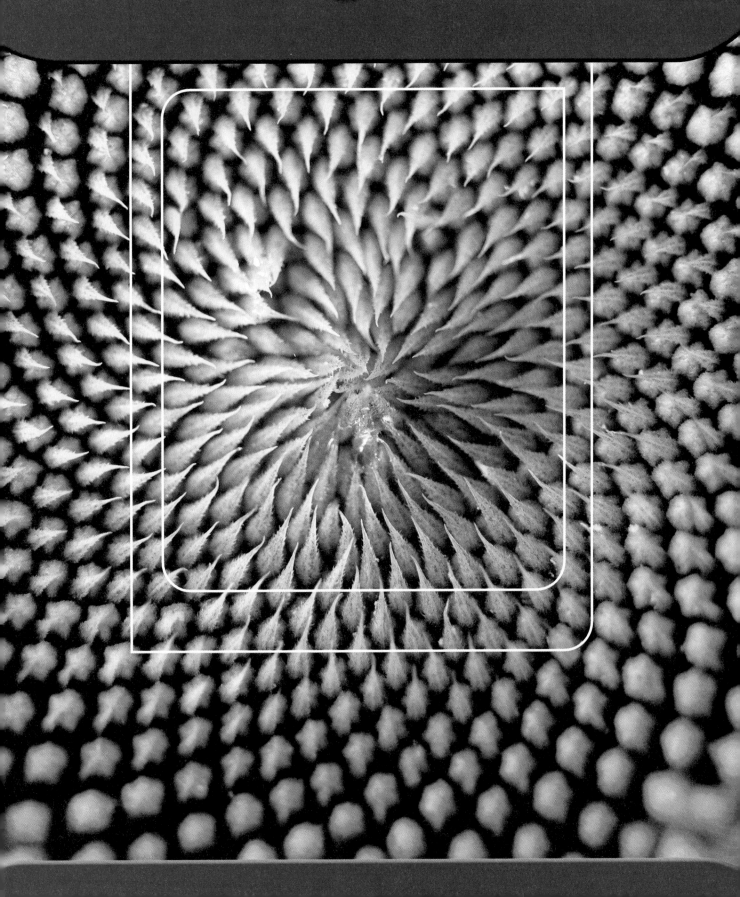

Algebra

The origins of algebra can be traced back thousands of years to ancient Egypt and Babylon. (For examples of Egyptian mathematics, see the margin for a photograph and description of the Rhind Papyrus.) In early times, algebra involved solving number problems for which one or more of the numbers was not known. The objective was to find these missing numbers, called the "unknowns." Words were used for the unknowns, as in "Five plus some number equals eight."

Then, in the late 1500s, François Viète began using letters to stand for unknown quantities, as in $5 + x = 8$. Viète's invention made solving number problems much easier and led to an explosion of discoveries in mathematics and science that has continued into modern times.

Most people think of algebra as a high school subject, mostly involving symbols and equations. (An equation is a number sentence that contains an equals sign.) You have studied algebra since first grade, when you had to find the missing number in simple equations such as $5 + \square = 9$. As you studied more mathematics, the equations became more complicated. But the basic problem has remained the same: to find the missing number in an equation.

Letters, blanks, or other symbols that are used to stand for unknown numbers are called **variables.** Variables are also used in several other ways.

Variables Can Be Used to State the Properties of the Number System

Properties of a number system are rules that are true for all numbers. For example, the commutative property of addition states that for all numbers a and b, $a + b = b + a$. You were introduced to this property in first grade as the "turn-around" shortcut to help you memorize addition facts.

Many additional properties are listed on pages 104–106. All of these properties are stated by using variables.

Did You Know?

Our word *algebra* comes from the Arabic word *al-jabru,* which means "restoration." Algebra was known as the "science of restoration and balancing."

The Rhind Papyrus was written in Egypt almost 4,000 years ago. It is divided into three parts: arithmetic problems, geometry problems, and a section that includes area and volume problems. Some of the problems used algebra, but words were used for the unknowns instead of letters or symbols.

Variables Can Be Used to Express Rules or Functions

Function machines and "What's My Rule?" tables have rules that tell you how to get the "out" numbers from the "in" numbers. These rules can be written using variables. For example, a function machine might have the rule "double and add 1." This rule can be written using variables, as $y = 2 * x + 1$ or as $y = 2x + 1$. The rule can also be graphed on a coordinate grid, as shown at the right.

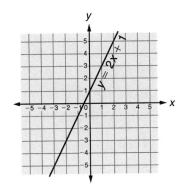

Variables Can Be Used in Formulas

Formulas are used in everyday life, in science, in business, and in many other situations as a compact way to express relationships. For example, the formula $d = r * t$ expresses the relationship between a distance (d), the rate at which one covers the distance (r), and the time (t) it takes to cover the distance.

miles (d)	50	100	150	?
hours (t)	1	2	3	4

$d = r * t$ $r = 50$ miles per hour

Variables Can Be Used in Computers and Calculators

Variables are used in computer spreadsheets, making it possible to evaluate formulas quickly and efficiently. Spreadsheets are useful for making predictions based on trends. Variables are also used in writing computer programs. *(See the margin.)* These computer programs are made up of a series of "commands" similar to number sentences that contain variables.

Some calculators, especially graphing calculators, use variables to name calculator key functions and to specify procedures.

```
PROGRAM SIMPLE
   ! A SAMPLE PROGRAM
   IMPLICIT NONE
   REAL :: A,B,C,TOTAL
   A = 2.0
   B = 3.0
   C = 4.0
   TOTAL = A + B + C
   PRINT 'TOTAL IS',TOTAL
END PROGRAM SIMPLE
```

A simple computer program

Check Your Understanding

For each problem, write a number sentence using a letter for the unknown.

1. Twice some number equals 40.

2. Some number, decreased by 9, equals 11.

Solve each problem. Use the formula $d = r * t$.

3. How far will a car travel in $2\frac{1}{2}$ hours if its average speed is 55 miles per hour?

4. Joan reads 3 pages in 2 minutes. How long will it take her to read $5\frac{1}{2}$ pages?

Check your answers on page 423.

Algebraic Expressions

Consider the following statement: "Marcia read four more books last year than Gina." You can't tell how many books Marcia read unless you know the number of books Gina read. There are many possibilities:

◆ If Gina read 8 books, then Marcia read 12 books.

◆ If Gina read 27 books, then Marcia read 31 books, and so on.

One way to represent the number of books Marcia read is to write an **algebraic expression** with a variable that represents the number of books Gina read. For example, if G represents the number of books Gina read, then $G + 4$ represents the number of books Marcia read.

Note

An **expression** uses operation symbols (such as $+$, $-$, $*$, and $/$) to combine numbers and variables. Some expressions like $3 + 4$ and $(2 * 8) / 5$ do not contain variables. Other expressions like $3 + x$ and $(2 * 8) / C$ do contain variables and are called **algebraic expressions**.

Examples Write an algebraic expression to answer each question.

Statement	Algebraic Expression
Mary is 5 years older than her sister Carla. How old is Mary?	If Carla is C years old, then Mary is $C + 5$ years old.
Mrs. Roth weighs 30 pounds less than Mr. Roth. How much does Mrs. Roth weigh?	If Mr. Roth weighs R pounds, then Mrs. Roth weighs $R - 30$ pounds.
Anna bought a box of crayons with 8 crayons per box for each of her nieces. How many crayons did she buy altogether?	If Anna has N nieces, then she bought a total of $8 * N$ crayons.
Claude earned $6 an hour and was paid an additional $3.50 for his lunch. How much did Claude get paid?	If Claude worked H hours, he got paid $(6 * H) + 3.50$ dollars.

Check Your Understanding

Write an algebraic expression for each situation using the suggested variable.

1. Mark is m inches tall. If Audrey is 3 inches shorter than Mark, what is Audrey's height?

2. It takes Herman H minutes to do his homework. If it takes Sue twice as long, how long does it take her?

3. Dawn went on R rides at the amusement park. If the park charges a $3 admission fee and $0.50 per ride, how much did Dawn spend?

Check your answers on page 423.

Number Sentences

Number sentences are made up of **mathematical symbols.**

		Mathematical Symbols		
Digits	**Variables**	**Operation Symbols**	**Relation Symbols**	**Grouping Symbols**
0, 1, 2, 3, 4, 5, 6, 7, 8, 9	$n\ x\ y\ z$ $a\ b\ c\ d$ $C\ M\ P\ ?\ \square$	$+$ plus $-$ minus $\times$ or $*$ times $/$ or $\div$ divided by	$=$ is equal to $\neq$ is not equal to $<$ is less than $>$ is greater than $\leq$ is less than or equal to $\geq$ is greater than or equal to	( ) parentheses [ ] brackets

A number sentence must contain numbers (or variables) and a relation symbol. It may or may not contain operation symbols and grouping symbols.

Number sentences that contain the $=$ symbol are called **equations.** Number sentences that contain any one of the symbols $\neq$, $<$, $>$, $\leq$, or $\geq$ are called **inequalities.**

If a number sentence does not contain variables, then it is always possible to tell whether it is **true** or **false.**

Examples

Equations

$3 + 3 = 8$ False $3 + 3 = 6$, not 8

$(24 + 3) / 9 = 3$ True $27 / 9 = 3$

$100 = 9^2 + 9$ False $9^2 + 9 = 90$, and 90 is not equal to 100

Inequalities

$-27 * 4 > 42$ False $-27 * 4 = -108$, which is not greater than 42.

$\frac{4}{5} - \frac{2}{3} < \frac{1}{2}$ True $\frac{4}{5} - \frac{2}{3} = \frac{2}{15}$, and $\frac{2}{15}$ is less than $\frac{1}{2}$.

$27 \neq 72$ True   27 is not equal to 72.

$19 < 19$ False   19 is not less than itself.

$16 * 4 \geq 80 \div 3$ True   64 is greater than or equal to $26\frac{2}{3}$.

Check Your Understanding

True or false?

1. $32 - 14 = 18$

2. $4 * 7 < 30$

3. $0 = \frac{5}{5}$

4. $25 + 5 \leq 5 * 6$

5. $50 - 12 = 7 * 2^2$

6. $84 \neq 84$

Check your answers on page 423.

Parentheses

The meaning of a number sentence or expression is not always clear. You may not know which operation to do first. But you can use parentheses to make the meaning clear. When there are parentheses in a number sentence or expression, *the operations inside the parentheses are always done first.*

Examples Evaluate. $(24 - 6) * 2 = ?$

The parentheses tell you to subtract $24 - 6$ first.	$(24 - 6) * 2$
	$18 * 2$
Then multiply by 2.	36

$(24 - 6) * 2 = 36$

Evaluate $24 - (6 * 2)$.

The parentheses tell you to multiply $6 * 2$ first.	$24 - (6 * 2)$
	$24 - 12$
Then subtract.	12

$24 - (6 * 2) = 12$

Example Evaluate $(3 * C) + H$ if $C = 4$ and $H = 5$.

Replace C with 4 and replace H with 5.
The result is $(3 * 4) + 5$.

The parentheses tell you to multiply $3 * 4$ first.	$(3 * 4) + 5$
Then add.	$12 + 5$
	17

If $C = 4$ and $H = 5$, then $(3 * C) + H$ is 17.

Note

To **evaluate** something is to find out what it is worth.

To evaluate an algebraic expression (like the last example), first replace each variable with its value and then carry out the operations.

Open Sentences

In some number sentences, one or more numbers may be missing. In place of each missing number is a letter, a question mark, or some other symbol. These number sentences are called **open sentences.** A symbol used in place of a missing number is called a **variable.**

For most open sentences, you can't tell whether the sentence is true or false until you know which number replaces the variable. For example, $5 + x = 12$ is an open sentence in which x stands for some number.

♦ If you replace x with 3 in $5 + x = 12$, you get the number sentence $5 + 3 = 12$, which is false.

♦ If you replace x with 7 in $5 + x = 12$, you get the number sentence $5 + 7 = 12$, which is true.

Note

Some open sentences are always true. $9 + y = y + 9$ is true if you replace y with any number.

Some open sentences are always false. $C - 1 > C$ is false if you replace C with any number.

If a number used in place of a variable makes the number sentence true, this number is called a **solution** of the open sentence. For example, the number 7 is a solution of the open sentence $5 + x = 12$ because the number sentence $5 + 7 = 12$ is true. Finding the solution(s) of an open number sentence is called **solving** the number sentence.

Many simple equations have just one solution, but inequalities may have many solutions. For example, 9, 3.5, $2\frac{1}{2}$, and -8 are all solutions of the inequality $x < 10$. In fact, $x < 10$ has infinitely many solutions—any number that is less than 10 is a solution.

Number Models

In *Everyday Mathematics,* a number sentence or an expression that describes some situation is called a **number model.** Often, two or more number models can fit a given situation. Suppose, for example, that you had $20, spent $8.50, and ended up with $11.50. The number model $20 − $8.50 = $11.50 fits this situation. The number model $20 = $8.50 + $11.50 also fits.

> **Note**
>
> Number models are not always number sentences. The expression $20 − $8.50 is not a number sentence, but it is another model that fits the situation.

Number models can be useful in solving problems. For example, the problem "Juan is saving for a bicycle that costs $119. He has $55. How much more does he need?" can be modeled by "$119 = $55 + x" or by "$119 − $55 = x." The first of these number models suggests counting up to find how much more Juan needs; the second suggests subtracting to find the answer.

Other kinds of mathematical models are discussed in the section on problem solving.

Check Your Understanding

Find the solution of each equation.

1. $6 + c = 20$

2. $42 = 6 * z$

3. $(2 * f) + 5 = 26$

Write a number model that fits each problem.

4. Hunter used a $20 bill to pay for a CD that cost $12.49. How much change did he get?

5. Eve earns $10 a week baby-sitting. How many weeks will it take her to earn $90?

Check your answers on page 423.

Inequalities

An **inequality** is a number sentence that contains one of these symbols: $\neq$, $<$, $>$, $\geq$, or $\leq$. An inequality that contains a variable is an open sentence. Any number substituted for the variable that makes the inequality a true number sentence is called a solution of the inequality.

Many open inequalities have an infinite number of solutions; therefore it is usually impossible to list all the solutions. Instead, the set of solutions, called the **solution set,** is either described or shown on a number-line graph.

Example Describe and graph the solution set of $x + 3 > 10$.

The inequality $x + 3 > 10$ is an open sentence.
100 is a solution of $x + 3 > 10$ because $100 + 3 > 10$ is true.
2 is not a solution of $x + 3 > 10$ because $2 + 3 > 10$ is not true.

Any number that is less than or equal to 7 is clearly not a solution of $x + 3 > 10$.
For example, 6 is not a solution of $x + 3 > 10$ because $6 + 3$ is not greater than 10.

Any number greater than 7 is a solution of $x + 3 > 10$.
The graph of the solution set of $x + 3 > 10$ looks like this:

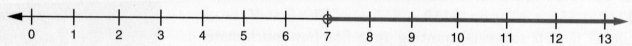

The shaded part of the number line tells you that any number greater than 7 is a solution (for example, 7.1, 8, 10.25). Notice the open circle at 7. This tells you that 7 is not part of the solution set.

Example Graph the solution set of $y - 3 \leq 1$.

Any number less than or equal to 4 is a solution. For example, 4, 2.1, −6 are all solutions. Notice the shaded circle at 4. This tells you that 4 is one of the solutions.

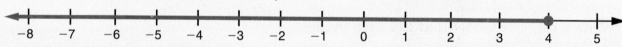

Check Your Understanding

Describe the solution set of the inequalities in Problems 1 and 2. Graph the solution sets in Problems 3 and 4.

1. $n - 6 < 2$ 2. $6 + c > 6$ 3. $y \leq 8$ 4. $-3 + p \geq 0$

Check your answers on page 423.

Formulas

A **formula** is a way of expressing a relationship between quantities.
(A **quantity** is a number with a unit, usually a measurement or a
count.) The quantities in a formula are represented by variables.

> **Example** What is the formula for the area of a parallelogram?
>
> The variable A stands for the area, b for the length of the base,
> and h for the height of the parallelogram.
>
> The formula for the area of a parallelogram is $A = b * h$.

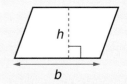

Variables in Formulas

Often the symbol for a variable is the first letter of the quantity
it represents.

> **Examples** Write each formula using variables.
>
> Area of parallelogram = base length * height $A = b * h$
> circumference of circle = pi * diameter of circle $c = \pi * d$

Note

The multiplication symbol
($*$ or $\times$) can be left out
when a variable is
multiplied by another
variable or by a number.
For example, the formula
for the area of a circle is
often written $A = \pi r^2$.
This is the same as
$A = \pi * r^2$.

A letter variable can have different meanings depending on
whether it is a capital or a lowercase (small) letter.

> **Example** The area of the shaded region in the figure at the right
> can be found by using the formula $A = S^2 - s^2$.
>
> Note that S stands for the length of the side of the larger square and
> s stands for the length of the side of the smaller square.

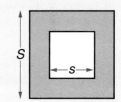

Evaluating a Formula

To **evaluate** a formula means to find the value of one variable in
the formula when the values of the other variables are given.

> **Example** Evaluate the formula for the area of the shaded region in the figure above
> when $S = 4$ cm and $s = 2$ cm.
>
> $A = S^2 - s^2$
> $A = (4 \text{ cm})^2 - (2 \text{ cm})^2$
> $A = 16 \text{ cm}^2 - 4 \text{ cm}^2 = 12 \text{ cm}^2$
>
> The area of the shaded region is 12 cm^2.

Units in Formulas

It is important that the **units** (hours, inches, meters, and so on) in a formula are consistent. For example, an area formula will not give a correct result if one measurement is in millimeters and another is in centimeters.

Example Find the area of the rectangle.

7 mm

5 cm

First, you need to change millimeters to centimeters or centimeters to millimeters. Then, use the formula $A = b * h$.

Change centimeters to millimeters: 5 cm = 50 mm $A = 7 \text{ mm} * 50 \text{ mm} = 350 \text{ mm}^2$
Change millimeters to centimeters: 7 mm = 0.7 cm $A = 0.7 \text{ cm} * 5 \text{ cm} = 3.5 \text{ cm}^2$

The area of the rectangle is 350 mm^2 or 3.5 cm^2.

If something travels at a constant speed, the distance formula is $d = r * t$. d is the distance traveled; t is the travel time; and r is the rate of travel (speed). If the value of r is in miles per hour, the value of t should be in hours. If the value of r is in meters per second, t should be in seconds.

Did You Know?

If an object is dropped over an open space, then $d = 16 * t^2$ and $s = 32 * t$. t is the time in seconds since the object started falling; d is the distance traveled (in feet); and s is the speed of the object (in feet per second).

Example Find the distance if $r = 50$ miles per hour and $t = 2$ hours. Use the formula $d = r * t$.

If $r = 50$ miles per hour and $t = 2$ hours, then
$d = 50$ miles per hour $* 2$ hours $= 100$ miles.

Check Your Understanding

A   B

16 inches

D   3 feet   C

1. Find the area of rectangle *ABCD*.
2. If a car is traveling 20 feet per second, how far will it travel in one minute?
3. If $S = 9$ meters and $s = 3$ meters, what is the area of the shaded region in the diagram at the right?

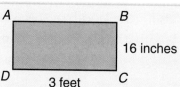

Express each of the following relationships with a formula.

4. The volume of a cone (V) is equal to $\frac{1}{3}$ of the area of the base (B) times the height of the cone (h).

5. The interest (i) earned on $1,000 deposited in a savings account is equal to $1,000 times the rate of interest (r) times the length of time the money is left in the account (t).

Check your answers on page 423.

Order of Operations

In many everyday situations, the order in which things are done is important. When you bake a cake, for example, you crack the eggs before adding them to the batter. In mathematics, too, many operations should be done in a certain order.

Rules for the Order of Operations

1. Do the operations inside **parentheses.**
   Follow rules 2–4 when you are working inside parentheses.
2. Calculate all expressions with **exponents.**
3. **Multiply** and **divide** in order from left to right.
4. **Add** and **subtract** in order from left to right.

Some people remember the order of operations by memorizing this sentence:

Please **E**xcuse **M**y **D**ear **A**unt **S**ally.

Parentheses **E**xponents **M**ultiplication **D**ivision **A**ddition **S**ubtraction

Example Evaluate. $5 * 4 - 6 * 3 + 2 = ?$

Multiply first.	$5 * 4 - 6 * 3 + 2$
Subtract next.	$20 - 18 + 2$
Then add.	$2 + 2$
	4

$5 * 4 - 6 * 3 + 2 = 4$

Example Evaluate $5^2 + (3 * 4 - 2)/5$.

Clear parentheses first.	$5^2 + (3 * 4 - 2)/5$
Calculate exponents next.	$5^2 + 10/5$
Divide.	$25 + 10/5$
Then add.	$25 + 2$
	27

$5^2 + (3 * 4 - 2)/5 = 27$

Check Your Understanding

Evaluate each expression.

1. $28 - 15/3 + 8$
2. $1 + (5 * 10)/4$
3. $10 * 6/2 - 30$
4. $10 * (12/6 + 4)/12 + 1$

Check your answers on page 423.

The Distributive Property

You have been using the **distributive property** for years, probably without knowing it.

For example, when you solve 40 * 57 with partial products, you think of 57 as 50 + 7 and multiply each part by 40.

The distributive property says: 40 * (50 + 7) = (40 * 50) + (40 * 7).

The distributive property can be illustrated by finding the area of a rectangle.

$$
\begin{array}{r}
57 \\
* \, 40 \\
\hline
40 * 50 = 2000 \\
40 * 7 = 280 \\
\hline
40 * 57 = 2280
\end{array}
$$

Example Show how the distributive property works by finding the area of the rectangle in two different ways.

3 cm

4 cm   2 cm

Method 1 Find the total width of the rectangle and multiply that by the height.

$A = 3 \text{ cm} * (4 \text{ cm} + 2 \text{ cm})$
$ = 3 \text{ cm} * 6 \text{ cm}$
$ = 18 \text{ cm}^2$

Method 2 Find the area of each smaller rectangle, and then add these areas.

$A = (3 \text{ cm} * 4 \text{ cm}) + (3 \text{ cm} * 2 \text{ cm})$
$ = 12 \text{ cm}^2 + 6 \text{ cm}^2$
$ = 18 \text{ cm}^2$

Both methods show that the area of the rectangle is 18 cm².

$3 * (4 + 2) = (3 * 4) + (3 * 2)$.

This is an example of the distributive property of multiplication over addition.

The distributive property of multiplication over addition can be stated in two ways:

$a * (x + y) = (a * x) + (a * y)$

$(x + y) * a = (x * a) + (y * a)$

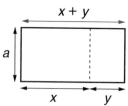

The distributive property of multiplication over subtraction can also be stated in two ways:

$a * (x - y) = (a * x) - (a * y)$

$(x - y) * a = (x * a) - (y * a)$

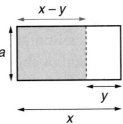

Example Show how the distributive property of multiplication over subtraction works by finding the area of the shaded part of the rectangle in two different ways.

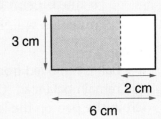

Method 1 Multiply the width of the shaded rectangle by its height.

$$A = 3 \text{ cm} * (6 \text{ cm} - 2 \text{ cm})$$
$$= 3 \text{ cm} * 4 \text{ cm}$$
$$= 12 \text{ cm}^2$$

Method 2 Subtract the area of the unshaded rectangle from the entire area of the whole rectangle.

$$A = (3 \text{ cm} * 6 \text{ cm}) - (3 \text{ cm} * 2 \text{ cm})$$
$$= 18 \text{ cm}^2 - 6 \text{ cm}^2$$
$$= 12 \text{ cm}^2$$

Both methods show that the area of the shaded part of the rectangle is 12 cm^2.
$3 * (6 - 2) = (3 * 6) - (3 * 2)$

This is an example of the distributive property of multiplication over subtraction.

Check Your Understanding

Use the distributive property to solve the problems.

1. $6 * (100 + 40)$ **2.** $(35 - 15) * 6$ **3.** $4 * (80 - 7)$

4. Use a calculator to verify that $1.23 * (456 + 789) = (1.23 * 456) + (1.23 * 789)$.

Check your answers on page 423.

Pan-Balance Problems and Equations

If two different kinds of objects are placed in the pans of a balance so that the pans balance, then you can find the weight of one kind of object in terms of the other kind of object.

Example

The pan-balance at the right has 2 balls and 6 marbles in one pan, and 1 ball and 8 marbles in the other pan. How many marbles weigh as much as 1 ball?

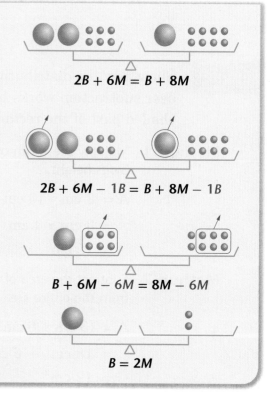

$$2B + 6M = B + 8M$$

Step 1:

If 1 ball is removed from each pan, the pan-balance will remain balanced. One ball and 6 marbles will be left in the pan on the left and 8 marbles will be left in the pan on the right.

$$2B + 6M - 1B = B + 8M - 1B$$

Step 2:

If 6 marbles are removed from each pan, the pan-balance will remain balanced. One ball will be left in the pan on the left and 2 marbles will be left in the pan on the right.

$$B + 6M - 6M = 8M - 6M$$

1 ball weighs as much as 2 marbles.

$$B = 2M$$

When solving a pan-balance problem, the pans must balance after each step. If you *always do the same thing to the objects in both pans,* then the pans will remain balanced. For example, you might remove the same number of the same kind of object from each pan as in Step 1 and Step 2 of the example above.

You can think of pan-balance problems as models for equations. Suppose that B stands for the weight of 1 ball and M stands for the weight of 1 marble. The pan-balance problem in the example above can then be expressed by the equation $2B + 6M = B + 8M$.

Some other ways to do the same thing to the objects in both pans:

1. Double the number of each kind of object in each pan.

2. Remove half of each kind of object from each pan.

Check Your Understanding

Write an equation for each pan-balance problem.

1.

2.

Check your answers on page 423.

A Systematic Method for Solving Equations

Many equations with just one unknown can be solved using only addition, subtraction, multiplication, and division. If the unknown appears on both sides of the equals sign, you must change the equation to an equation with the unknown appearing on one side only. You may also have to change the equation to one with all the constants on the other side of the equals sign.

Note

A **constant** is just a number, such as 3 or 7.5 or π. Constants don't change, or vary, the way variables do.

The operations you use to solve an equation are similar to the operations you use to solve a pan-balance problem. Remember— you must *always perform the same operation on both sides of the equals sign.*

Example   Solve $3y + 10 = 7y - 6$

Step	Operation	Equation
1. Remove the unknown term (the variable term) from the left side of the equation.	Subtract 3y from each side. (S 3y)	$\begin{aligned} 3y + 10 &= 7y - 6 \\ -3y\phantom{{}+10} & \phantom{{}={}} -3y\phantom{{}-6} \\ \hline 10 &= 4y - 6 \end{aligned}$
2. Remove the constant term from the right side of the equation.	Add 6 to both sides. (A 6)	$\begin{aligned} 10 &= 4y - 6 \\ +6 & \phantom{{}={}} +6 \\ \hline 16 &= 4y \end{aligned}$
3. Change the 4y term to a 1y term. (Remember, 1y and 1 * y and y all mean the same thing.)	Divide both sides by 4. (D 4)	$\begin{aligned} 16 &= 4y \\ 16/4 &= 4y/4 \\ 4 &= y \end{aligned}$

Check: Substitute the solution, 4, for y in the original equation:

$$
\begin{aligned}
3y + 10 &= 7y - 6 \\
3 * 4 + 10 &= 7 * 4 - 6 \\
12 + 10 &= 28 - 6 \\
22 &= 22
\end{aligned}
$$

Since $22 = 22$ is true, the solution, 4, is correct.

So, $y = 4$.

Each step in the above example produced a new equation that looks different from the original equation. But even though these equations look different, they all have the same solution (which is 4). Equations that have the same solution are called **equivalent equations.**

Like terms are terms that have exactly the same unknown or unknowns. The terms $4x$ and $2x$ are like terms because they both contain x. The terms 6 and 15 are like terms because they both contain no variables; 6 and 15 are both constants.

If an equation has parentheses, or if the unknown or constants appear on both sides of the equals sign, here is how you can **simplify** it.

♦ If an equation has parentheses, use the distributive property or other properties to write an equation without parentheses.

♦ If an equation has two or more like terms on one side of the equals sign, combine the like terms. To **combine like terms** means to rewrite the sum or difference of like terms as a single term. For example, $4y + 7y = 11y$ and $4y - 7y = -3y$.

Example   Solve $5(b + 3) - 3b + 5 = 4(b - 1)$.

Operation	Equation
1. Use the distributive property to remove the parentheses.	$5b + 15 - 3b + 5 = 4b - 4$
2. Combine like terms.	$2b + 20 = 4b - 4$
3. Subtract $2b$ from both sides. (S $2b$)	$\begin{aligned} 2b + 20 &= 4b - 4 \\ -2b &= -2b \\ \hline 20 &= 2b - 4 \end{aligned}$
4. Add 4 to both sides. (A 4)	$\begin{aligned} 20 &= 2b - 4 \\ +4 &= +4 \\ \hline 24 &= 2b \end{aligned}$
5. Divide both sides by 2. (D 2)	$\begin{aligned} 24 &= 2b \\ 24/2 &= 2b/2 \\ 12 &= b \end{aligned}$

Note

Reminder:
$5(b + 3)$ means the same as $5 * (b + 3)$.

Check Your Understanding

1. Check that 12 is the solution of the equation in the example above.

Solve.

2. $5x - 7 = 1 + 3x$ **3.** $5 * (s + 12) = 10 * (3 - s)$ **4.** $3(9 + b) = 6(b + 3)$

Check your answers on page 423.

"What's My Rule?" Problems

Imagine a machine that works like this: When a number (the *input,* or "in" number) is dropped into the machine, the machine changes the number according to a rule. A new number (the *output,* or "out" number) comes out the other end.

This machine adds 5 to any "in" number. Its rule is "+ 5."

- ♦ If 4 is dropped in, 9 comes out.
- ♦ If 7 is dropped in, 12 comes out.
- ♦ If 53 is dropped in, 58 comes out.
- ♦ If −6 is dropped in, −1 comes out.

"In" and "out" numbers can be displayed in table form as shown at the right.

To solve a "What's My Rule?" problem, you need to find the missing information. In the following examples, the solutions (the missing information) appear in color.

4
↓

Rule

+5

↓
9

in	out
x	$x + 5$
4	9
7	12
53	58
−6	−1

Examples

Find the "out" numbers.

Rule: subtract 7 from "in"

in	out	
z	$z - 7$	
9	2	$9 - 7 = 2$
27	20	$27 - 7 = 20$

Find the "in" numbers.

Rule: multiply "in" by 2

in	out	
w	$w * 2$	
4	8	$4 * 2 = 8$
24	48	$24 * 2 = 48$

Find the rule.

Rule: raise "in" to the second power

in	out	
r	r^2	
2	4	$2^2 = 4$
5	25	$5^2 = 25$

Check Your Understanding

Solve these "What's My Rule?" problems.

1. *Rule:* divide "in" by 3

in	out
n	$n / 3$
9	
36	

2. *Rule:* subtract 4 from "in"

in	out
k	$k - 4$
	7
	24

3. *Rule:* ?

in	out
x	
4	120
10	300

Check your answers on page 423.

Rules, Tables, and Graphs

Many problems can be solved by showing relationships between variables with rules, tables, or graphs.

Example Sara earns $6 per hour. Use a rule, a table, and a graph to find how much Sara earns in $2\frac{1}{2}$ hours.

Rule: If h stands for the number of hours Sara works, and e stands for her earnings, then $e = \$6 * h$.

$$e = \$6 * h = \$6 * 2\frac{1}{2}$$
$$= \$15$$

Table:

Time (hours)	Earnings ($)
h	$6 * h$
0	0
1	6
2	12
3	18

Think of $2\frac{1}{2}$ hours as 2 hours + $\frac{1}{2}$-hour. For 2 hours, Sara earns $12. For $\frac{1}{2}$-hour, Sara earns half of $6, or $3. In all, Sara earns $12 + $3 = $15. Or, note that $2\frac{1}{2}$ hours is halfway between 2 hours and 3 hours, so her earnings are halfway between $12 and $18, which is $15.

Graph:

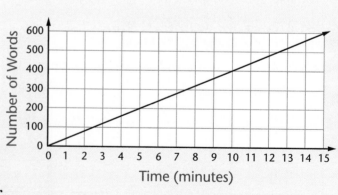

Time (hours)

To draw the graph, plot several of the number pairs from the table. For example, plot (0,0), (1,6), (2,12), and (3,18). Plot each number pair as a point on the coordinate grid. The straight line connecting these points is the graph.

To use the graph, first find $2\frac{1}{2}$ hours on the horizontal axis. Then go straight up to the line for Sara's earnings. Turn left and go across to the vertical axis. You will end up at the same answer as you did when you used the table, $15.

Sara earns $15 in $2\frac{1}{2}$ hours.

Check Your Understanding

1. Christie types about 40 words per minute. Use the graph to find how many words she can type in 12.5 minutes.

2. Daniel earns $4.50 an hour. He worked 7 hours. Use the rule to find how much he earned. (e stands for earnings; h stands for the number of hours worked.) *Rule:* $e = \$4.50 * h$.

Check your answers on page 423.

Independent and Dependent Variables

Formulas and the rules in "What's My Rule?" problems often use equations to relate two or more variables. For example:

- The formula for the area A of a square with a given side length s is $A = s^2$.

- An equation for the number of miles m a car travels in a given number of hours h at a constant speed of 45 miles per hour is $45 * h = m$.

- An equation for the "What's My Rule?" function machine at right is $y = x + 5$.

Each of these equations has an independent variable and a dependent variable. An **independent variable** is one whose value does not rely on any other variable. A **dependent variable** is one whose value depends on the value of another variable.

Example   Identify the independent and dependent variables in the equations above.

In $A = s^2$, the area depends on the side length. So s is independent and A is dependent.	In $45 * h = m$, the miles traveled depends on the number of hours. So h is independent and m is dependent.	In $y = x + 5$, the "out" variable depends on the value of the "in" variable. So x is independent and y is dependent.

The situation for which an equation is written determines the independent variable or variables. The independent variables in the area and distance equations are the given values: side length s for area and hours h for distance. In "What's My Rule?" problems, the independent variable is the "in" number.

The next example shows that sometimes it is not clear which variable or variables are independent.

Example   Judd wants to build an 84 square foot rectangular deck w feet wide and l feet long. So $84 = l * w$. Which variable in this equation is the independent variable?

Because Judd can choose a value for either l or w, both are independent.

Once a value for either l or w is chosen, the other variable is dependent on that value to make the equation true.

By convention, values of independent variables in equations are listed in the left column of a table and graphed on the horizontal axis of a graph.

$A = s^2$

s	A
1	1
2	4
3	9

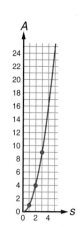

$45 * h = m$

h	m
1	45
2	90
3	135

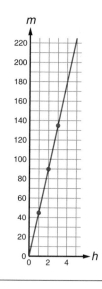

$y = x + 5$

x	y
1	6
2	7
3	8

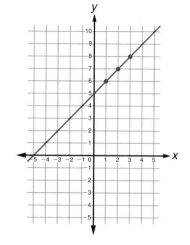

It is possible to have more than one independent variable in an equation. The formula for the volume V of a cylinder with base area B and height h is $V = B * h$. Both B and h are independent variables and V is dependent on both.

Check Your Understanding

Identify the independent and dependent variables in each situation. Then write an equation relating them.

1. The rule in a "What's My Rule?" problem is "multiply by $\frac{1}{2}$".

2. The volume V of a sphere with radius r

Check your answers on page 425.

Problem Solving

Mathematical Modeling

A **mathematical model** is something mathematical that describes something in the real world. A sphere, for example, is a model of a volleyball. The formula $d = (60 \text{ miles/hour}) * t$ is a model for the distance a car travels in t hours at 60 miles per hour. The graph at the right is a model for the cost of renting a bicycle at B & H Rentals, which charges $10 for the first hour and $2.50 for each additional half hour or less.

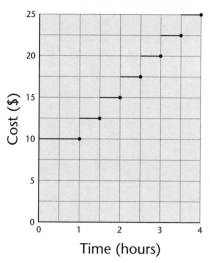

Cost of renting a bicycle at B & H Rentals

You have used mathematical models to solve problems for many years. Beginning in Kindergarten, you have solved problems using physical models such as counters, base-10 blocks, pattern blocks, and straws and connectors. In later grades, you learned to use other models such as situation diagrams, graphs, and number models to solve problems. As you continue studying mathematics, you will learn to make and use more powerful mathematical models.

Everyday Mathematics has many different kinds of problems. Some problems ask you to find something. Other problems ask you to make something. When you get older, you will often be asked to prove things, which means giving good reasons why they are true or correct.

Problems that ask you to find something	Problems that ask you to make something
• Five out of 8 students in Lane school play a musical instrument. A total of 140 students play an instrument. How many students attend Lane School? • What are the missing numbers? 1, 3, 6, 10, __, 21, 28, __, __	• Use pattern-block triangles and squares to make a semi-regular tessellation. • Use a compass and straightedge to construct a parallelogram that includes this angle:

Problems you already know how to solve offer good practice. But the problems that will help you learn the most are the ones you can't solve right away. Learning to be a good problem solver means learning what to do when you don't know what to do.

A Guide for Solving Number Stories

Learning to solve problems is the main reason for studying mathematics. One way you learn to solve problems is by solving number stories. A **number story** is a story with a problem that can be solved with arithmetic.

1. Understand the problem.
- ♦ Read the problem. Can you retell it in your own words?
- ♦ What do you want to find out?
- ♦ What do you know?
- ♦ Do you have all the information needed to solve the problem?

2. Plan what to do.
- ♦ Is the problem like one you solved before?
- ♦ Is there a pattern you can use?
- ♦ Can you draw a picture or a diagram?
- ♦ Can you write a number model or make a table?
- ♦ Can you use counters, base-10 blocks, or some other tool?
- ♦ Can you estimate the answer and check if you are right?

3. Carry out the plan.
- ♦ After you decide what to do, do it. Be careful.
- ♦ Make a written record of what you do.
- ♦ Answer the question.

4. Look back.
- ♦ Does your answer make sense?
- ♦ Does your answer agree with your estimate?
- ♦ Can you write a number model for the problem?
- ♦ Can you solve the problem in another way?

Note

Understanding the problem is an important step. Good problem solvers make sure they really understand the problem.

Note

Sometimes it's easy to know what to do. Other times you need to be creative.

Check Your Understanding

Use the Guide for Solving Number Stories to help you solve the following problems. Explain your thinking at each step, and explain your answers.

1. There are 20 students in Mr. Khalid's sixth grade class. Two out of 8 have no brothers or sisters. How many students have no brothers or sisters?

2. Your school ran a weekend car wash and raised $315. Your class received $\frac{3}{7}$ of this total. How much did your class receive?

Check your answers on page 423.

A Problem-Solving Diagram

Over the years, you have developed good skills for solving number stories. But problem solving is much more than just solving number stories. Problems from everyday life, science, and business are often more complicated than the number stories you solve in school. Sometimes the steps in the Guide for Solving Number Stories may not be helpful.

The diagram below shows another way to think about problem solving. This diagram is more complicated than a list, but it shows more accurately what people do when they solve problems in science and business. The arrows connecting the boxes are meant to show that you don't always do things in the same order.

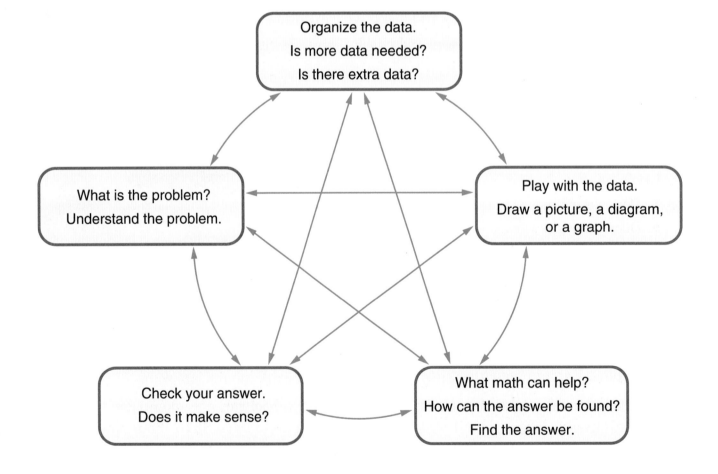

Using the diagram on the previous page as you solve problems may help you to be a better problem solver. Here are some things to try for each of the boxes in the diagram. These are suggestions to try, not rules that must be followed. They include most of the strategies that good problem solvers use.

♦ *What is the problem?* Try to understand the problem. Can you retell it in your own words? What do you want to find out? What do you know? Try to imagine what an answer might look like.

♦ *Organize the data.* Study the data you have and organize it in a list or in some other way. Look for more data if you need it. Get rid of any data you don't need.

♦ *Play with the data.* Try drawing a picture, a diagram, or a graph. Can you write a number model? Can you model the problem with counters or blocks?

♦ *What math can help?* Can you use arithmetic, geometry, algebra, or other mathematics to find the answer? Do the math. Use a calculator when it makes sense to do so. Label the answer with units.

Business people brainstorming

♦ *Check your answer.* Does it make sense? Compare your answer to a friend's answer. Review the problem to see whether the question you are trying to answer is the right question. Try the answer in the problem. If you need to, go back to the strategies listed above. Can you solve the problem another way?

Check Your Understanding

Use the problem-solving diagram to help you decide how you would answer the following question:

Walking at a normal pace, about how many steps would you take in 1 hour?

Check your answer on page 423.

Interpreting a Remainder in Division

Some number stories are solved by dividing whole numbers. You may need to decide what to do when there is a nonzero remainder.

Here are three possible choices:

♦ Ignore the remainder. Use the quotient as the answer.

♦ Round the quotient up to the next whole number.

♦ Rewrite the remainder as a fraction or decimal. Use this fraction or decimal as part of the answer.

To rewrite a remainder as a fraction:

1. Make the remainder the *numerator* of the fraction.

2. Make the divisor the *denominator* of the fraction.

Problem	Answer	Remainder Rewritten as a Fraction	Answer Written as a Mixed Number	Answer Written as a Decimal
375 / 4	93 R3	$\frac{3}{4}$	$93\frac{3}{4}$	93.75

Examples

• Suppose 3 people share 20 counters equally. How many counters will each person get?

$20 / 3 \rightarrow 6$ R2

Ignore the remainder. Use the quotient as the answer.

Each person will have 6 counters. Two counters are left over.

• Suppose 20 photos are placed in a photo album. How many pages are needed if 3 photos can fit on a page?

$20 / 3 \rightarrow 6$ R2

Round the quotient up to the next whole number. The album will have 6 pages filled and another page only partially filled.

So, 7 pages are needed.

• Suppose 3 friends share a 20-inch string of licorice. How long is each piece if the friends receive equal shares?

$20 / 3 \rightarrow 6$ R2

The answer, 6 R2, shows that if each friend receives 6 inches of licorice, 2 inches remain to be divided. Divide this 2-inch remainder into $\frac{1}{3}$-inch pieces. Each friend receives two $\frac{1}{3}$-inch pieces, or $\frac{2}{3}$ inch.

1 in.   1 in.

$\frac{1}{3}$ in. $\frac{1}{3}$ in. $\frac{1}{3}$ in. $\frac{1}{3}$ in. $\frac{1}{3}$ in. $\frac{1}{3}$ in.

The remainder (2) has been rewritten as a fraction ($\frac{2}{3}$). Use this fraction as part of the answer.

Each friend will get a $6\frac{2}{3}$-inch piece of licorice.

Making Estimates

An **estimate** is an answer that should be close to an exact answer.

♦ Sometimes you must estimate because it is impossible to know the exact answer. For example, weather forecasters must estimate when they predict the weather.

♦ Sometimes you estimate because finding an exact answer is not practical. For example, you could estimate the number of books in your school library instead of actually counting them.

♦ Sometimes you estimate because finding an exact answer is not worth the trouble. For example, you might estimate the cost of several items at the store, to be sure you have enough money.

If you choose to estimate an answer, you must decide how accurate your estimate should be. Keep in mind that an estimate is *not* a guess. A guess is just an opinion that may not be based on much real information.

One kind of rough estimate is called a **magnitude estimate.** When making a magnitude estimate, ask yourself "Is the answer in the tens? Hundreds? Thousands?" and so on. You can use magnitude estimates to check answers or to judge whether information you read or hear makes sense.

You can use **rounded numbers** to make both magnitude estimates and other more accurate estimates.

Did You Know?

The New York Public Library is the largest public library in the world. In 2005, it had a collection of 6.6 million items and was growing at the rate of 10,000 items per week.

By 2010, it is likely that the library will hold *at least* 8.5 million items, but *not more than* 10 million items.

reading room of the
New York Public Library

Example Mr. Huber has started a new job at a salary of $2,957 a month. He expects to get a raise after 6 months. About how much can he expect to earn in the first 6 months?

To estimate the answer, round $2,957 to $3,000 and multiply the rounded number by 6.

Since 6 * $3,000 = $18,000, he can expect to earn a little under $18,000 in the first 6 months on the job.

When a range of possible values is more useful than just one estimated value, make an **interval estimate.** An interval estimate consists of two numbers. One number is less than the exact value, and the other number is greater than the exact value. For example: "This compact car weighs *between* 2,400 and 3,000 pounds."

Rounding Numbers

When you **round** a number, you adjust the number to make it easier to work with. Often numbers are rounded to the nearest multiple of 10, 100, 1,000, and so on.

In many situations, exact numbers are not needed. For example, Jupiter revolves around the sun every 4,332.6 Earth days, and Saturn revolves around the sun every 10,759.2 Earth days. To compare the length of a year on these two planets, round the numbers to the nearest thousand. A year on Jupiter is about 4,000 Earth days, and a year on Saturn is about 11,000 Earth days. Using the rounded numbers, you can see that a year on Saturn is almost 3 times as long as a year on Jupiter.

If you use rounded numbers to estimate results of operations such as addition, subtraction, multiplication, and division— always round the numbers first, *before* estimating. The examples show the steps to follow in rounding a number.

Did You Know?

Neptune takes 60,189 days to orbit the sun. That is almost 165 Earth years.

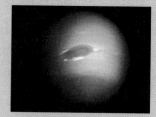

Neptune seen from Voyager 2

Examples

1. Round 3,548 to the nearest hundred.
2. Round 86,721 to the nearest thousand.
3. Round 2,595 to the nearest ten.
4. Round 4.563 to the nearest tenth.

	Step 1: Find the digit in the place you are rounding to.	**Step 2:** Rewrite the number, replacing all digits to the right of this digit with zeros. This is the **lower number**.	**Step 3:** Add 1 to the digit in the place you are rounding to. If the sum is 10, write 0 and add 1 to the digit to its left. This is the **higher number**.	**Step 4:** Is the number you are rounding closer to the lower number or to the higher number?	**Step 5:** Round to the closer of the two numbers. If it is halfway between the lower and the higher number, round to the higher number.
1.	3,548	3,500	3,600	lower number	3,500
2.	86,721	86,000	87,000	higher number	87,000
3.	2,595	2,590	2,600	halfway	2,600
4.	4.563	4.500	4.600	higher number	4.600 = 4.6

Check Your Understanding

Round 35,481.746 to the nearest:

1. hundred
2. ten thousand
3. thousand
4. hundredth

Check your answers on page 424.

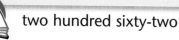

Venn Diagrams

A **Venn diagram** is a picture that uses circles or other closed curves to show relationships between sets.

Example
A regular deck of playing cards includes 12 face cards (Jacks, Queens, and Kings) and 40 non-face cards.

The Venn diagram shows this information in two non-overlapping circles. Each card is either a face card or it is not; it cannot be both. Adding the numbers in each section shows the deck has 52 cards in it. (12 + 40 = 52)

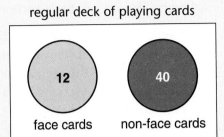
regular deck of playing cards

Example
Ms. Waller teaches one music class and one art class. She has 21 students in her music class and 14 students in her art class. Four of the students are in both her music and art classes.

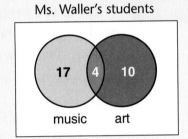
Ms. Waller's students

The Venn diagram shows this information in two overlapping circles. The overlapping part of the diagram represents the students who are in both her music and art classes. The numbers 17 and 10 in the diagram may be obtained by subtracting the overlap from each of the total class sizes:

21 music students − 4 music and art students = 17 students taking music only

14 art students − 4 music and art students = 10 students taking art only

The Venn diagram shows that there are 17 + 4 + 10 = 31 *different* students in these two classes.

Example
Mr. Dean's students recorded the types of TV shows they viewed last week.

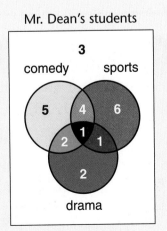
Mr. Dean's students

The Venn diagram for this situation uses three overlapping circles. Three of the students said they had not watched TV. To show this, the number "3" is outside all of the circles. The numbers in the different sections of the Venn diagram may be used to collect other interesting facts:

no TV: 3 students   sports, comedy, and drama: 1 student
comedy only: 5 students   sports and comedy: 4 + 1 = 5 students
drama only: 2 students   sports and drama: 1 + 1 = 2 students
sports only: 6 students   comedy and drama: 2 + 1 = 3 students

There is a total of 1 + 1 + 2 + 2 + 3 + 4 + 5 + 6 = 24 students in the class.

Venn diagrams can be useful in solving probability problems.

Example What is the probability that the spinner will land on red?

The spinner has 20 equal sections. Each section has the same chance of being landed on. Each section is likely to come up $\frac{1}{20}$ of the time.

The Venn diagram for this situation uses two circles that do not overlap. It shows that 7 of the sections are red. Each of these sections has a $\frac{1}{20}$ chance of coming up. So the total probability of landing on a red section is $7 * \frac{1}{20} = \frac{7}{20}$.

20-section spinner

13 blue   7 red

Example What is the probability that the spinner above will land on an odd number? A red section with an odd number? A blue section with an even number?

20-section spinner

6

7 3 4

odd   red

The Venn diagram for this situation uses two circles that overlap. It shows that 10 sections have an odd number. (Three of these sections are red, and 7 are not red.) Since each section has a $\frac{1}{20}$ chance of coming up, the total probability of landing on an odd-numbered section is $10 * \frac{1}{20} = \frac{10}{20}$, or $\frac{1}{2}$.

The diagram shows that there are 3 spinner sections in the overlap. These 3 sections are red with an odd number. Since each section has a $\frac{1}{20}$ chance of coming up, the total probability of landing on a red section with an odd number is $3 * \frac{1}{20} = \frac{3}{20}$.

The number "6" in the diagram is printed *outside* both circles. This shows that 6 sections do *not* have odd numbers and are *not* red. These sections are blue and have even numbers. Since each section has a $\frac{1}{20}$ chance of coming up, the total probability of landing on a blue section with an even number is $6 * \frac{1}{20} = \frac{6}{20}$.

Check Your Understanding

The spinner above shows the prime numbers 2, 3, 5, 7, 11, 13, 17, and 19.

1. Copy and complete the Venn diagram.

2. What is the chance of landing on a red section that does not have a prime number?

3. What is the chance of landing on a section that has a prime number or is red, or both?

20-section spinner

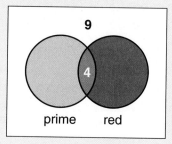

9

4

prime   red

Check your answers on page 424.

Calculators

About Calculators

You have used calculators to help you learn to count. Now you can use them for working with whole numbers, fractions, decimals, and percents.

As with any mathematical tool or strategy, you need to think about when and how to use a calculator. It can help you compute quickly and accurately when you have many problems to do in a short time. Calculators can help you compute with very large and very small numbers that may be hard to do in your head or with pencil and paper. Whenever you use a calculator, estimation should be part of your work. Always ask yourself if the number in the display makes sense.

There are many different kinds of calculators. **Four-function calculators** do little more than add, subtract, multiply, and divide whole numbers and decimals. More advanced **scientific calculators** let you find powers and reciprocals, and perform some operations with fractions. After elementary school, you may use **graphic calculators** that draw graphs, find data landmarks, and do even more complicated mathematics.

There are many calculators that work well with *Everyday Mathematics*. If the instructions in this book don't work for your calculator, or the keys on your calculator are not explained, you should refer to the directions that came with your calculator, or you can ask your teacher for help.

Calculator A

Calculator B

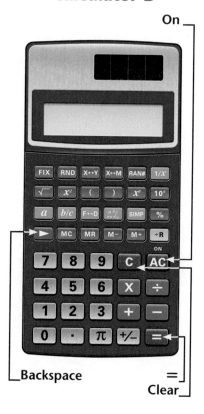

Basic Operations

You must take care of your calculator. Dropping it, leaving it in the sun, or other kinds of carelessness may break it or make it less reliable.

Many four-function and scientific calculators use light cells for power. If you press the ON key and see nothing on the display, hold the front of the calculator toward a light or a sunny window for a moment and then press ON again.

Entering and Clearing

Pressing a key on a calculator is called **keying in** or **entering.** In this book, calculator keys, except numbers and decimal points, are shown in rectangular boxes: $+$, $=$, $\times$, and so on. A set of instructions for performing a calculation is called a **key sequence.**

The simplest key sequences turn the calculator on and enter or clear numbers or other characters. These keys are labeled on the photos and summarized below.

Note

Calculators have two kinds of memory. **Short-term memory** is for the last number entered. The keys with an "M" are for **long-term memory** and are explained on pages 292–298.

Calculator A	Key Sequence	Purpose
	(On/Off)	Turn the display on.
	(Clear) and (On/Off) together	Clear the display and the short-term memory.
	(Clear)	Clear only the display.
	(←)	Clear the last digit.

Calculator B	Key Sequence	Purpose
	ON [AC]	Turn the display on.
	[AC]	Clear the display and the short-term memory.
	[C]	Clear a number if you make a mistake while entering a calculation.
	[▶]	Clear the last digit.

Always clear both the display and the memory each time you turn the calculator on.

Many calculators have a backspace key that will clear the last digit or digits you entered without re-entering the whole number.

Example Enter 123.444. Change it to 123.456.

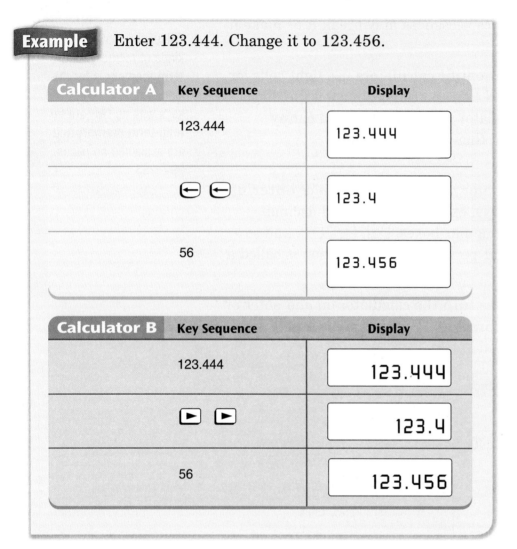

Try using the backspace key on your calculator.

Order of Operations and Parentheses

When you use a calculator for basic arithmetic operations, you enter the numbers and operations and press $\boxed{=}$ or $\boxed{\text{Enter}}$ to see the answer.

Note

Examples for arithmetic calculations are given in earlier sections of this book.

Try your calculator to see if it follows the rules for the order of operations. Key in 5 $\boxed{+}$ 6 $\boxed{\times}$ 2 $\boxed{=}$.

♦ If your calculator follows the order of operations, it will display 17.

♦ If it does not follow the order of operations, it will probably do the operations in the order they were entered: adding and then multiplying, displaying 22.

If you want the calculator to do operations in an order different from the order of operations, use the parentheses keys $\boxed{(}$ and $\boxed{)}$.

Examples Evaluate. $7 - (2 + 1)$

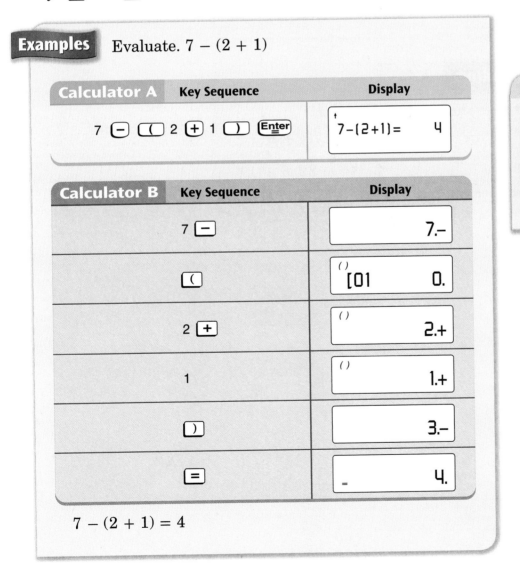

Note

If you see a tiny up or down arrow on the calculator display, you can use the up or down arrows to scroll the screen.

$7 - (2 + 1) = 4$

Calculators

Sometimes expressions are shown without all of the multiplication signs. *Remember to press the multiplication key even when it is not shown.*

Example   Evaluate. $9 - 2(1 + 2)$

Calculator A   **Key Sequence**   **Display**

9 ⊖ 2 ⊗ (1 ⊕ 2) Enter $9-2\times(1+2)=\ \ 3$

Calculator B   **Key Sequence**   **Display**

9 ⊖ 2 ⊗ (1 ⊕ 2) ⊜ $=$ 3.

$9 - 2(1 + 2) = 3$

Check Your Understanding

Use your calculator to evaluate each expression.

1. $79 - (4 + 8)$
2. $92 - 7(4 + 8)$
3. $3(6 + 7.6) - 29$
4. $(46 - 22)/6 + 19$

Check your answers on page 424.

Negative Numbers

How you enter a negative number depends on your calculator. You will use the change sign key, either ⌊+⁄−⌋ or ⌊(−)⌋ depending on your calculator. Both keys change the sign of the number.

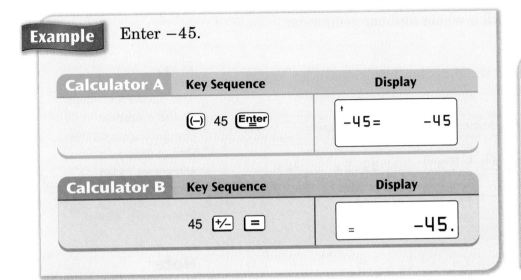

Example   Enter −45.

Calculator A	Key Sequence	Display
	⌊(−)⌋ 45 ⌊Enter⌋	⁺−45= −45

Calculator B	Key Sequence	Display
	45 ⌊+⁄−⌋ ⌊=⌋	= −45.

Note

If the number on the display is positive, it becomes negative after you press ⌊+⁄−⌋. If the number on the display is negative, it becomes positive after pressing ⌊+⁄−⌋. Keys like this are called **toggles**.

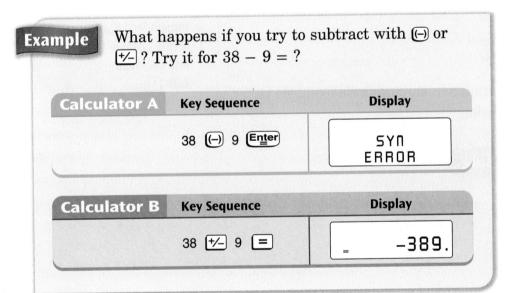

Example   What happens if you try to subtract with ⌊(−)⌋ or ⌊+⁄−⌋ ? Try it for 38 − 9 = ?

Calculator A	Key Sequence	Display
	38 ⌊(−)⌋ 9 ⌊Enter⌋	SYN ERROR

Calculator B	Key Sequence	Display
	38 ⌊+⁄−⌋ 9 ⌊=⌋	= −389.

Note

"SYN" is short for "syntax," which means the ordering and meaning of keys in a sequence.

Note

If you try to subtract using ⌊+⁄−⌋ on this calculator, it just changes the sign of the first number and adds the digits of the second number to it.

Division with Remainders

The answer to a division problem with whole numbers does not always result in whole number answers. When this happens, most calculators display the answer as a decimal. Some calculators also have a second division key that displays the whole number quotient with a whole number remainder.

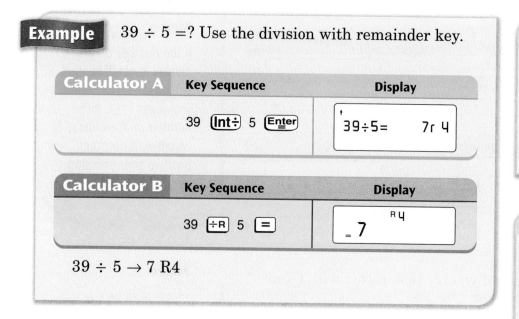

Example   39 ÷ 5 =? Use the division with remainder key.

Calculator A   Key Sequence   Display

39 (Int÷) 5 (Enter)   39÷5=   7r 4

Calculator B   Key Sequence   Display

39 (÷R) 5 (=)   = 7 R4

39 ÷ 5 → 7 R4

Note

"Int" stands for "integer" on this calculator. Use (Int÷) because this kind of division is sometimes called "integer division."

Note

(÷R) means "divide with remainder." You can also divide positive fractions and decimals with (÷R).

Try the division with remainder in the previous example to see how your calculator works.

Check Your Understanding

Divide with remainder.

1. 122 ÷ 7   **2.** 273 ÷ 13   **3.** 22,222 ÷ 54

Check your answers on page 424.

Fractions and Percent

Some calculators let you enter, rewrite, and do operations with fractions. Once you know how to enter a fraction, you can add, subtract, multiply, or divide them just like whole numbers and decimals.

Entering Fractions and Mixed Numbers

Most calculators that let you enter fractions use similar key sequences. For proper fractions, always start by entering the numerator. Then press a key to tell the calculator to begin writing a fraction.

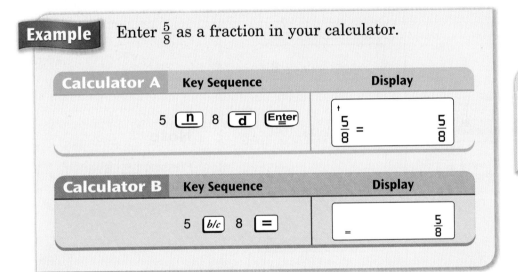

Example Enter $\frac{5}{8}$ as a fraction in your calculator.

Calculator A	Key Sequence	Display
	5 [n] 8 [d] [Enter]	$\frac{5}{8} = \frac{5}{8}$

Calculator B	Key Sequence	Display
	5 [b/c] 8 [=]	$\frac{5}{8}$

Note

Pressing [d] after you enter the denominator is optional.

To enter a mixed number, enter the whole number part and then press a key to tell the calculator what you did.

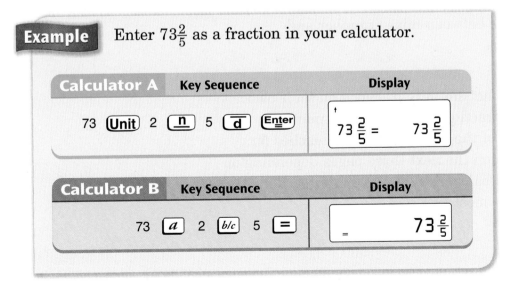

Example Enter $73\frac{2}{5}$ as a fraction in your calculator.

Calculator A	Key Sequence	Display
	73 [Unit] 2 [n] 5 [d] [Enter]	$73\frac{2}{5} = 73\frac{2}{5}$

Calculator B	Key Sequence	Display
	73 [a] 2 [b/c] 5 [=]	$73\frac{2}{5}$

Try entering a mixed number on your calculator.

The keys to convert between mixed numbers and improper fractions are similar on all fraction calculators.

Example   Convert $\frac{45}{7}$ to a mixed number with your calculator. Then change it back.

Calculator A	Key Sequence	Display
	45 [n] 7 [d] [Enter]	$\frac{45}{7} = \quad 6\frac{3}{7}$
	[U$\frac{n}{d}$↔$\frac{n}{d}$]	$\frac{45}{7}$
	[U$\frac{n}{d}$↔$\frac{n}{d}$]	$6\frac{3}{7}$

Note

Pressing [Enter] is *not* optional in this key sequence.

Calculator B	Key Sequence	Display
	45 [b/c] 7 [=]	$\frac{45}{7}$
	[a b/c ↔d/c]	$6\frac{3}{7}$
	[a b/c ↔d/c]	$\frac{45}{7}$

Note

Pressing [=] is optional in this key sequence.

Both [U$\frac{n}{d}$↔$\frac{n}{d}$] and [a b/c ↔d/c] toggle between mixed number and improper fraction notation.

Simplifying Fractions

Ordinarily, calculators do not simplify fractions on their own. The steps for simplifying fractions are similar for many calculators, but the order of the steps varies. Approaches for two calculators are shown on the next three pages depending on the keys you have on your calculator. Read the approaches for the calculator having keys most like yours.

Simplifying Fractions on Calculator A

This calculator lets you simplify a fraction in two ways. Each way divides the numerator and denominator by a common factor. The first approach uses (Simp) to automatically divide by the smallest common factor, and (Fac) to display the factor.

Calculator A

(Simp) simplifies a fraction by a common factor.

(Fac) displays the common factor used to simplify a fraction.

$\frac{N}{D} \to \frac{n}{d}$ means that the fraction shown is not in simplest form.

Example Convert $\frac{18}{24}$ to simplest form using smallest common factors.

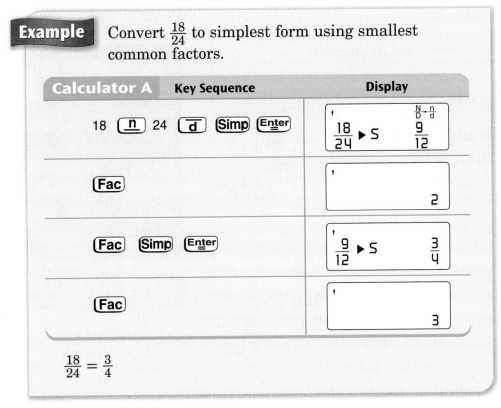

Calculator A	Key Sequence	Display
	18 (n) 24 (d) (Simp) (Enter)	$\frac{18}{24} \blacktriangleright S \quad \frac{9}{12}$
	(Fac)	2
	(Fac) (Simp) (Enter)	$\frac{9}{12} \blacktriangleright S \quad \frac{3}{4}$
	(Fac)	3

$\frac{18}{24} = \frac{3}{4}$

In the second approach, you can simplify the fraction in one step by telling the calculator to divide by the greatest common factor of the numerator and the denominator.

Example Convert $\frac{18}{24}$ to simplest form in one step by dividing the numerator and the denominator by their greatest common factor, 6.

Calculator A	Key Sequence	Display
	18 (n) 24 (d) (Simp) 6 (Enter)	$\frac{18}{24} \blacktriangleright S6 \quad \frac{3}{4}$
	(Fac)	6

$\frac{18}{24} = \frac{3}{4}$

Note

Pressing (Fac) toggles between the display of the factor and the display of the fraction.

Simplifying Fractions on Calculator B

Calculators like the one shown here let you simplify fractions in three different ways. Each way divides the numerator and denominator by a common factor. The first approach uses $=$ to give the simplest form in one step. The word Simp in the display means that the fraction shown is not in simplest form.

Calculator B

$\boxed{\text{SIMP}}$ simplifies a fraction by a common factor.

Press $\boxed{=}$ $\boxed{=}$ to display the fraction.

Example Convert $\frac{18}{24}$ to simplest form in one step.

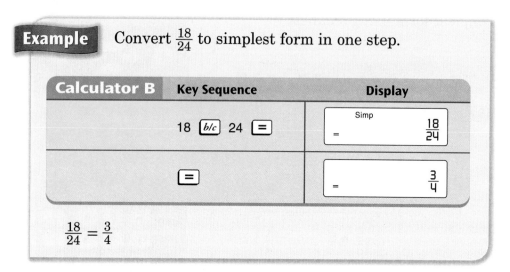

Calculator B	Key Sequence	Display
	18 $\boxed{b/c}$ 24 $\boxed{=}$	Simp = $\frac{18}{24}$
	$\boxed{=}$	= $\frac{3}{4}$

$$\frac{18}{24} = \frac{3}{4}$$

If you enter a fraction that is already in simplest form, you will not see Simp on the display. The one-step approach does not tell you the common factor as the next two approaches do using $\boxed{\text{SIMP}}$.

Example Convert $\frac{18}{24}$ to simplest form using smallest common factors.

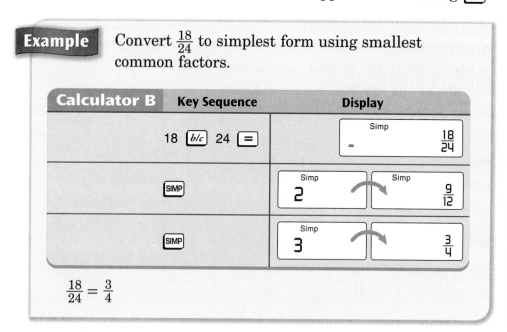

Calculator B	Key Sequence	Display
	18 $\boxed{b/c}$ 24 $\boxed{=}$	Simp = $\frac{18}{24}$
	$\boxed{\text{SIMP}}$	Simp 2 → Simp $\frac{9}{12}$
	$\boxed{\text{SIMP}}$	Simp 3 → $\frac{3}{4}$

$$\frac{18}{24} = \frac{3}{4}$$

Note

Each time you press $\boxed{\text{SIMP}}$ in the smallest-common-factor approach you briefly see the common factor, then the simplified fraction. This can be done without pressing $\boxed{=}$ first.

In the last approach to simplifying fractions with this type of calculator you tell it what common factor to divide by. If you use the greatest common factor on the numerator and the denominator, you can simplify the fraction in one step.

Calculator A

% divides a number by 100.

Example   Convert $\frac{18}{24}$ to simplest form by dividing the numerator and the denominator by their greatest common factor, 6.

Calculator B	Key Sequence	Display
	18 [b/c] 24 [=]	Simp = $\frac{18}{24}$
	6 [SIMP]	Simp 6 → $\frac{3}{4}$

$$\frac{18}{24} = \frac{3}{4}$$

Note

If you enter a number that is not a common factor of the numerator and the denominator, you will get an error symbol "E" in the display with the unchanged fraction.

Try simplifying the fractions in the previous examples to see how your calculator works.

Percent

The calculators shown here, and many others, have a **%** key, but it is likely that they work differently. The best way to learn what your calculator does with percents is to read its manual.

Calculator B

% finds *a* percent of *b*.

Most calculators include [%] to solve "percent of" problems.

Example   Calculate 25% of 180.

Using the first calculator, you can multiply 180 and 25% in either order, so both ways are shown.

Calculator A	Key Sequence	Display
	180 [×] 25 [%] (Enter)	180x25%= 45
	25 [%] [×] 180 (Enter)	25%x180= 45

Calculator B	Key Sequence	Display
	180 [×] 25 [%]	45.

25% of 180 is 45.

You can change percents to decimals with [%].

Examples   Display 85%, 250%, and 1% as decimals.

Calculator A	Key Sequence	Display
	85 [%] (Enter)	85%= 0.85
	250 [%] (Enter)	250%= 2.5
	1 [%] (Enter)	1%= 0.01

Calculator B	Key Sequence	Display
	1 [×] 85 [%]	0.85
	1 [×] 250 [%]	2.5
	1 [×] 1 [%]	0.01

Note

To convert percents to decimals on this calculator, you calculate a percent of 1, like in the previous example.

85% = 0.85; 250% = 2.5; 1% = 0.01

You can also use ⬚% to convert percents to fractions.

On many calculators, first change the percent to a decimal as in the previous examples, then use ⬚F↔D to change to a fraction.

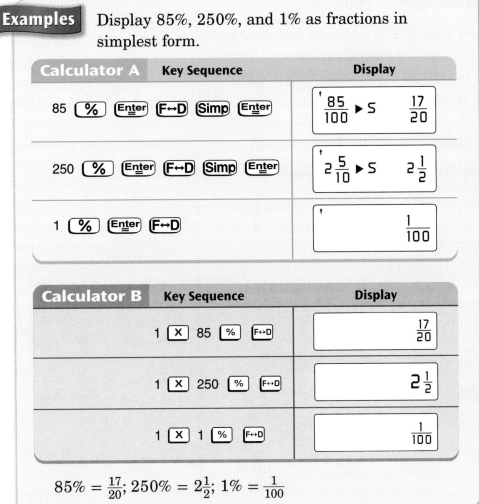

Examples Display 85%, 250%, and 1% as fractions in simplest form.

Calculator A	Key Sequence	Display
85 ⬚% ⬚Enter ⬚F↔D ⬚Simp ⬚Enter		$\frac{85}{100}$ ▸ S $\frac{17}{20}$
250 ⬚% ⬚Enter ⬚F↔D ⬚Simp ⬚Enter		$2\frac{5}{10}$ ▸ S $2\frac{1}{2}$
1 ⬚% ⬚Enter ⬚F↔D		$\frac{1}{100}$

Note
You may need to use ⬚SIMP to simplify.

Calculator B	Key Sequence	Display
1 ⬚X 85 ⬚% ⬚F↔D		$\frac{17}{20}$
1 ⬚X 250 ⬚% ⬚F↔D		$2\frac{1}{2}$
1 ⬚X 1 ⬚% ⬚F↔D		$\frac{1}{100}$

Note
This calculator simplifies automatically.

$85\% = \frac{17}{20}; 250\% = 2\frac{1}{2}; 1\% = \frac{1}{100}$

Try displaying some percents as fractions and decimals on your calculator.

Fraction/Decimal/Percent Conversions

Conversions of fractions to decimals and percents can be done on any calculator. For example, to rename $\frac{3}{5}$ as a decimal, simply enter 3 $\boxed{\div}$ 5 $\boxed{=}$. The display will show 0.6. To rename a decimal as a percent, just multiply by 100.

Conversions of decimals and percents to fractions can only be done on calculators that have special keys for fractions. Such calculators also have keys to change a fraction to its decimal equivalent or a decimal to an equivalent fraction.

Example Convert $\frac{3}{8}$ to a decimal and back to a fraction.

Calculator A	Key Sequence	Display
	3 $\boxed{n}$ 8 $\boxed{d}$ $\boxed{\text{Enter}}$	$\frac{3}{8} = \quad \frac{3}{8}$
	$\boxed{F \leftrightarrow D}$	0.375
	$\boxed{F \leftrightarrow D}$	$\frac{N \rightarrow n}{D \rightarrow d}$ $\frac{375}{1000}$
	$\boxed{\text{Simp}}$ $\boxed{\text{Enter}}$	$\frac{375}{1000} \blacktriangleright S \quad \frac{75}{200}$
	$\boxed{\text{Simp}}$ $\boxed{\text{Enter}}$	$\frac{75}{200} \blacktriangleright S \quad \frac{15}{40}$
	$\boxed{\text{Simp}}$ $\boxed{\text{Enter}}$	$\frac{15}{40} \blacktriangleright S \quad \frac{3}{8}$

Note

$\boxed{F \leftrightarrow D}$ toggles between fraction and decimal notation.

Calculator B	Key Sequence	Display
	3 $\boxed{b/c}$ 8	$\frac{3}{8}$
	$\boxed{F \leftrightarrow D}$	0.375
	$\boxed{F \leftrightarrow D}$	$\frac{3}{8}$

$\frac{3}{8} = 0.375$

See how your calculator changes fractions to decimals.

The tables below show examples of various conversions. Although only one key sequence is shown for each conversion, there are often other key sequences that work as well.

Conversion	Starting Number	Calculator A Key Sequence	Display
Fraction to decimal	$\frac{3}{5}$	3 [n] 5 [d] (Enter) (F↔D)	↑ 0.6
Decimal to fraction	0.125	.125 (Enter) (F↔D)	↑ $\frac{125}{1000}$
Decimal to percent	0.75	.75 (▶%) (Enter)	↑ 0.75▶% 75%
Percent to decimal	125%	125 (%) (Enter)	↑ 125%= 1.25
Fraction to percent	$\frac{5}{8}$	5 [n] 8 [d] (▶%) (Enter)	↑ $\frac{5}{8}$▶% 62.5%
Percent to fraction	35%	35 (%) (Enter) (F↔D)	↑ $\frac{35}{100}$

Conversion	Starting Number	Calculator B Key Sequence	Display
Fraction to decimal	$\frac{3}{5}$	3 [b/c] 5 [F↔D]	0.6
Decimal to fraction	0.125	.125 [F↔D]	$\frac{1}{8}$
Decimal to percent	0.75	.75 [×] 100 [=]	= 75.
Percent to decimal	125%	1 [×] 125 [%]	1.25
Fraction to percent	$\frac{5}{8}$	5 [b/c] 8 [F↔D] [×] 100 [=]	= 62.5
Percent to fraction	35%	1 [×] 35 [%] [F↔D]	$\frac{7}{20}$

Check Your Understanding

Use your calculator to convert between fractions, decimals, and percents.

1. $\frac{3}{16}$ to a decimal
2. 0.185 to a fraction
3. 0.003 to a percent
4. 723% to a decimal
5. $\frac{9}{32}$ to a percent
6. 68% to a fraction

Check your answers on page 424.

Other Operations

Your calculator can do more than simple arithmetic with whole numbers, fractions, and decimals. Every calculator does some things that other calculators cannot, or does them in different ways. See your owner's manual or ask your teacher to help you explore these things. The following pages explain some other things that many calculators can do.

Rounding

All calculators can round decimals. Decimals must be rounded to fit on the display. For example, if you key in 2 ÷ 3 =

♦ Calculator A shows 11 digits and rounds to the nearest value: 0.6666666667.

♦ Calculator B shows 8 digits and rounds down to 0.6666666.

Try 2 ÷ 3 = on your calculator to see how big the display is and how it rounds.

All scientific calculators have a FIX key to set, or **fix,** the place value of decimals on the display. Fixing always rounds to the nearest value.

> **Note**
>
> To turn off fixed rounding on a calculator, press FIX ⋅ .

> **Note**
>
> On this calculator you can fix the decimal places to the left of the decimal point, but are limited to the right of the decimal point to thousandths (0.001).

Examples Clear your calculator and fix it to round to tenths. Round each number 1.34; 812.79; and 0.06 to the nearest tenth.

Calculator A	Key Sequence	Display
	Clear Fix 0.1	↑ Fix ◄
	1.34 Enter	↑ Fix 1.34 = 1.3
	812.79 Enter	↑ Fix 812.79 = 812.8
	.06 Enter	↑ Fix .06 = 0.1

1.34 rounds to 1.3; 812.79 rounds to 812.8; 0.06 rounds to 0.1.

Examples Clear your calculator and fix it to round to tenths. Round each number 1.34; 812.79; and 0.06 to the nearest tenth.

Note

This calculator only lets you fix places to the right of the decimal point.

Calculator B	Key Sequence	Display
	AC FIX	0−7
	1	FIX 0.0
	1.34 =	FIX = 1.3
	812.79 =	FIX = 812.8
	.06 =	FIX = 0.1

1.34 rounds to 1.3; 812.79 rounds to 812.8; 0.06 rounds to 0.1.

Note

You can fix either calculator to round without clearing the display first. It will round the number on the display.

Check Your Understanding

Use your calculator to round to the indicated place.

1. 0.67 to tenths

2. 427.88 to ones

3. 4384.4879 to thousandths

4. 0.7979 to hundredths

Check your answers on page 424.

two hundred eighty-three **283**

Fixing the display to round to hundredths is helpful for solving problems about dollars and cents.

Example One CD costs $11.23 and another costs $14.67. Set your calculator to round to the nearest cent and calculate the total cost of the CDs.

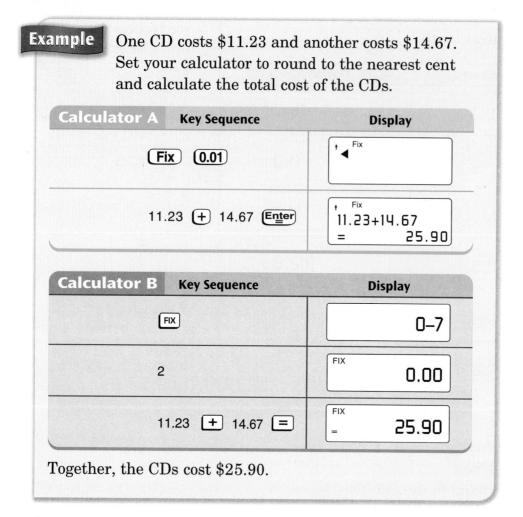

Together, the CDs cost $25.90.

On most calculators, if you find the total in the Example above with the "fix" turned off, the display reads 25.9. To show the answer in dollars and cents, fix the display to round to hundredths and you will see 25.90.

Powers, Reciprocals, and Square Roots

Powers of numbers can be calculated on all scientific calculators. Look at your calculator to see which key it has for finding powers of numbers.

♦ The key may look like x^y and is read "x to the y."

♦ The key may look like $\boxed{\wedge}$, and is called a **caret.**

To compute a number to a negative power, be sure to use the change-sign key $\boxed{(-)}$ or $\boxed{+/-}$, not the subtraction key $\boxed{-}$.

Calculator A

$\boxed{\wedge}$ finds powers and reciprocals.

$\boxed{\sqrt{}}$ finds square roots.

Examples Find the values of 3^4 and 5^{-2}.

Calculator A	Key Sequence	Display
	3 $\boxed{\wedge}$ 4 $\boxed{\text{Enter}}$	3 ^ 4 = 81
	5 $\boxed{\wedge}$ $\boxed{(-)}$ 2 $\boxed{\text{Enter}}$	5 ^ -2 = 0.04

Note

If you press $\boxed{(-)}$ after the 2, you will get an error message.

Calculator B

$\boxed{x^y}$ finds powers.

$\boxed{1/x}$ finds reciprocals.

Calculator B	Key Sequence	Display
	3 $\boxed{x^y}$ 4 $\boxed{=}$	= 81.
	5 $\boxed{x^y}$ 2 $\boxed{+/-}$ $\boxed{=}$	= 0.04

Note

If you press $\boxed{+/-}$ before the 2, it will change the sign of the 5 and display the result of $(-5)^2 = 25$.

$3^4 = 81;\ 5^{-2} = 0.04$

$\boxed{\sqrt{}}$ finds square roots.

$\boxed{x^2}$ is a shortcut to square numbers.

Calculators

Most scientific calculators have a reciprocal key [1/x]. On all scientific calculators you can find a reciprocal of a number by raising it to the −1 power.

Examples Find the reciprocals of 25 and $\frac{2}{3}$.

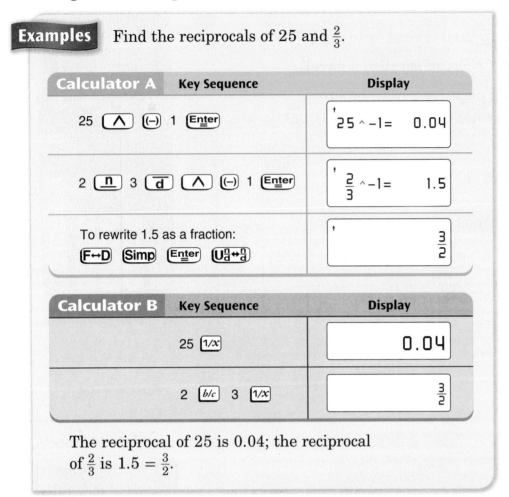

Calculator A	Key Sequence	Display
	25 [∧] [(−)] 1 [Enter]	25 ^ −1 = 0.04
	2 [n] 3 [d] [∧] [(−)] 1 [Enter]	$\frac{2}{3}$ ^ −1 = 1.5
	To rewrite 1.5 as a fraction: [F↔D] [Simp] [Enter] [U$\frac{n}{d}$↔$\frac{n}{d}$]	$\frac{3}{2}$

Calculator B	Key Sequence	Display
	25 [1/x]	0.04
	2 [b/c] 3 [1/x]	$\frac{3}{2}$

The reciprocal of 25 is 0.04; the reciprocal of $\frac{2}{3}$ is $1.5 = \frac{3}{2}$.

Note

You don't need to press the last key in the final step if your calculator is set to keep fractions in improper form. See the owner's manual for details.

Almost all calculators have a [√] key for finding square roots. It depends on the calculator whether you press [√] before or after entering a number.

Examples Find the square roots of 25 and 10,000.

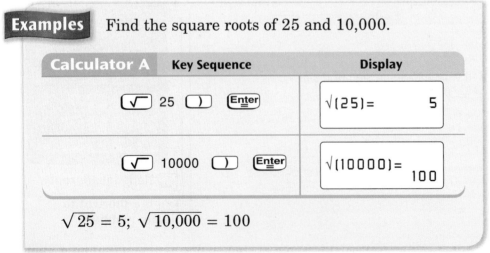

Calculator A	Key Sequence	Display
	[√] 25 [)] [Enter]	√(25) = 5
	[√] 10000 [)] [Enter]	√(10000) = 100

Note

On this calculator you have to "close" the square root by pressing [)] after the number.

$\sqrt{25} = 5$; $\sqrt{10{,}000} = 100$

Examples Find the square roots of 25 and 10,000.

Calculator B	Key Sequence	Display
	25 $\sqrt{}$	5.
	10000 $\sqrt{}$	100.

$$\sqrt{25} = 5; \sqrt{10,000} = 100$$

This calculator uses the caret ^ to display scientific notation.

Try finding the square roots of a few numbers.

Scientific Notation

Scientific notation is a way of writing very large or very small numbers. A number in scientific notation is shown as a product of a number between 1 and 10 and a power of 10. In scientific notation, the 9,000,000,000 bytes of memory on a 9-gigabyte hard drive is written $9 * 10^9$. On scientific calculators, numbers with too many digits to fit on the display are automatically shown in scientific notation like the bottom calculator in the margin.

Different calculators use different symbols for scientific notation. Your calculator may display raised exponents of 10, although most do not. Since the base of the power is always 10, most calculators leave out the 10 and simply put a space between the number and the exponent.

This calculator shows $9 * 10^9$.

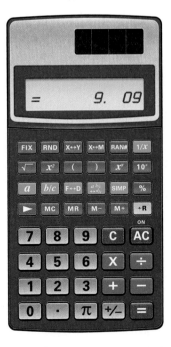

two hundred eighty-seven

Examples Convert $7 * 10^4$, $4.35 * 10^5$ and $8 * 10^{-3}$ to decimal notation.

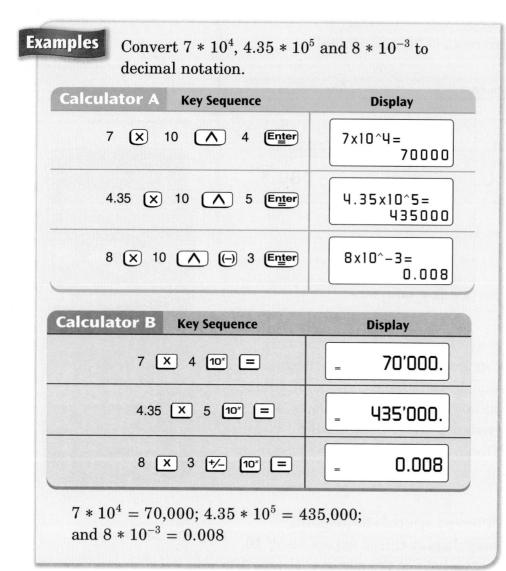

Calculator A	Key Sequence	Display
	7 [×] 10 [∧] 4 (Enter)	7x10^4 = 70000
	4.35 [×] 10 [∧] 5 (Enter)	4.35x10^5 = 435000
	8 [×] 10 [∧] [(−)] 3 (Enter)	8x10^−3 = 0.008

Calculator B	Key Sequence	Display
	7 [×] 4 [10ˣ] [=]	= 70'000.
	4.35 [×] 5 [10ˣ] [=]	= 435'000.
	8 [×] 3 [+/−] [10ˣ] [=]	= 0.008

Note

Neither calculator displays large numbers in decimal notation with a comma like you do with pencil and paper. One uses an apostrophe; the other uses no symbol at all.

$7 * 10^4 = 70{,}000$; $4.35 * 10^5 = 435{,}000$;
and $8 * 10^{-3} = 0.008$

How does your calculator display large numbers?

Check Your Understanding

Use your calculator to convert the following to standard notation:

1. $7.2 * 10^{-4}$ **2.** $8.3 * 10^7$ **3.** $3.726 * 10^{-6}$ **4.** $-3.4 * 10^{-5}$

Check your answers on page 424.

Calculators have different limits to the numbers they can display without scientific notation.

Note

A calculation resulting in a number larger than the limit is automatically displayed in scientific notation.

Example Write 123,456 * 654,321 in scientific notation. Then write the product in decimal notation.

Calculator A	Key Sequence	Display
	123456 $\times$ 654321 (Enter)	8.078x10^10

The product is $8.078 * 10^{10} = 80,780,000,000$.

Calculator B	Key Sequence	Display
	123456 $\times$ 654321 (=)	8.0779 10

The product is $8.0779 * 10^{10} = 80,779,000,000$.

Note

Using (FIX) to round answers does not affect scientific notation on either calculator.

Example Write 1 * 2 * 3 * 4 * 5 * 6 * 7 * 8 * 9 * 10 * 11 * 12 * 13 * 14 * 15 in scientific notation.

Calculator A	Key Sequence	Display
	1 $\times$ 2 $\times$ 3 $\times$ 4 $\times$ 5 $\times$ 6 $\times$ 7 $\times$ 8 $\times$ 9 $\times$ 10 $\times$ 11 $\times$ 12 $\times$ 13 $\times$ 14 $\times$ 15 (Enter)	1.308x10^12

The product is $1.308 * 10^{12}$, or 1,308,000,000,000.

Calculator B	Key Sequence	Display
	1 $\times$ 2 $\times$ 3 $\times$ 4 $\times$ 5 $\times$ 6 $\times$ 7 $\times$ 8 $\times$ 9 $\times$ 10 $\times$ 11 $\times$ 12 $\times$ 13 $\times$ 14 $\times$ 15 (=)	1.3076 12

The product is $1.3076 * 10^{12}$, or 1,307,600,000,000.

Check Your Understanding

Write in scientific notation.

1. 995 * 7 * 54 * 65 * 659 * 807 * 468

2. 956 * 859 * 760 * 862

3. 527 * 32 * 987 * 424 * 77 * 145 * 195

4. 15^9 * 13 * 996 * 558

5. The number of different 5-card hands that can be drawn from a standard deck of 52 cards is: 52 * 51 * 50 * 49 * 48. How many hands is this in scientific notation? In decimal notation?

Check your answers on page 424.

Pi (π)

The formulas for the circumference and area of circles involve **pi (π)**. Pi is a number that is a little more than 3. The first nine digits of pi are 3.14159265. All scientific calculators have a pi key $\boxed{\pi}$ that gives an approximate value in decimal form. A few calculators display an exact value using the π symbol.

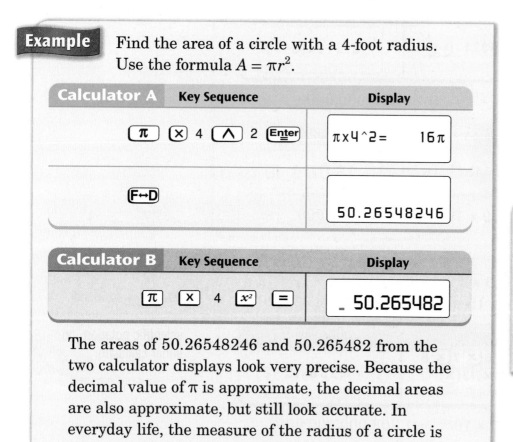

Example Find the area of a circle with a 4-foot radius. Use the formula $A = \pi r^2$.

Calculator A	Key Sequence	Display
	$\boxed{\pi}$ $\boxed{\times}$ 4 $\boxed{\wedge}$ 2 $\boxed{\text{Enter}}$	$\pi \times 4^2 = \quad 16\pi$
	$\boxed{\text{F}\leftrightarrow\text{D}}$	50.26548246

Calculator B	Key Sequence	Display
	$\boxed{\pi}$ $\boxed{\times}$ 4 $\boxed{x^2}$ $\boxed{=}$	₌ 50.265482

Note

You can set the number of decimal places on your calculator's display to show 50 by pressing either $\boxed{\text{Fix}}$ $\boxed{1.}$ or $\boxed{\text{Fix}}$ 0 depending on the calculator.

The areas of 50.26548246 and 50.265482 from the two calculator displays look very precise. Because the decimal value of π is approximate, the decimal areas are also approximate, but still look accurate. In everyday life, the measure of the radius of a circle is probably approximate, and giving an area to 6 or 8 decimal places does not make sense.

So a good approximation of the area of the 4-foot radius circle is about 50 square feet.

Example Find the circumference of a circle with a 15-centimeter diameter to the nearest tenth of a centimeter. Use the formula $C = \pi d$.

Calculator A	Key Sequence	Display
	(Fix) (0.1)	↑ Fix
	(π) (×) 15 (Enter)	↑ Fix π×15= 15π
	(F↔D)	↑ Fix 47.1

Calculator B	Key Sequence	Display
	(FIX) 1	FIX 0.0
	(π) (×) 15 (=)	FIX 47.1 =

The circumference is about 47.1 centimeters.

Note

When you are finished, remember to turn off the fixed rounding by pressing (FIX) (·).

Check Your Understanding

1. Find the area of a circle with a 70-foot radius. Display your answer to the nearest square foot.
2. To the nearest tenth of a centimeter, find the circumference of a circle with a 36.7-centimeter diameter.

Check your answers on page 424.

Using Calculator Memory

Many calculators let you save a number in **long-term memory** using keys with "M" on them. Later on, when you need the number, you can recall it from memory. Most calculators display an "M" or similar symbol when there is a number other than 0 in the memory.

Memory Basics

There are two main ways to enter numbers into long-term memory. Some calculators, including most 4-function calculators, have the keys in the table on page 294. If your calculator does not have at least the [M+] and [M-] keys, see the examples on this page.

Memory on Calculator A

One way that calculators can put numbers in memory is using a key to **store** a value. On the first calculator the store key is [▶M] and only works on numbers that have been entered into the display with [Enter].

Calculator A	Key Sequence	Purpose
	(MR/MC) (MR/MC)	Clear the long-term memory. This should always be the first step to any key sequence using the memory. Afterward, there will be no "M" in the display. This tells you there is no number in memory.
	(▶M) (Enter)	Store the number entered in the display in memory.
	(MR/MC)	Recall the number stored in memory and show it in the display.

Calculator A

(▶M) stores the displayed number in memory.

(MR/MC) recalls and displays the number in memory. Press it twice to clear memory.

Note

If you press (MR/MC) more than twice, you will recall and display the 0 that is now in memory. Press (Clear) to clear the display.

The following example first shows what happens if you don't enter a number before trying to store it.

Example Store 25 in memory and recall it to show that it was saved.

Calculator A	Key Sequence	Display
	(MR/MC) (MR/MC) (Clear)	
	25 (▶M) (Enter)	MEM ERROR

Oops. Start again. First, press (Clear) twice.

Calculator A	Key Sequence	Display
	(MR/MC) (MR/MC) (Clear)	
	25 (Enter) (▶M) (Enter)	M 25= 25
	(Clear)	M
	(MR/MC)	M 25

If your calculator is like this one, try the Check Your Understanding problem on page 294.

Memory on Calculator B

Calculators put a 0 into memory when [MC] is pressed. To store a single number in a cleared memory, simply enter the number and press [M+].

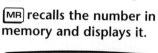

[MC] clears the memory.

[MR] recalls the number in memory and displays it.

Calculator B	Key Sequence	Purpose
	[MC]	Clear the long-term memory. This should always be the first step to any key sequence using the memory. Afterward, there will be no "M" in the display. This tells you there is no number in memory.
	[MR]	Recall the number stored in memory and show it in the display.
	[M+]	Add the number on the display to the number in memory.
	[M−]	Subtract the number on the display from the number in memory.

[M−] subtracts the number on the display from the number in memory.

[M+] adds the number on the display to the number in memory.

Example Store 25 in memory and recall it to show that it was saved.

Calculator B	Key Sequence	Display
	[AC] [MC]	0.
	25 [M+]	M 25.
	[AC]	M 0.
	[MR]	M 25.

Note

When this calculator turns off, the display clears, but a value in memory is *not* erased.

Check Your Understanding

Store π in the long-term memory. Clear the display. Then compute the area A of a circle whose radius r is 16 feet, without pressing the π key. ($A = \pi r^2$)

Check your answer on page 424.

Using Memory in Problem Solving

A common use of memory in calculators is to solve problems
that have two or more steps in the solution.

Example   Compute a 15% tip on a $25 bill. Store the tip in
the memory, then find the total bill.

Calculator A	Key Sequence	Display
	(MR/MC) (MR/MC) (Clear)	
	15 (%) (×) 25 (Enter)	15%×25= 3.75
	(▶M) (Enter)	M 15%×25= 3.75
	25 (+) (MR/MC) (Enter)	M 25+3.75= 28.75

Note

Always be sure to clear
the memory after solving
one problem and
before beginning
another.

Calculator B	Key Sequence	Display
	(AC) (MC)	0.
	25 (×) 15 (%) (=)	3.75
	(M+)	ᴹ 3.75
	25 (+) (MR) (=)	ᴹ = 28.75.

Calculator B	Key Sequence	Display
	(AC) (MC)	0.
	25 (×) 15 (%) (+)	28.75

Note

The second solution
shows how this calculator
solves the problem by
using memory
automatically.

The total bill is $28.75.

Check Your Understanding

Compute a 15% tip on an $85 bill. Then find the total bill.

Check your answer on page 424.

Example Marguerite ordered the following food at the food court: 2 hamburgers at $1.49 each and 3 hot dogs at $0.89 each. How much change will she receive from a $10 bill?

Calculator A	Key Sequence	Display
	(MR/MC) (MR/MC) (Clear)	
	2 (×) 1.49 (Enter) (▶M) (Enter)	ᴹ 2×1.49= 2.98
	3 (×) .89 (Enter) (▶M) (+)	ᴹ 3×.89= 2.67
	10 (−) (MR/MC) (Enter)	ᴹ 10−5.65= 4.35

Note

The key sequence (▶M) (+) is a shortcut to add the displayed number to memory. Similarly, (▶M) (−) subtracts a number from memory.

Calculator B	Key Sequence	Display
	(AC) (MC)	0.
	2 (×) 1.49 (=) (M+)	ᴹ 2.98
	3 (×) .89 (=) (M+)	ᴹ 2.67
	10 (−) (MR) (=)	ᴹ = 4.35

Marguerite will receive $4.35 in change.

Example Mr. Beckman bought 2 adult tickets at $8.25 each and 3 child tickets at $4.75 each. He redeemed a $5 gift certificate. How much did he pay for the tickets?

Note

If you fix the rounding to hundredths, all the values will be displayed as dollars and cents.

Calculator A	Key Sequence	Display
(MR/MC) (MR/MC) (Clear)		
2 (×) 8.25 (Enter) (►M) (Enter)		ᴹ 2x8.25= 16.5
3 (×) 4.75 (Enter) (►M) (+)		ᴹ 3x4.75= 14.25
(MR/MC) (−) 5 (Enter)		ᴹ 30.75-5= 25.75

Calculator B	Key Sequence	Display
(AC) (MC)		0.
2 (×) 8.25 (=) (M+)		ᴹ 16.5
3 (×) 4.75 (=) (M+)		ᴹ 14.25
(MR) (−) 5 (=)		ᴹ = 25.75

Mr. Beckman paid $25.75 for the tickets.

Example Juan bought the following tickets to a baseball game for himself and 6 friends: 2 bleacher seats at $15.25 each and 5 mezzanine seats at $27.50 each. If everyone intends to split the costs evenly seven ways how much does each person owe Juan?

Note

If you fix the rounding to hundredths, all the values will be displayed as dollars and cents.

Calculator A	Key Sequence	Display
(MR/MC) (MR/MC) (Clear)		
2 (×) 15.25 (Enter) (▶M) (Enter)		M 2×15.25= 30.5
5 (×) 27.50 (Enter) (▶M) (+)		M 5×27.50= 137.5
(MR/MC) (÷) 7 (Enter)		M 168÷7= 24

Calculator B	Key Sequence	Display
(AC) (MC)		0.
2 (×) 15.25 (=) (M+)		M 30.5
5 (×) 27.50 (=) (M+)		M 137.5
(MR) (÷) 7 (=)		M = 24.

Each friend owes Juan $24.00 for the tickets.

Check Your Understanding

1. How much would 2 flags and 4 banners cost if flags cost $24.25 each and banners cost $10.50 each?

2. How much would it cost to take a family of 2 adults and 4 children to a matinee if tickets cost $6.25 for adults and $4.25 for children?

Check your answers on page 424.

Skip Counting on a Calculator

In earlier grades, you may have been using a 4-function calculator to skip-count.

Recall that the program needs to tell the calculator:

1. What number to count by;
2. Whether to count up or down;
3. What number to start at;
4. When to count.

Here's how to program each calculator.

(Op1) and (Op2) allow you to program and repeat operations.

> **Example** Starting at 3, count by 7s on this calculator.

Calculator A		
Purpose	**Key Sequence**	**Display**
Tell the calculator to count up by 7. (Op1) is programmed to do any operation with any number that you enter between presses of (Op1).	(Op1) (+) 7 (Op1)	Op1 +7
Tell the calculator to start at 3 and do the first count.	3 (Op1)	Op1 3+7 1 10
Tell the calculator to count again.	(Op1)	Op1 10+7 2 17
Keep counting by pressing (Op1).	(Op1)	Op1 17+7 3 24

To count back by 7, begin with (Op1) (−) 7 (Op1).

Note

You can use (Op2) to define a second constant operation. (Op2) works in exactly the same way as (Op1).

Note

The number in the lower left corner of the display shows how many counts you have made.

Example Starting at 3, count by 7s on this calculator.

Calculator B		
Purpose	**Key Sequence**	**Display**
Tell the calculator to count up by 7. The "K" on the display means you have successfully programmed the "constant," as the count-by number is sometimes called.	7 $\boxed{+}$ $\boxed{+}$	K 7.+
Tell the calculator to start at 3 and do the first count.	3 $\boxed{=}$	K 10.+
Tell the calculator to count again.	$\boxed{=}$	K 17.+
Keep counting by pressing $\boxed{=}$.	$\boxed{=}$	K 24.+

To count back by 7, begin with 7 $\boxed{-}$ $\boxed{-}$.

Check Your Understanding

Use your calculator to do the following counts. Write five counts each.

1. Starting at 11, count on by 7s.
2. Starting at 120, count back by 13s.

Check your answers on page 424.

Games

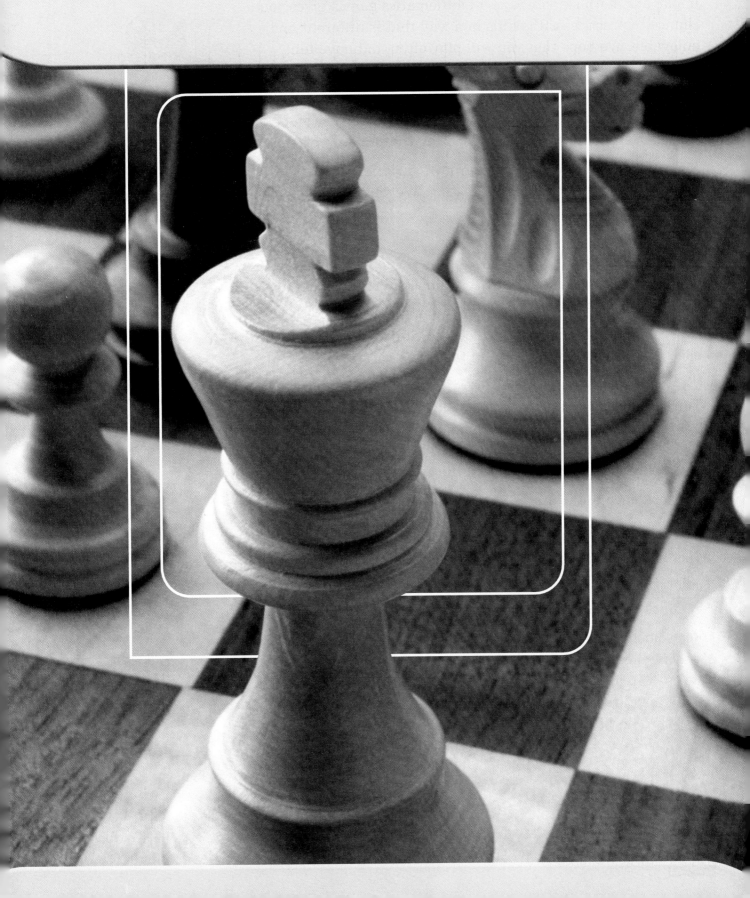

Games

Throughout the year, you will play games that help you practice important math skills. Playing mathematics games gives you a chance to practice math skills in a way that is different and enjoyable. We hope that you will play often and have fun!

In this section of your *Student Reference Book*, you will find the directions for many games. The numbers in most games are generated randomly. This means that the games can be played over and over without repeating the same problems.

Many students have created their own variations to these games to make them more interesting. We encourage you to do this too.

Materials

You need a deck of number cards for many of the games. You can use an Everything Math Deck, a deck of regular playing cards, or make your own deck out of index cards.

An Everything Math Deck includes 54 cards. There are 4 cards each for the numbers 0–10. And there is 1 card for each of the numbers 11–20.

You can also use a deck of regular playing cards after making a few changes. A deck of playing cards includes 54 cards (52 regular cards, plus 2 jokers). To create a deck of number cards, use a permanent marker to mark the cards in the following way:

♦ Mark each of the 4 aces with the number 1.

♦ Mark each of the 4 queens with the number 0.

♦ Mark the 4 jacks and 4 kings with the numbers 11 through 18.

♦ Mark the 2 jokers with the numbers 19 and 20.

For some games you will have to make a gameboard, a score sheet, or a set of cards that are not number cards. The instructions for doing this are included with the game directions. More complicated gameboards and card decks are available from your teacher.

Algebra Election

Materials ☐ 1 set of *Algebra Election* Cards
(*Math Journal 1*, Activity Sheets 3 and 4)

☐ 1 Electoral Vote Map
(*Math Masters*, pp. 434 and 435)

☐ 1 six-sided die

☐ 4 counters

☐ 1 calculator

Players 2 teams, each with 2 players

Skill Variable substitution, solving equations

Object of the game Players move their counters on a map of the United States. For each state or the District of Columbia (D.C.) that a player lands on, the player tries to win that state's electoral votes by solving a problem. The first team to collect 270 or more votes wins the election. Winning-team members become President and Vice President.

Algebra Election Cards
(sample cards)

Find: x squared x to the fourth power $\frac{1}{x}$	Find *n*. (*Hint: n* could be a negative number.) $1,000 + n = x$ $1,000 + n = -x$
Insert parentheses in $\frac{1}{10} * x - 2$ so that its value is greater than 0 and less than 4.	Find *n*. (*Hint: n* could be a negative number.) $n + 10 = x$ $n - 10 = x$
$T = B - (2 * \frac{H}{1,000})$ If $B = 80$ and $H = 100x$, what does *T* equal?	Find *n*. $n = (2 * x) / 10$ $n + 1 = (2 * x)$
Tell whether each is true or false. $10 * x > 100$ $\frac{1}{2} * x * 100 < 10^3$ $x^3 * 1,000 > 4 * 10^4$	Which number is this? $x * 10^2$ $x * 10^5$

Directions

1. Each player puts a counter on Iowa.

2. One member of each team rolls the die. The team with the higher roll goes first.

3. Alternate turns between teams and partners: Team 1, Player 1; Team 2, Player 1; Team 1, Player 2; Team 2, Player 2.

4. Shuffle the *Algebra Election* Cards. Place them writing-side down in a pile on the table.

5. The first player rolls the die. The result is the number of moves the player must make from the current state. Each new state counts as one move. Moves can be in any direction as long as they pass between states that share a common border.
Exceptions: Players can get to and from Alaska by way of Washington state and to and from Hawaii by way of California. Once a player has been in a state, the player may not return to that state on the same turn.

6. The player moves their counter and takes the top *Algebra Election* Card. The state's number of electoral votes is substituted for the variable *x* in the problem on the card. The map names how many electoral votes the state has. The player solves the problem and offers an answer. The other team checks the answer with a calculator.

7. If the answer is correct, the player's team wins the state's electoral votes. They do the following:

◆ Write the state's name and its electoral votes on a piece of scratch paper.

◆ Write their first initials in pencil on the state to show that they have won it.

Once a state is won, it is out of play. The opposing team may land on the state, but they cannot get its votes.

Electoral Vote Map

NOTE: Alaska and Hawaii are not drawn to scale.

8. If the player does not solve the problem correctly, the state remains open. Players may still try to win its votes.

9. The next player rolls the die, moves his or her counter, and repeats Steps 6–8.

10. The first team to get at least 270 votes wins the election.

11. When all the *Algebra Election* Cards have been used, shuffle the deck and use it again.

12. Each player begins a turn from the last state he or she landed on.

Notes

◆ "A state" means "a state or the District of Columbia (D.C.)."

◆ Partners may discuss the problem with one another. Each player, however, has to answer the problem on his or her own.

◆ If a player does not want to answer an *Algebra Election* Card, the player may say "Pass" and draw another card. A player may "Pass" 3 times during a game.

◆ If an *Algebra Election* Card contains several problems, a player must answer all the questions on a card correctly to win a state's votes.

◆ Suggested strategy: Look at the map to see which states have the most votes, then work with your partner to win those states.

Variations

1. Agree on a time limit for answering problems.

2. Give one extra point if the player can name the capital of the state landed on.

> **Note**
>
> A shorter version of the game can be played by going through all 32 *Algebra Election* Cards just once. The team with more votes at that time is the winner.

Angle Tangle

Materials ☐ 1 protractor

☐ 1 straightedge

☐ several blank sheets of paper

Players 2

Skill Estimating and measuring angle size

Object of the game To estimate angle sizes accurately and have the lower total score.

Directions

In each round:

1. Player 1 uses a straightedge to draw an angle on a sheet of paper.

2. Player 2 estimates the degree measure of the angle.

3. Player 1 measures the angle with a protractor. Players agree on the measure.

4. Player 2's score is the difference between the estimate and the actual measure of the angle. (The difference will be 0 or a positive number.)

5. Players trade roles and repeat Steps 1–4.

Players add their scores at the end of five rounds. The player with the lower total score wins the game.

Example

	Player 1			**Player 2**		
	Estimate	**Actual**	**Score**	**Estimate**	**Actual**	**Score**
Round 1	120°	108°	12	50°	37°	13
Round 2	75°	86°	11	85°	87°	2
Round 3	40°	44°	4	15°	19°	4
Round 4	60°	69°	9	40°	56°	16
Round 5	135°	123°	12	150°	141°	9
Total score			48			44

Player 2 has the lower total score. Player 2 wins the game.

Build-It

Materials ☐ 1 *Build-It* Card Deck (*Math Masters*, p. 427)

☐ 1 *Build-It* Gameboard for each player
(*Math Masters*, p. 428)

Players 2

Skill Comparing and ordering fractions

Object of the game To be the first player to arrange 5 fraction cards in order from smallest to largest.

Directions

1. Shuffle the fraction cards. Deal 1 card number-side down on each of the 5 spaces on the 2 *Build-It* gameboards.

2. Put the remaining cards number-side down for a **draw pile.** Turn the top card over and place it number-side up in a **discard pile.**

3. Players turn over the 5 cards on their gameboards. Do not change the order of the cards at any time during the game.

4. Players take turns. When it is your turn:

   ♦ Take either the top card from the draw pile or the top card from the discard pile.

   ♦ Decide whether to keep this card or put it on the discard pile.

   ♦ If you keep the card, it must replace 1 of the 5 cards on your *Build-It* gameboard. Put the replaced card on the discard pile.

5. If all the cards in the draw pile are used, shuffle the discard pile. Place them number-side down in a draw pile. Turn over the top card to start a new discard pile.

6. The winner is the first player to have all 5 cards on his or her gameboard in order from the smallest fraction to the largest.

Build-It Gameboard

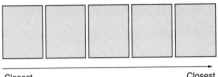

Closest to 0 → Closest to 1

Build-It Card Deck

$\frac{5}{9}$	$\frac{1}{3}$	$\frac{11}{12}$	$\frac{1}{12}$
$\frac{7}{12}$	$\frac{3}{8}$	$\frac{1}{4}$	$\frac{1}{5}$
$\frac{2}{3}$	$\frac{3}{7}$	$\frac{4}{7}$	$\frac{3}{4}$
$\frac{3}{5}$	$\frac{4}{5}$	$\frac{7}{9}$	$\frac{5}{6}$

Games

Credits/Debits Game (Advanced Version)

Materials
- ☐ 1 complete deck of number cards
- ☐ 1 penny
- ☐ 1 *Credits/Debits* Game Recording Sheet for each player (*Math Masters*, p. 431)

Players 2

Skill Addition and subtraction of positive and negative numbers

Object of the game To have more money after adding and subtracting credits and debits.

Recording Sheet

	Start	Change — Addition or Subtraction	Change — Credit or Debit	End, and next start
1	+$10			
2				
3				
4				
5				
6				
7				
8				
9				
10				

Directions

You are an accountant for a business. Your job is to keep track of the company's **current balance.** The current balance is also called the **bottom line.**

1. Shuffle the deck and lay it number-side down between the players. The black-numbered cards are the "credits," and the blue- or red-numbered cards are the "debits."

2. The heads side of the coin tells you to **add** a credit or debit to the bottom line. The tails side of the coin tells you to **subtract** a credit or debit from the bottom line.

3. Each player begins with a bottom line of +$10.

4. Players take turns. On your turn, do the following:
   - ♦ Flip the coin. This tells you whether to add or subtract.
   - ♦ Draw a card. The card tells you what amount in dollars (positive or negative) to add or subtract from your bottom line. Red or blue numbers are negative numbers. Black numbers are positive numbers.
   - ♦ Record the results in your table.

Scoring

After 10 turns each, the player with more money is the winner of the round. If both players have negative dollar amounts, the player whose amount is closer to 0 wins.

Examples
Max has a "Start" balance of $4. His coin lands heads up and he records + in the "Addition or Subtraction" column.

He draws a red 7 and records −$7 in the "Credit or Debit" column. Max adds: $4 + (−$7) = −$3. He records −$3 in the "End" balance column and also in the "Start" column on the next line.

Beth has a "Start" balance of −$20. Her coin lands tails up, which means subtract. She draws a black 9 (+$9). She subtracts: −$20 − (+$9) = −$29. Her "End" balance is −$29.

Divisibility Dash

Materials ☐ number cards 0–9 (4 of each)

☐ number cards: 2, 3, 5, 6, 9, and 10 (2 of each)

Players 2 or 3

Skill Recognizing multiples, using divisibility tests

Object of the game To discard all cards.

Directions

1. Shuffle the divisor cards and place them number-side down on the table. Shuffle the draw cards and deal 8 cards to each player. Place the remaining draw cards number-side down next to the divisor cards.

2. For each round, turn the top divisor card number-side up. Players take turns. When it is your turn:

   ♦ Use the cards in your hand to make 2-digit numbers that are multiples of the divisor card. Make as many 2-digit numbers that are multiples as you can. A card used to make one 2-digit number may not be used again to make another number.

   ♦ Place all the cards you used to make 2-digit numbers in a discard pile.

   ♦ If you cannot make a 2-digit number that is a multiple of the divisor card, you must take a card from the draw pile. Your turn is over.

3. If a player disagrees that a 2-digit number is a multiple of the divisor card, that player may challenge. Players use the divisibility test for the divisor card value to check the number in question. Any numbers that are not multiples of the divisor card must be returned to the player's hand.

4. If the draw pile or divisor cards have all been used, they can be reshuffled and put back into play.

5. The first player to discard all of his or her cards is the winner.

> **Note**
>
> The number cards 2, 3, 5, 6, 9, and 10 (2 of each) are the **divisor cards**.
>
> The number cards 0–9 (4 of each) are the **draw cards**. This set of cards is also called the **draw pile**.

Example   Andrew's cards: `1` `2` `5` `5` `7` `8`   Divisor card: `3`

Andrew uses his cards to make 2 numbers that are multiples of 3: `1` `5` `7` `5`

He discards these 4 cards and holds the 2 and 8 for the next round of play.

Doggone Decimal

Materials ☐ number cards 0–9 (4 of each)

☐ 4 index cards labeled 0.1, 1, 10, and 100

☐ 2 counters per player (to use as decimal points)

☐ 1 calculator for each player

Players 2

Skill Estimating products of whole numbers and decimals

Object of the game To collect more number cards.

Directions

1. One player shuffles the number cards and deals 4 cards to each player.

2. The other player shuffles the index cards, places them number-side down, and turns over the top card. The number that appears (0.1, 1, 10, or 100) is the **target number.**

3. Using 4 number cards and 2 decimal-point counters, each player forms 2 numbers. Each number must have 2 digits and a decimal point.

   ♦ Players try to form 2 numbers whose product is as close as possible to the target number.

   ♦ The decimal point can go anywhere in a number—for example:

4. Each player computes the product of their numbers using a calculator.

5. The player whose product is closer to the target number takes all 8 number cards.

6. Four new number cards are dealt to each player and a new target number is turned over. Repeat Steps 3–5 using the new target number.

7. The game ends when all the target numbers have been used.

8. The player with more number cards wins the game. In the case of a tie, reshuffle the index cards and turn over a target number. Play one tie-breaking round.

Example The target number is 10.

Briana is dealt 1, 4, 8, and 8. She forms the numbers 8.8 and 1.4.

Evelyn is dealt 2, 3, 6, and 9. She forms the numbers 2.6 and 3.9.

Briana's product is 12.32 and Evelyn's is 10.14.

Evelyn's product is closer to 10. She wins the round and takes all 8 cards.

Exponent Ball

Materials ☐ 1 *Exponent Ball* Gameboard (*Math Masters*, p. 436)

☐ 1 six-sided die

☐ 1 counter

☐ 1 calculator

Players 2

Skill Converting exponential notation to standard notation, comparing probabilities

Object of the game To score more points in 4 turns.

Directions

1. The game is similar to U.S. football. Player 1 first puts the ball (counter) on one of the 20-yard lines. The objective is to reach the opposite goal line, 80 yards away. A turn consists of 4 chances to advance the counter to the goal line and score.

2. The first 3 chances must be runs on the ground. To run, Player 1 rolls the die twice. The first roll names the **base,** the second roll names the **exponent.** For example, rolls of 5 and 4 name the number $5^4 = 625$.

3. Player 1 calculates the value of the roll and uses Table 1 on the gameboard page to find how far to move the ball forward (+) or backward (−).

4. If Player 1 does not score in the first 3 chances, he or she may choose to run or kick on the fourth chance. To kick, the player rolls the die once and multiplies the number shown by 10. The result is the distance the ball travels (Table 2 on the gameboard).

5. If the ball reaches the goal line on a run, the player scores 7 points. If the ball reaches the goal line on a kick, the player scores 3 points.

6. If the ball does not reach the goal line in 4 chances, Player 1's turn ends. Player 2 starts where Player 1 stopped and moves toward the opposite goal line.

7. If Player 1 scores, Player 2 puts the ball on one of the 20-yard lines and follows the directions above.

8. Players take turns. A round consists of 4 turns for each player. The player with more points wins.

Exponent Ball Gameboard

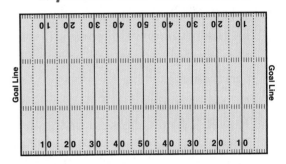

Table 1: Runs		
Value of Roll	**Move Ball**	**Chances of Gaining on the Ground**
1	−15 yd	−15 yards: 1 out of 6 or about 17%
2 to 6	+10 yd	10 yards or more: 5 out of 6 or about 83%
8 to 81	+20 yd	20 yards or more: 4 out of 6 or about 67%
in the 100s	+30 yd	30 yards or more: 13 out of 36 or about 36%
in the 1,000s	+40 yd	40 yards or more: 7 out of 36 or about 19%
in the 10,000s	+50 yd	50 yards: 1 out of 18 or about 6%

Table 2: Kicks		
Value of Roll	**Move Ball**	**Chances of Kicking**
1	+10 yd	10 yards or more: 6 out of 6 or 100%
2	+20 yd	20 yards or more: 5 out of 6 or about 83%
3	+30 yd	30 yards or more: 4 out of 6 or about 67%
4	+40 yd	40 yards or more: 3 out of 6 or about 50%
5	+50 yd	50 yards or more: 2 out of 6 or about 33%
6	+60 yd	60 yards: 1 out of 6 or about 17%

Note

If a backward move should carry the ball behind the goal line, the ball (counter) is put on the goal line.

Factor Captor

Materials   ☐ 1 calculator for each player

☐ paper and pencil for each player

☐ 1 *Factor Captor* Grid—either Grid 1 or Grid 2 (*Math Masters*, pp. 437 and 438)

☐ coin-size counters (48 for Grid 1; 70 for Grid 2)

Players   2

Skill   Finding factors of a number

Object of the game   To have the higher total score.

Directions

1. To start the first round, Player 1 chooses a 2-digit number on the number grid, covers it with a counter, and records the number on scratch paper. This is Player 1's score for the round.

2. Player 2 covers all of the factors of Player 1's number. Player 2 finds the sum of the factors and records it on scratch paper. This is Player 2's score for the round.

A factor may be covered only once during a round.

3. If Player 2 missed any factors, Player 1 can cover them with counters and add them to his or her score.

4. In the next round, players switch roles. Player 2 chooses a number that is not covered by a counter. Player 1 covers all factors of that number.

5. Any number that is covered by a counter is no longer available and may not be used again.

6. The first player in a round may not cover a number that is less than 10, unless no other numbers are available.

7. Play continues with players trading roles after each round, until all numbers on the grid have been covered. Players then use their calculators to find their total scores. The player with the higher score wins the game.

Grid 1
(Beginning Level)

1	2	2	2	2	2
2	3	3	3	3	3
3	4	4	4	4	5
5	5	5	6	6	7
7	8	8	9	9	10
10	11	12	13	14	15
16	18	20	21	22	24
25	26	27	28	30	32

Grid 2
(Advanced Level)

1	2	2	2	2	3
3	3	3	3	4	4	4
4	5	5	5	5	6	6
6	7	7	8	8	9	9
10	10	11	12	13	14	15
16	17	18	19	20	21	22
23	24	25	26	27	28	30
32	33	34	35	36	38	39
40	42	44	45	46	48	49
50	51	52	54	55	56	60

Example

Round 1: James covers 27 and scores 27 points. Emma covers 1, 3, and 9, and scores 1 + 3 + 9 = 13 points.

Round 2: Emma covers 18 and scores 18 points. James covers 2, 3, and 6, and scores 2 + 3 + 6 = 11 points. Emma covers 9 with a counter, because 9 is also a factor of 18. Emma adds 9 points to her score.

First to 100

Materials ☐ one set of *First to 100* Problem Cards
(*Math Journal* 2, Activity Sheets 5 and 6)

☐ 2 six-sided dice

☐ 1 calculator

Players 2 to 4

Skill Variable substitution, solving equations

Object of the game To solve problems and be the first player to collect at least 100 points.

Directions

1. Shuffle the Problem Cards and place them word-side down on the table.

2. Players take turns. When it is your turn:

   ♦ Roll 2 dice and find the product of the numbers.

   ♦ Turn over the top Problem Card and substitute the product for the variable x in the problem on the card.

   ♦ Solve the problem mentally or use paper and pencil. Give the answer. (You have 3 chances to use a calculator to solve difficult problems during a game.) Other players check the answer with a calculator.

   ♦ If the answer is correct, you win the number of points equal to the product that was substituted for the variable x. Some Problem Cards require 2 or more answers. In order to win any points, you must answer all parts of the problem correctly.

   ♦ Put the used Problem Card at the bottom of the deck.

3. The first player to get at least 100 points wins the game.

Example Alice rolls a 5 and a 6. The product is 30.

She turns over a Problem Card: $20 * x = ?$
She substitutes 30 for x and answers 600.

The answer is correct. Alice wins 30 points.

First to 100 Problem Cards

How many inches are there in x feet? How many centimeters are there in x meters? **1**	How many quarts are there in x gallons? **2**	What is the smallest number of x's you can add to get a sum greater than 100? **3**	Is $50 * x$ greater than 1,000? Is $\frac{x}{10}$ less than 1? **4**
$\frac{1}{2}$ of $x = ?$ $\frac{1}{10}$ of $x = ?$ **5**	$1 - x = ?$ $x + 998 = ?$ **6**	If x people share 1,000 stamps equally, how many stamps will each person get? **7**	What time will it be x minutes from now? What time was it x minutes ago? **8**
It is 102 miles to your destination. You have gone x miles. How many miles are left? **9**	What whole or mixed number equals x divided by 2? **10**	Is x a prime or a composite number? Is x divisible by 2? **11**	The time is 11:05 A.M. The train left x minutes ago. What time did the train leave? **12**
Bill was born in 1939. Freddy was born the same day, but x years later. In what year was Freddy born? **13**	Which is larger: $2 * x$ or $x + 50$? **14**	There are x rows of seats. There are 9 seats in each row. How many seats are there in all? **15**	Sargon spent x cents on apples. If she paid with a $5 bill, how much change should she get? **16**
The temperature was 25°F. It dropped x degrees. What was the new temperature? **17**	Each story in a building is 10 ft high. If the building has x stories, how tall is it? **18**	Which is larger: $2 * x$ or $\frac{100}{x}$? **19**	$20 * x = ?$ **20**
Name all the whole-number factors of x. **21**	Is x an even or an odd number? Is x divisible by 9? **22**	Shalanda was born on a Tuesday. Linda was born x days later. On what day of the week was Linda born? **23**	Will had a quarter plus x cents. How much money did he have in all? **24**
Find the perimeter and area of this square. x cm $\times$ x cm **25**	What is the median of these weights? 5 pounds 21 pounds x pounds What is the range? **26**	What is the median of these weights? **27**	$x^2 = ?$ 50% of $x^2 = ?$ **28**
$(3x + 4) - 8 = ?$ **29**	x out of 100 students voted for Ruby. Is this more than 25%, less than 25%, or exactly 25% of the students? **30**	There are 200 students at Wilson School. $x\%$ speak Spanish. How many students speak Spanish? **31**	People answered a survey question either Yes or No. $x\%$ answered Yes. What percent answered No? **32**

Frac-Tac-Toe

2-4-5-10 Frac-Tac-Toe

Materials ☐ number cards 0–10 (4 of each)

☐ 1 *Frac-Tac-Toe* Number-Card Board (*Math Masters,* p. 439)

☐ 1 *2-4-5-10 Frac-Tac-Toe* (Decimals) gameboard (*Math Masters,* p. 444)

☐ counters (2 colors), or pennies (one player using heads, the other using tails)

☐ 1 calculator

Players 2

Skill Renaming fractions as decimals and percents

Object of the game To cover 3 squares in a row, in any direction (horizontal, vertical, diagonal).

Advance Preparation Separate the cards into 2 piles on the Number-Card Board—a numerator pile and a denominator pile. For a *2-4-5-10* game, place 2 each of the 2, 4, 5, and 10 cards in the denominator pile. All other cards are placed on the numerator pile.

Shuffle the cards in each pile. Place the piles number-side down in the left-hand spaces. When the numerator pile is completely used, reshuffle that pile, and place it number-side down in the left-hand space. When the denominator pile is completely used, turn it over and place it number-side down in the left-hand space without reshuffling it.

Directions

1. Players take turns. When it is your turn:
   ◆ Turn over the top card from each pile to form a fraction (numerator card above denominator card).
   ◆ Try to match the fraction shown with one of the grid squares on the gameboard. If a match is found, cover that grid square with your counter and your turn is over. If no match is found, your turn is over.

2. To change the fraction shown by the cards to a decimal, players may use either a calculator or the *Table of Decimal Equivalents for Fractions* on page 374.

Frac-Tac-Toe
Number-Card Board

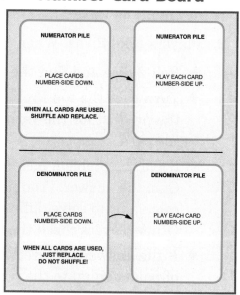

gameboard for *2-4-5-10*
Frac-Tac-Toe **(Decimals)**

>1.0	0 or 1	>2.0	0 or 1	>1.0
0.1	0.2	0.25	0.3	0.4
>1.5	0.5	>1.5	0.5	>1.5
0.6	0.7	0.75	0.8	0.9
>1.0	0 or 1	>2.0	0 or 1	>1.0

Examples

The cards show the fraction $\frac{4}{5}$. The player may cover the 0.8 square, unless that square has already been covered.

The cards show the fraction $\frac{0}{5}$. The player may cover any 1 of the 4 squares labeled "0 or 1" that has not already been covered.

The cards show the fraction $\frac{4}{2}$. The player may cover any square labeled "> 1.0" or "> 1.5" that has not already been covered. The player may not cover a square labeled "> 2.0," because $\frac{4}{2}$ is equal to, but not greater than, 2.0.

3. The first player to cover 3 squares in a row in any direction (horizontal, vertical, diagonal) is the winner of the game.

Variation

Play a version of the *2-4-5-10* game using the percent gameboard shown at the right. Use *Math Masters*, page 445.

gameboard for *2-4-5-10*
***Frac-Tac-Toe* (Percents)**

>100%	0% or 100%	>200%	0% or 100%	>100%
10%	20%	25%	30%	40%
>100%	50%	>200%	50%	>100%
60%	70%	75%	80%	90%
>100%	0% or 100%	>200%	0% or 100%	>100%

2-4-8 and 3-6-9 Frac-Tac-Toe

Play the *2-4-8* or the *3-6-9* version of the game. Gameboards for the different versions are shown below.

♦ For a *2-4-8* game, place 2 each of the 2, 4, and 8 cards in the denominator pile. Use *Math Masters,* page 440 or 442.

♦ For a *3-6-9* game, place 2 each of the 3, 6, and 9 cards in the denominator pile. Use *Math Masters,* page 441 or 443.

2-4-8 Frac-Tac-Toe (Decimals)

>2.0	0 or 1	>1.5	0 or 1	>2.0
1.5	0.125	0.25	0.375	1.5
>1.0	0.5	0.25 or 0.75	0.5	>1.0
2.0	0.625	0.75	0.875	2.0
>2.0	0 or 1	1.125	0 or 1	>2.0

2-4-8 Frac-Tac-Toe (Percents)

>200%	0% or 100%	>150%	0% or 100%	>200%
150%	$12\frac{1}{2}\%$	25%	$37\frac{1}{2}\%$	150%
>100%	50%	25% or 75%	50%	>100%
200%	$62\frac{1}{2}\%$	75%	$87\frac{1}{2}\%$	200%
>200%	0% or 100%	$112\frac{1}{2}\%$	0% or 100%	>200%

3-6-9 Frac-Tac-Toe (Decimals)

>1.0	0 or 1	$0.\overline{1}$	0 or 1	>1.0
$0.1\overline{6}$	$0.\overline{2}$	$0.\overline{3}$	0.3	$0.\overline{4}$
>2.0	$0.\overline{5}$	>1.0	$0.\overline{6}$	>2.0
$0.\overline{6}$	0.7	$0.8\overline{3}$	$0.\overline{8}$	$1.\overline{3}$
>1.0	0 or 1	$1.\overline{6}$	0 or 1	>1.0

3-6-9 Frac-Tac-Toe (Percents)

>100%	0% or 100%	11.1%	0% or 100%	>100%
$16\frac{2}{3}\%$	22.2%	$33\frac{1}{3}\%$	33.3%	44.4%
>200%	55.5%	>100%	66.6%	>200%
$66\frac{2}{3}\%$	77.7%	$83\frac{1}{3}\%$	88.8%	$133\frac{1}{3}\%$
>100%	0% or 100%	$166\frac{2}{3}\%$	0% or 100%	>100%

Fraction Action, Fraction Friction

Materials ☐ 1 *Fraction Action, Fraction Friction* Card Deck
(*Math Masters*, p. 446)

☐ 1 or more calculators

Players   2 or 3

Skill   Estimating sums of fractions

Object of the game To collect a set of fraction cards with a sum as close as possible to 2, without going over 2.

Fraction Action, Fraction Friction Card Deck

$\frac{1}{2}$	$\frac{1}{3}$	$\frac{2}{3}$	$\frac{1}{4}$
$\frac{3}{4}$	$\frac{1}{6}$	$\frac{1}{6}$	$\frac{5}{6}$
$\frac{1}{12}$	$\frac{1}{12}$	$\frac{5}{12}$	$\frac{5}{12}$
$\frac{7}{12}$	$\frac{7}{12}$	$\frac{11}{12}$	$\frac{11}{12}$

Directions

1. Shuffle the deck and place it number-side down on the table between the players.

2. Players take turns.

   ♦ On each player's first turn, he or she takes a card from the top of the pile and places it number-side up on the table.

   ♦ On each of the player's following turns, he or she announces one of the following:

   "Action" This means that the player wants an additional card. The player believes that the sum of the fraction cards he or she already has is *not* close enough to 2 to win the hand. The player thinks that another card will bring the sum of the fractions closer to 2, without going over 2.

   "Friction" This means that the player does not want an additional card. The player believes that the sum of the fraction cards he or she already has *is* close enough to 2 to win the hand. The player thinks there is a good chance that taking another card will make the sum of the fractions greater than 2.

Once a player says "Friction," he or she cannot say "Action" on any turn after that.

3. Play continues until all players have announced "Friction" or have a set of cards whose sum is greater than 2. The player whose sum is closest to 2 without going over 2 is the winner of that round. Players may check each other's sums on their calculators.

4. Reshuffle the cards and begin again. The winner of the game is the first player to win 5 rounds.

Fraction Capture

Materials ☐ 1 *Fraction Capture* Gameboard (*Math Masters*, p. 447)

☐ 2 six-sided dice

Players   2

Skill   Addition of fractions, finding equivalent fractions

Object of the game To capture more squares on the *Fraction Capture* gameboard.

Directions

1. Player 1 rolls the 2 dice and makes a fraction with the numbers that come up. The number on either die can be the denominator. The number on the other die becomes the numerator.

   A fraction equal to a whole number is NOT allowed. For example, if a player rolls 3 and 6, the fraction cannot be $\frac{6}{3}$, because $\frac{6}{3}$ equals 2. If the 2 dice show the same number, the player rolls again.

2. Player 1 initials sections of one or more gameboard squares to show the fraction formed. This **claims** the sections for the player.

Figure 1

Figure 2

> **Examples**
> - A player rolls a 4 and a 3 and makes $\frac{3}{4}$. The player claims three $\frac{1}{4}$ sections by initialing them. (Figure 1)
>
> - Equivalent fractions can be claimed. If a player rolls a 1 and a 2 and makes $\frac{1}{2}$, the player can initial one $\frac{1}{2}$ section of a square, or two $\frac{1}{4}$ sections, or three $\frac{1}{6}$ sections. (Figure 2)
>
> - The fraction may be split between squares. A player can show $\frac{5}{4}$ by claiming $\frac{4}{4}$ on one square and $\frac{1}{4}$ on another square. (Figure 3)

3. Players take turns. If a player forms a fraction, but cannot claim enough sections to show that fraction, no sections can be claimed and the player's turn is over.

4. A player **captures** a square when that player has claimed sections making up **more than** $\frac{1}{2}$ of the square. If each player has initialed $\frac{1}{2}$ of a square, no one has captured that square.

5. Blocking is allowed. For example, if Player 1 initials $\frac{1}{2}$ of a square, Player 2 may initial the other half, so that no one can capture the square.

6. Play ends when all of the squares have either been captured or blocked. The winner is the player who has captured more squares.

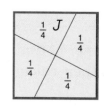

Figure 3

Fraction/Whole Number Top-It

Materials ☐ number cards 1–10 (4 of each)

☐ 1 calculator (optional)

Players 2 to 4

Skill Multiplication of whole numbers and fractions, converting improper fractions to mixed numbers

Object of the game To collect the most cards.

Directions

1. Shuffle the cards and place them number-side down on the table.

2. Each player turns over 3 cards. The card numbers are used to form 1 whole number and 1 fraction.

   ◆ The first card drawn is placed number-side up on the table. This card number is the whole number.

   ◆ The second and third cards drawn are used to form a fraction and are placed number-side up next to the first card. The fraction that these cards form must be less than or equal to 1.

3. Each player calculates the product of their whole number and fraction and calls it out as a mixed number. The player with the largest product takes all the cards. Players may use a calculator to compare their products.

4. In case of a tie for the largest product, each tied player repeats Steps 2 and 3. The player with the largest product takes all the cards from both plays.

5. The game ends when there are not enough cards left for each player to have another turn. The player with the most cards wins.

Example

Amy turns over a 3, a 9, and a 5, in that order.
Roger turns over a 7, a 2, and an 8, in that order.

Amy's product is $3 * \frac{5}{9} = \frac{15}{9}$.

Roger's product is $7 * \frac{2}{8} = \frac{14}{8}$.

$\frac{15}{9} = \frac{5}{3} = 1\frac{2}{3}$ $\qquad$ $\frac{14}{8} = \frac{7}{4} = 1\frac{3}{4}$

Roger's product is larger, so he takes all of the cards.

Fraction/Whole Number Top-It (continued)

Advanced Version

Each player turns over 4 cards and forms 1 fraction from their first 2 cards and a second fraction from their last 2 cards. (All fractions must be less than or equal to 1.) Each player calculates the product of their fractions, and the player with the largest product takes all the cards.

Example

Kenny turns over a 2, a 1, a 4, and an 8, in that order.
Liz turns over a 2, a 3, a 5, and a 5, in that order.

Kenny's product is $\frac{1}{2} * \frac{4}{8} = \frac{4}{16}$.

Liz's product is $\frac{2}{3} * \frac{5}{5} = \frac{10}{15}$.

$\frac{4}{16} = \frac{1}{4}$ $\frac{10}{15} = \frac{2}{3}$

Liz's product is larger, so she takes all the cards.

Kenny

Liz

Getting to One

Materials ☐ 1 calculator

Players 2

Skill Estimation

Object of the game To correctly guess a mystery number in as few tries as possible.

Directions

1. Player 1 chooses a mystery number that is between 1 and 100.

2. Player 2 guesses the mystery number.

3. Player 1 uses a calculator to divide Player 2's guess by the mystery number. Player 1 then reads the answer in the calculator display. If the answer has more than 2 decimal places, only the first 2 decimal places are read.

4. Player 2 continues to guess until the calculator result is 1. Player 2 keeps track of the number of guesses.

5. When Player 2 has guessed the mystery number, players trade roles and follow Steps 1–4 again. The player who guesses their mystery number in the fewest number of guesses wins the round. The first player to win 3 rounds wins the game.

> **Note**
>
> For a decimal number, the places to the right of the decimal point with digits in them are called *decimal places*. For example, 4.06 has 2 decimal places, 123.4 has 1 decimal place, and 0.780 has 3 decimal places.

Example Player 1 chooses the mystery number 65.

Player 2 guesses: 45. Player 1 keys in: 45 ÷ 65 =.
Answer: 0.69 Too small.

Player 2 guesses: 73. Player 1 keys in: 73 ÷ 65 =.
Answer: 1.12 Too big.

Player 2 guesses: 65. Player A keys in: 65 ÷ 65 =.
Answer: 1. Just right!

Advanced Version
Allow mystery numbers up to 1,000.

Greedy

Materials ☐ 1 *Greedy* Score Sheet for each player
(*Math Masters,* p. 453)

☐ 1 six-sided die

Players group of 4–8 students, or whole class
group leader

Skill Developing a game strategy

Object of the game To have the highest total score.

Directions

1. Each player records his or her own score on a
separate score sheet.

2. To begin a round, all players stand up.

3. The group leader rolls the die twice. Each
player's score is equal to the sum of the
numbers rolled.

4. Before rolling the die again, the group leader
asks whether any player would like to sit
down. Players who sit down keep the score they
have and record it as the score for that round
on their score sheet.

5. If any players remain standing, the group
leader rolls the die twice.

♦ If either roll is a 2, all players standing record a score of
0 for that round. The round is over.

♦ If neither roll is a 2, each player standing calculates their
new score by adding the sum of the numbers rolled to
their last score.

6. Repeat Steps 4 and 5 until no players remain standing.
The round is over.

7. Begin again with Step 2 and play another round. Play
continues for 6 rounds.

8. The player with the largest total score for 6 rounds wins
the game.

Greedy **Score Sheet**

Round	Total Score
1	
2	
3	
4	
5	
6	
Total Score	

High-Number Toss

Materials ☐ 1 six-sided die

☐ 1 sheet of paper for each player

Players 2

Skill Place value, exponential notation

Object of the game To make the largest numbers possible.

Directions

1. Each player draws 4 blank lines on a sheet of paper to record the numbers that come up on the rolls of the die.

   Player 1: _____ _____ _____ | _____

   Player 2: _____ _____ _____ | _____

2. Player 1 rolls the die and writes the number on any of his or her 4 blank lines. It does not have to be the first blank—it can be any of them. *Keep in mind that the larger number wins!*

3. Player 2 rolls the die and writes the number on one of his or her blank lines.

4. Players take turns rolling the die and writing the number 3 more times each.

5. Each player then uses the 4 numbers on his or her blanks to build a number.

   ♦ The numbers on the first 3 blanks are the first 3 digits of the number the player builds.

   ♦ The number on the last blank tells the number of zeros that come after the first 3 digits.

6. Each player reads his or her number. (See the place-value chart below.) The player with the larger number wins the round. The first player to win 4 rounds wins the game.

Note

If you don't have a die, you can use a deck of number cards. Use all cards with the numbers 1 through 6. Instead of rolling the die, draw the top card from the facedown deck.

Hundred Millions	Ten Millions	Millions	,	Hundred Thousands	Ten Thousands	Thousands	,	Hundreds	Tens	Ones

Example

		First three digits			Number of zeros	
Player 1:	1	3	2		6	= 132,000,000 (132 million)
Player 2:	3	5	6		4	= 3,560,000 (3 million, 560 thousand)

Player 1 wins.

High-Number Toss: Decimal Version

Materials ☐ number cards 0–9 (4 of each)

☐ 1 scorecard for each player (*Math Masters*, p. 455)

Players 2

Skill Decimal place value, subtraction, and addition

Object of the game To make the largest decimal numbers possible.

Scorecard

Game 1	
Round 1	Score
0. __ __ __	_____
Round 2	
0. __ __ __	_____
Round 3	
0. __ __ __	_____
Round 4	
0. __ __ __	_____
Total:	_____

Directions

1. Each player makes a scorecard like the one at the right. Players fill out their own scorecards at each turn.

2. Shuffle the cards and place them number-side down on the table.

3. In each round:

   ♦ Player 1 draws the top card from the deck and writes the number on any of the 3 blanks on the scorecard. It need not be the first blank—it can be any of them.

   ♦ Player 2 draws the next card from the deck and writes the number on one of his or her blanks.

   ♦ Players take turns doing this 2 more times. The player with the larger number wins the round.

4. The winner's score for a round is the difference between the two players' numbers. (Subtract the smaller number from the larger number.) The loser scores 0 points for the round.

Example

Player 1: 0 . __7__ __6__ __3__

Player 2: 0 . __9__ __2__ __1__

Since 0.921 − 0.763 = 0.158, Player 2 scores 0.158 point for the round. Player 1 scores 0 points.

5. Players take turns starting a round. At the end of 4 rounds, they find their total scores. The player with the larger total score wins the game.

Landmark Shark

Materials ☐ 1 complete deck of number cards

☐ 1 each of Range, Median, and Mode *Landmark Shark* Cards for each player (*Math Masters,* p. 456)

☐ 1 *Landmark Shark* Score Sheet (*Math Masters,* p. 457)

Players   2 or 3

Skill   Finding the range, mode, median, and, mean

Object of the game To score the most points by finding data landmarks.

Directions

1. To play a round:

   ♦ The dealer shuffles the number cards and deals 5 cards number-side down to each player.

   ♦ Players put their cards in order from the smallest number to the largest.

   ♦ There are 3 ways a player may score points using their five cards:

Range: The player's score is the range of the 5 numbers.

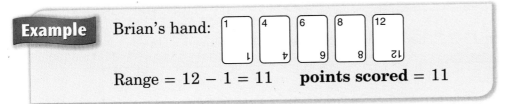

Example   Brian's hand:

Range = 12 − 1 = 11   **points scored = 11**

Median: The player's score is the median of the 5 numbers.

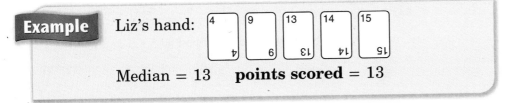

Example   Liz's hand:

Median = 13   **points scored = 13**

Mode: The player must have at least 2 cards with the same number. The player's score is found by multiplying the mode of the 5 numbers by the number of modal cards. If there is more than one mode, the player uses the mode that will produce the most points.

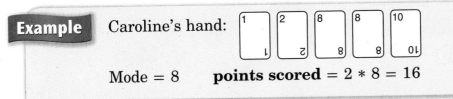

Example   Caroline's hand:

Mode = 8   **points scored = 2 * 8 = 16**

Landmark Shark (continued)

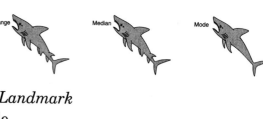

Range   Median   Mode

2. Each player decides which landmark will yield the highest score for their hand. A player indicates their choice by placing 1 of the 3 *Landmark Shark* cards (Range, Median, or Mode) on the table.

3. Players can try to improve their scores by exchanging up to 3 of their cards for new cards from the deck. However, the *Landmark Shark* card stays the same.

Examples **Brian's hand:**

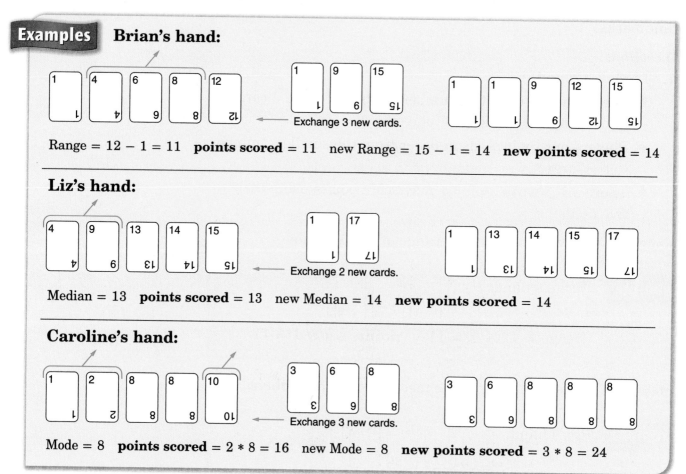

Range = 12 − 1 = 11   **points scored = 11**   new Range = 15 − 1 = 14   **new points scored = 14**

Liz's hand:

Median = 13   **points scored = 13**   new Median = 14   **new points scored = 14**

Caroline's hand:

Mode = 8   **points scored = 2 * 8 = 16**   new Mode = 8   **new points scored = 3 * 8 = 24**

4. Players lay down their cards and record their points scored on the score sheet.

5. **Bonus Points:** Each player calculates the *mean* of their card numbers, to the nearest tenth. Each player's score for the round is the sum of their points scored plus any bonus points.

		Player 1	Player 2	Player 3
Round 1:	Points Scored			
	Bonus Points			
	Round 1 Score			

6. Repeat Steps 1–5 for each round. The winner is the player with the highest total after 5 rounds.

Mixed-Number Spin

Materials ☐ 1 *Mixed-Number Spin* Record Sheet (*Math Masters*, p. 459)

☐ 1 *Mixed-Number* Spinner (*Math Masters*, p. 458)

☐ 1 large paper clip

Players 2

Skill Addition and subtraction of fractions and mixed numbers, solving inequalities

Object of the game To complete 10 number sentences that are true.

Directions

1. Each player writes his or her name in one of the boxes on the Record Sheet.

2. Players take turns. When it is your turn:

   ♦ Anchor the paper clip to the center of the spinner with the point of your pencil and use your other hand to spin the paper clip.

   ♦ Once the paper clip stops, write the fraction or mixed number it most closely points to on any one of the blanks below your name.

3. The first player to complete 10 true number sentences is the winner.

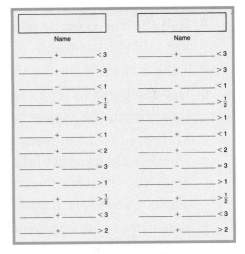

Mixed-Number Spin
Record Sheet

Example Ella has filled in 2 blanks in different sentences.

On her next turn, Ella spins $1\frac{3}{8}$. She has 2 choices for where she can write this mixed number. She can place it on a line where there are 2 blanks.

Or, she can use it to form the true number sentence $1\frac{3}{8} - 1\frac{1}{8} < \frac{1}{2}$.

She cannot use it on the first line because $2 + 1\frac{3}{8}$ is not < 3.

Ella	Ella	Ella
name	name	name
$\underline{\quad 2 \quad} + \underline{\qquad} < 3$	$\underline{\quad 2 \quad} + \underline{\qquad} < 3$	$\underline{\quad 2 \quad} + \underline{\qquad} < 3$
$\underline{\qquad} + \underline{\qquad} > 3$	$\underline{\qquad} + \underline{\qquad} > 3$	$\underline{\qquad} + \underline{\qquad} > 3$
$\underline{\qquad} - \underline{\qquad} < 1$	$\underline{1\frac{3}{8}} - \underline{\qquad} < 1$	$\underline{\qquad} - \underline{\qquad} < 1$
$\underline{\qquad} - \underline{1\frac{1}{8}} < \frac{1}{2}$	$\underline{\qquad} - \underline{1\frac{1}{8}} < \frac{1}{2}$	$\underline{1\frac{3}{8}} - \underline{1\frac{1}{8}} < \frac{1}{2}$

Multiplication Bull's-Eye

Materials ☐ number cards 0–9 (4 of each)

☐ 1 six-sided die

☐ 1 calculator

Players 2

Skill Estimating products of 2- and 3-digit numbers

Object of the game To score more points.

Directions

1. Shuffle the deck and place it number-side down on the table.

2. Players take turns. When it is your turn:

   ♦ Roll the die. Look up the target range of the product in the table at the right.

   ♦ Take 4 cards from the top of the deck.

   ♦ Use the cards to try to form 2 numbers whose product falls within the target range. **Do not use a calculator.**

   ♦ Multiply the 2 numbers on your calculator to determine whether the product falls within the target range. If it does, you have hit the bull's-eye and score 1 point. If it doesn't, you score 0 points.

   ♦ Sometimes it is impossible to form 2 numbers whose product falls within the target range. If this happens, you score 0 points for that turn.

3. The game ends when each player has had 5 turns.

4. The player scoring more points wins the game.

Number on Die	Target Range of Product
1	500 or less
2	501–1,000
3	1,001–3,000
4	3,001–5,000
5	5,001–7,000
6	more than 7,000

Example

Tom rolls a 3, so the target range of the product is from 1,001 to 3,000. He turns over a 5, a 7, a 2, and a 9.

Tom uses estimation to try to form 2 numbers whose product falls within the target range—for example, 97 and 25.

He then finds the product on the calculator: 97 * 25 = 2,425.

Since the product is between 1,001 and 3,000, Tom has hit the bull's-eye and scores 1 point.

Some other possible winning products from the 5, 7, 2, and 9 cards are: 25 * 79, 27 * 59, 9 * 257, and 2 * 579.

Name That Number

Materials ☐ 1 complete deck of number cards

Players 2 or 3

Skill Naming numbers with expressions

Object of the game To collect the most cards.

Directions

1. Shuffle the deck and deal 5 cards to each player. Place the remaining cards number-side down on the table between the players. Turn over the top card and place it beside the deck. This is the **target number** for the round.

2. Players try to match the target number by adding, subtracting, multiplying, or dividing the numbers on as many of their cards as possible. A card may only be used once.

3. Players write their solutions on a sheet of paper. When players have written their best solutions:

   ♦ Each player sets aside the cards they used to match the target number.

   ♦ Each player replaces the cards they set aside by drawing new cards from the top of the deck.

   ♦ The old target number is placed on the bottom of the deck.

   ♦ A new target number is turned over, and another round is played.

4. Play continues until there are not enough cards left to replace all of the players' cards. The player who has set aside the most cards wins the game.

Example   Target number: 16

Player 1's cards:

Some possible solutions:

$10 + 8 - 2 = 16$ (3 cards used)

$7 * 2 + 10 - 8 = 16$ (4 cards used)

$8/2 + 10 + 7 - 5 = 16$ (all 5 cards used)

The player sets aside the cards used to make a solution and draws the same number of cards from the top of the deck.

Polygon Capture

Materials ☐ 1 set of *Polygon Capture* Pieces
(*Math Masters*, p. 470)

☐ 1 set of *Polygon Capture* Property Cards
(*Math Masters*, p. 471)

Players 2, or two teams of 2

Skill Properties of polygons

Object of the game To collect more polygons.

Directions

1. Write the letter A on the back of each Property Card that mentions "angle." Write the letter S on the back of each Property Card that mentions "side."

2. Spread the polygons out on the table. Shuffle the Property Cards and sort them writing-side down into A-card and S-card piles.

3. Players take turns. When it is your turn:

   ◆ Draw the top card from each pile of Property Cards.

   ◆ Take all of the polygons that have **both** of the properties shown on the Property Cards in your hand.

   ◆ If there are no polygons with both properties, draw one additional Property Card—either an A- or an S-card. Look for polygons that have this new property and one of the properties already drawn. Take these polygons.

   ◆ At the end of a turn, if you have not captured a polygon that you could have taken, the other player may name and capture it.

4. When all the Property Cards in either pile have been drawn, shuffle *all* of the Property Cards. Sort them writing-side down into A-card and S-card piles. Continue play.

5. The game ends when there are fewer than 3 polygons left.

6. The winner is the player who has captured more polygons.

Polygon Capture **Pieces**

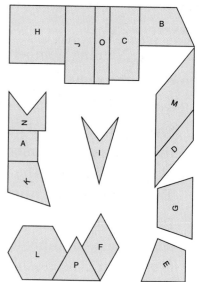

Polygon Capture
Property Cards
(writing-side up)

There is only one right angle.	There are one or more right angles.	All angles are right angles.	There are no right angles.
There is at least one acute angle.	At least one angle is more than 90°.	All angles are right angles.	There are no right angles.
All opposite sides are parallel.	Only one pair of sides is parallel.	There are no parallel sides.	All sides are the same length.
All opposite sides are parallel.	Some sides have the same length.	All opposite sides have the same length.	**Wild Card:** Pick your own side property.

Example Liz has these Property Cards: "All angles are right angles" and "All sides are the same length." She can take all the squares (polygons A and H). Liz has "captured" these polygons.

Scientific Notation Toss

Materials □ 2 six-sided dice

Players 2

Skill Converting from scientific notation to standard notation

Object of the game To create the largest number written in scientific notation.

Directions

When a player rolls 2 dice, either number may be used to name a power of 10, such as 10^2 or 10^4. The other number is used to multiply that power of 10.

Examples

A 5 and a 4 are rolled.	A 2 and a 3 are rolled.
Either $4 * 10^5$ or $5 * 10^4$ can be written.	Either $2 * 10^3$ or $3 * 10^2$ can be written.

1. Each player rolls the dice 3 times and writes each result in scientific notation (as shown above).

2. Players convert their numbers from scientific notation to standard notation. Then they order the numbers from largest to smallest.

3. Players compare lists. The player who has the largest number wins. In case of a tie, they roll a fourth time.

Examples

Ann rolls: 2 and 4   5 and 3   1 and 6

writes: $2 * 10^4$ $3 * 10^5$ $1 * 10^6$

$= 2 * 10,000$ $= 3 * 100,000$ $= 1 * 1,000,000$

$= 20,000$ $= 300,000$ $= 1,000,000$

orders: 1,000,000; 300,000; 20,000

Keith rolls: 5 and 5   2 and 1   4 and 3

writes: $5 * 10^5$ $1 * 10^2$ $3 * 10^4$

$= 5 * 100,000$ $= 1 * 100$ $= 3 * 10,000$

$= 500,000$ $= 100$ $= 30,000$

orders: 500,000; 30,000; 100

Ann's largest number is greater than Keith's largest number. Ann wins.

Solution Search

Materials ☐ 1 deck of *Solution Search* Cards
(*Math Masters*, p. 473)

☐ 1 complete deck of number cards

Players 3 or 4

Skill Solving inequalities

Object of the game To be the first player to discard all cards.

Directions

1. Shuffle the deck of *Solution Search* cards and place it number-side down in the center of the table.

2. Shuffle the deck of number cards and deal 8 cards to each player. Place the remainder of the deck number-side down in the center of the table. Any player may begin the first round.

3. The player who begins a round turns over the top *Solution Search* card.

4. The player who begins a round also takes the first turn in that round. Play continues in a clockwise direction. The round is over when each player has had 1 turn.

5. When it is your turn:

   ◆ Discard any one of your number cards that is a solution of the inequality on the *Solution Search* card.

   ◆ If you do not have a number card that is a solution, continue to draw from the deck of number cards until a solution card is drawn. Then discard that solution card.

6. The player who takes the last turn in a round begins the next round. Follow Steps 3–5 to play another round.

7. When all the *Solution Search* cards have been used, turn the pile number-side down. Without shuffling, take the next card. When no more number cards remain, shuffle the discard pile and place it number-side down. Continue play.

8. The winner is the first player to discard all of his or her cards.

Solution Search Cards

$q * 2 > 20$	$m < 3.5$	$y^2 < 5$	$x > 9$
$b < 6$	$5 \neq s$	$100 / k > 25$	$(9 * z) + 2 > 65$
$49 \leq p^2$	$r / 2 \geq 5$	$w - 3 < 2$	$-2 + a \geq 5$
$\sqrt{25} \leq t$	$10 < 50 / d$	$c * 7 \leq 14$	$81 > f^2$

Variation

2s and 7s are special cards. 2s are WILD. A player may choose to play a 2 card with its given value of 2, or a player may assign any value they wish to the 2 card. The value of the 7 card is always 7. However, if a player plays the 7 card, the next player loses their turn.

Spoon Scramble

Materials ☐ one set of *Spoon Scramble* Cards
(*Math Journal 1,* Activity Sheet 1 or 2)

☐ 3 spoons

Players 4

Skill Fraction, decimal, and percent multiplication

Object of the game To avoid getting all the letters in SPOONS.

Advance Preparation Use all 16 cards from Activity Sheet 1.
Or, use all 16 cards from Activity Sheet 2.

Directions

1. Place the spoons in the center of the table.
2. The dealer shuffles the cards and deals 4 cards number-side down to each player.
3. Players look at their cards. If a player has 4 cards of equal value, proceed to Step 5 below. Otherwise, each player chooses a card to discard and passes it, number-side down, to the player on the left.
4. Each player picks up the new card and repeats Step 3. The passing of the cards should proceed quickly.
5. As soon as a player has 4 cards of equal value, the player places the cards number-side up on the table and grabs a spoon.
6. The other players then try to grab one of the 2 remaining spoons. The player left without a spoon is assigned a letter from the word *SPOONS,* starting with the first letter. If a player incorrectly claims to have 4 cards of equal value, that player receives a letter instead of the player left without a spoon.
7. A new round begins. Players put the spoons back in the center of the table. The dealer shuffles and deals the cards (Step 2 above).
8. Play continues until 3 players each get all the letters in the word *SPOONS.* The remaining player is the winner.

Variations

♦ For 3 players: Eliminate one set of 4 equivalent *Spoon Scramble* Cards. Use only 2 spoons.
♦ Players can make their own deck of *Spoon Scramble* Cards. Each player writes 4 computation problems that have equivalent answers on 4 index cards. Check to be sure the players have all chosen different values.

Spoon Scramble **Cards 1**
(Activity Sheet 1)

$\frac{1}{7}$ of 42	$\frac{24}{4} * \frac{5}{5}$	$\frac{54}{9}$	$2\frac{16}{4}$
$\frac{1}{5}$ of 35	$\frac{21}{3} * \frac{4}{4}$	$\frac{56}{8}$	$4\frac{36}{12}$
$\frac{1}{8}$ of 64	$\frac{48}{6} * \frac{3}{3}$	$\frac{32}{4}$	$3\frac{25}{5}$
$\frac{1}{4}$ of 36	$\frac{63}{7} * \frac{6}{6}$	$\frac{72}{8}$	$5\frac{32}{8}$

Spoon Scramble **Cards 2**
(Activity Sheet 2)

$1 \div 2$	$\frac{35}{70}$	$\frac{1}{8} * 4$	0.5
$\frac{1}{3}$	$\frac{1}{6} * 2$	$33\frac{1}{3}\%$	$\frac{1}{2} - \frac{1}{6}$
$\frac{26}{13}$	$(\frac{6}{9} * \frac{9}{6}) * 2$	2	$4 * \frac{1}{2}$
$\frac{3}{4}$	$\frac{600}{800}$	0.75	$3 \div 4$

three hundred thirty-three

Spreadsheet Scramble

Materials ☐ *Spreadsheet Scramble* Game Mat (*Math Masters*, p. 475)

Players 2

Skill Addition of positive and negative numbers

Object of the game To score more points.

Directions

1. Player 1 uses the positive numbers 1, 2, 3, 4, 5, and 6. Player 2 uses the negative numbers -1, -2, -3, -4, -5, and -6.

2. Player 1 begins the game. Players take turns writing one of their numbers in a cell within the 3-by-4 rectangle outlined on the spreadsheet. Once a player has written a number, it cannot be used again.

3. After all 12 numbers have been used, fill in Total cells F2, F3, and F4 by adding each row across. For example, F2 = B2 + C2 + D2 + E2. Fill in Total cells B5, C5, D5, and E5 by adding each column down. For example, C5 = C2 + C3 + C4.

4. Seven cells show row and column totals: F2, F3, F4, B5, C5, D5, and E5. Player 1 gets one point for each cell that contains a positive number. Player 2 gets one point for each cell that contains a negative number. Neither player gets a point for a cell that contains 0. The player with more points wins.

Spreadsheet Scramble Game Mat

	A	B	C	D	E	F
1						Total
2						
3						
4						
5	Total					

Example

Game 1:

Player 1 gets 1 point each for cells F3, F4, and C5.
Player 2 gets 1 point each for F2 and E5.
Player 1 wins the game, 3 points to 2 points.

	A	B	C	D	E	F
1						Total
2		-1	-6	3	-5	-9
3		4	2	-4	6	$+8$
4		-3	5	1	-2	$+1$
5	Total	0	$+1$	0	-1	

Game 2:

Player 1 gets 1 point each for F2, F4, and C5.
Player 2 gets 1 point each for F3, B5, and E5.
The game is a tie, 3 points to 3 points.

	A	B	C	D	E	F
1						Total
2		-4	6	4	-5	$+1$
3		-1	-3	-6	-2	-12
4		3	5	2	1	$+11$
5	Total	-2	$+8$	0	-6	

3-D Shape Sort

Materials ☐ 1 set of *3-D Shape Sort* Shape Cards
(*Math Masters,* p. 477)

☐ 1 set of *3-D Shape Sort* Property Cards
(*Math Masters,* p. 476)

Players 2, or two teams of 2

Skill Properties of 3-D shapes

Object of the game To collect more Shape Cards.

Directions

1. Write the letters V/E on the back of each Property Card that mentions "vertex" or "edge." Write the letters S/F on the back of each Property Card that mentions "surface" or "face."

2. Spread out the Shape Cards writing-side up on the table. Shuffle the Property Cards and sort them writing-side down into V/E-card and S/F-card piles.

3. Players take turns. When it is your turn:

   ◆ Draw the top card from each pile of Property Cards.

   ◆ Take all the Shape Cards that have **both** of the properties shown on the Property Cards.

   ◆ If there are no Shape Cards with both properties, draw 1 additional Property Card—either a V/E Card or an S/F Card. Look for Shape Cards that have the new property and one of the properties drawn before. Take those Shape Cards.

   ◆ At the end of a turn, if you have not taken a Shape Card that you could have taken, the other player may name and take it.

4. When all of the Property Cards in either pile have been drawn, shuffle *all* of the Property Cards. Sort them writing-side down into V/E-card and S/F-card piles. Continue play.

5. The game ends when there are fewer than 3 Shape Cards left.

6. The winner is the player with more Shape Cards.

Shape Cards

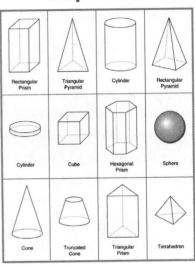

Property Cards
(writing-side up)

I have an even number of vertices.	I have no vertices.	I have at least 2 edges that are parallel to each other.	I have an odd number of edges.
One of my vertices is formed by an even number of edges.	I have at least one curved edge.	I have fewer than 6 vertices.	I have at least 2 edges that are perpendicular to each other.
All of my surfaces are polygons.	I have at least one face (flat surface).	I have at least one curved surface.	All of my faces are triangles.
All of my faces are regular polygons.	At least one of my faces is a circle.	I have at least one pair of faces that are parallel to each other.	**Wild Card:** Pick your own surface property.

Top-It Games

Materials ☐ number cards 1–10 (4 of each)

☐ 1 calculator (optional)

Players 2 to 4

Skill Multiplication and division facts

Object of the game To collect the most cards.

Multiplication Top-It

Directions

1. Shuffle the deck and place it number-side down.

2. Each player turns over 2 cards and calls out the product of the numbers. The player with the largest product takes all the cards. In case of a tie for the largest product, each tied player turns over 2 more cards and calls out the product of the numbers. The player with the largest product takes all the cards from both plays.

3. Check answers using a Multiplication Table or a calculator.

4. The game ends when there are not enough cards left for each player to have another turn. The player with the most cards wins.

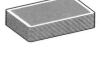

4 * 7 = 28

Variation

Use only the number cards 1–9. Each player turns over 3 cards, forms a 2-digit number, then multiplies the 2-digit number by the remaining number.

Division Top-It

Directions

1. Use only the number cards 1–9. Each player turns over 3 cards and uses them to generate a division problem as follows:

   ◆ Choose 2 cards to form the dividend.

   ◆ Use the remaining card as the divisor.

   ◆ Divide and drop any remainder.

2. The player with the largest quotient takes all the cards. In case of a tie for the largest quotient, repeat Step 1. The player with the largest quotient takes all the cards from both plays.

Advanced Version

Use only the number cards 1–9. Each player turns over 4 cards, chooses 3 of them to form a 3-digit number, then divides the 3-digit number by the remaining number.

Top-It Games with Positive and Negative Numbers

Materials ☐ 1 complete deck of number cards

☐ 1 calculator (optional)

Players 2 to 4

Skill Addition and subtraction of positive and negative numbers

Object of the game To collect the most cards.

Addition Top-It with Positive and Negative Numbers
Directions

The color of the number on each card tells you if a card is a positive number or a negative number.

- ♦ Black cards (spades and clubs) are *positive numbers*.

- ♦ Red cards (hearts and diamonds) or blue cards (Everything Math Deck) are *negative numbers*.

1. Shuffle the deck and place it number-side down.

2. Each player turns over 2 cards and calls out the sum of the numbers. The player with the largest sum takes all the cards.

3. In case of a tie, each tied player turns over 2 more cards and calls out the sum of the numbers. The player with the largest sum takes all the cards from both plays. If necessary, check answers with a calculator.

4. The game ends when there are not enough cards left for each player to have another turn. The player with the most cards wins.

Example

Lindsey turns over a red 3 and a black 6.

$-3 + 6 = 3$

Fred turns over a red 2 and a red 5.

$-2 + (-5) = -7$

$3 > -7$

Lindsey takes all 4 cards because 3 is greater than -7.

Variation

Each player turns over 3 cards and finds the sum.

Subtraction Top-It with Positive and Negative Numbers
Directions

The color of the number on each card tells you if a card is
a positive number or a negative number.

♦ Black cards (spades and clubs) are *positive numbers*.

♦ Red cards (hearts and diamonds) or blue cards
(Everything Math Deck) are *negative numbers*.

1. Shuffle the deck and place it number-side down.

2. Each player turns over 2 cards, one at a time, and subtracts
the second number from the first number. The player who
calls out the largest difference takes all the cards.

3. In case of a tie, each tied player turns over 2 more cards and
calls out the difference of the numbers. The player with the
largest answer takes all the cards from both plays. If
necessary, check answers with a calculator.

4. The game ends when there are not enough cards left for
each player to have another turn. The player with the most
cards wins.

Example

Lindsey turns over a black 2 first, then a red 3.

$+2 - (-3) = 5$

Fred turns over a red 5 first, then a black 8.

$-5 - (+8) = -13$

$5 > -13$

Lindsey takes all 4 cards because 5 is greater than -13.

Computer-generated art is created when a person develops
a computer program, runs the program, and then selects the
best works from all those generated by the computer. This is
quite different from **computer-aided art,** which involves
using computer software to create original art or to manipulate
images that have been loaded into the computer. There is
incredible variety in computer-generated art, both in the way
in which it is developed and in the types of images created.

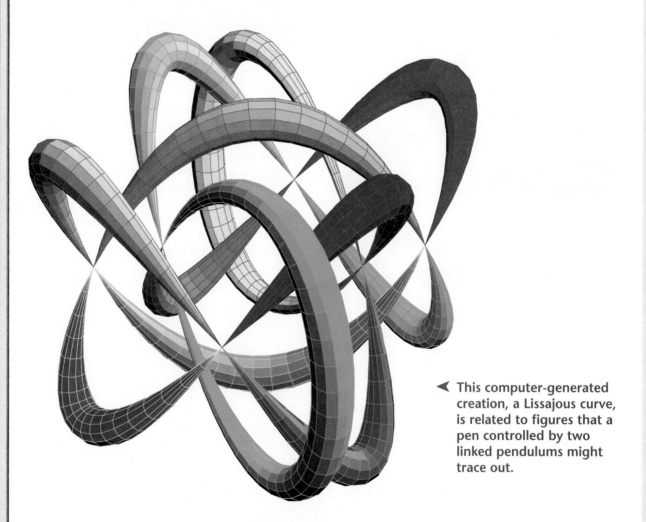

◀ This computer-generated
creation, a Lissajous curve,
is related to figures that a
pen controlled by two
linked pendulums might
trace out.

Optical Art

Computer-generated Optical Art uses the repetition of simple forms and colors to create the illusion of movement. Optical Art, also called *Op Art*, tricks the eye by creating a variety of visual effects such as foreground-background confusion and an exaggerated sense of depth.

This image was created by starting with the largest circle. Each successive circle is two units smaller in diameter than the previous one, and moved 45 degrees up and to the left of it. ➤

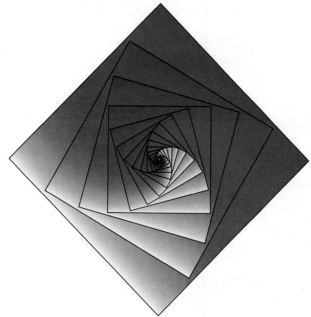

◀ This image was created by starting with the largest square. Successive squares were created so each new square was 90% the size of the one before it. Each new square was then rotated counter-clockwise and inscribed inside the preceding square. Finally, every second and third square was deleted so the squares would not touch each other.

This image was created by inscribing triangles inside triangles. Starting with the largest triangle, successive triangles were rotated 5 degrees clockwise and then reduced in size. ➤

Transformations

Any image that is drawn on the coordinate grid, or plane, is made up of points that can be named as ordered pairs. Some computer programs generate art by moving each ordered pair (x, y) on a coordinate grid to a new location (x', y') according to a rule. Such a rule is called a "transformation of the plane."

This image was created using a special type of rule, called a Mobius transformation, which moves circles into other circles. ➤

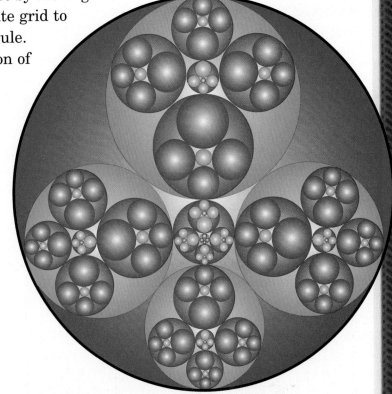

This image was created using rotation transformations. ▼

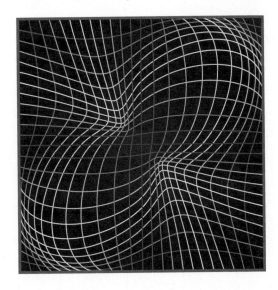

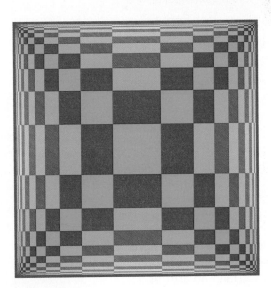

▲

The transformation that created this image moves grid lines on the plane to new locations. The original grid squares become rectangles of many different shapes and sizes.

Brushes and Paint

AARON, a computer system with its own painting robot, produces original art with brushes and paint. AARON was created by Harold Cohen, an artist and retired professor. Cohen believed that if he could teach a computer how to be an artist, rather than just an artistic tool, he would understand how humans learn to represent their world in drawings and pictures.

◄ To make a painting, AARON first uses artificial intelligence to create several drawings, or rough drafts. Then, Cohen selects one of the drawings and sends it to a computer that controls the robot painter.

AARON can only paint what it has been programmed to "know." For example, AARON knows the sizes of arms, legs, and other body parts. It also knows some rules for how a real body works. By following these rules, AARON will only attach body parts in ways that are physically possible. AARON won't attach an arm to a head, for example. ►

◄ AARON remembers everything it does in a painting. In fact, the robot spends at least half of its time building an internal "representation" of the image as it develops. AARON always needs to know where the lines are, where it has been painting, and what each little patch of a painting represents.

AARON's paintings have been exhibited ➤ in galleries and museums in the United States and Europe.

Repetition

One of the most exciting developments in computer-generated art is the use of iterations to create extremely detailed images. An **iteration** is a repetition of the same operation over and over again. The figures that result from infinitely many iterations are called **fractals.** A mathematician named Benoit Mandelbrot coined the word "fractal" from the Latin word frangere, which means "to break apart or fragment."

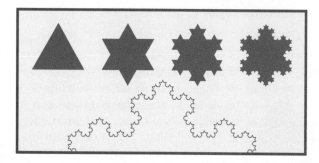

◀ In this series, each iteration adds smaller equilateral triangles, each with sides $\frac{1}{3}$ the length of the side of the larger triangle. After several iterations, the image looks like a shape known as the "Koch Snowflake."

This image was created by putting points through a special function machine that applies its rule over and over again. The color of a point in the image depends on what happens when the function machine applies its rule to that point infinitely many times. Look for patterns in this image that are repeated at different sizes. ▶

◀ Iteration can be used to create realistic and imaginary landscapes, including some that have been used in movies. Fractal landscapes are created using simple mathematical formulas that take up very little computer memory.

Art and Design
Activities

Perspective Drawing

Drawings are often used to represent real or imagined objects. Which of the drawings below do you think represents a cube?

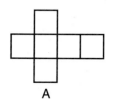
A

B

C

D

E

F

An ancient Egyptian drawing

A small child, an ancient Egyptian, or a modern French painter such as Marie Laurencin (1885–1956) might pick Figure A. It shows all six faces of the cube as squares—the way you might draw them after feeling the cube with your eyes closed. In a similar way, the ancient Egyptians showed the human figure from several different points of view at the same time.

Greek and Roman artists and mathematicians experimented with ways of making pictures that look like what we actually see. They noticed several things:

♦ An object moving away from the viewer appears to get smaller.

♦ Parallel lines moving away from the viewer seem to get closer.

After the Roman period, European artists returned to a flat style of painting. When they did try to show depth, they often made a drawing like Figure B above to represent a cube.

In the late Middle Ages, artists and scholars rediscovered the writings of the Greeks. They began their own experiments in representing 3-dimensional space. Their first attempts looked something like Figures C and D.

Many artists, especially those in Italy, continued to experiment with these ideas. They created more and more realistic images. In 1425, architect and engineer Filippo Brunelleschi (1377–1446) demonstrated what is now known as geometric **perspective** (also called linear perspective or Renaissance perspective). This method of drawing uses geometry to produce an illusion of depth, as if the flat surface of a drawing were a window looking onto a 3-dimensional scene.

Each part of the body was drawn from the direction that would make it most recognizable. The head, arms, and legs were drawn in profile (from the side), while the shoulders and chest were shown from the front.

Did You Know?

The term *perspective* is derived from the Latin *per* (meaning *through* or *forward*) and *specere* (meaning *to look*).

Brunelleschi discovered that to create a convincing illusion of three dimensions, objects should be drawn smaller when they are farther away from the viewer. Parallel lines moving away from the viewer should meet at a common point, called the **vanishing point.** Figure E on page 346 shows a cube drawn using this method. Later artists perfected a method using two vanishing points, as shown in Figure F on page 346.

In 1435, architect Leon Battista Alberti (1404–1472) wrote a book about painting that included detailed rules of perspective. Leonardo da Vinci (1452–1519) also experimented with perspective. Dissatisfied with Alberti's rules, he worked on a method of "natural" perspective that came closer to human vision. In his method, shapes were projected onto a curved surface, like the human eye, rather than onto the flat surface of a painting. These investigations and others that followed led to a branch of mathematics known as **projective geometry.**

The painting above was made in the twelfth century. It shows a flat style of painting and there is no illusion of depth. The ships in the distance (at the top of the painting) are the same size as the ones in the foreground (at the bottom of the painting).

This painting above, *A View of the Grand Canal* by Canaletto (1697–1768), shows a similar subject, this time using the perspective method to create an illusion of depth. The building and ships in the foreground are larger than those in the background. The lines in the buildings meet at a vanishing point (at the middle-right side of the painting).

When German artist Albrecht Dürer (1471–1528) visited Italy, he was exposed to new ideas circulating among artists and scholars. He later wrote a manual on painting that included the woodcut below. It shows an easy way to make a drawing with correct perspective. (Some of Dürer's best-known works are woodcuts, a type of print that is made in the same way as designs printed with a rubber stamp.)

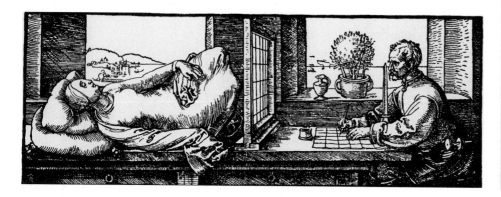

The artist looks through a grid made of threads set in a wooden frame. He copies what he sees onto a similar grid on paper. The object directly in front of the artist helps him keep his eye in the same position at all times.

Return to the question asked earlier: Which of the drawings below do you think represents a cube?

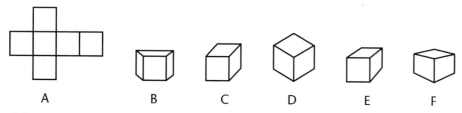

After reading the previous pages, your answer may be "They all do." Depending on your personal taste or how you want to use a drawing, any one of the above may be the "best" representation of a cube.

If you are interested in learning more about perspective drawing, look in an encyclopedia under topic headings such as *Mechanical Drawing, Greek Art, Human Perception, Medieval Art, Perspective, Projection, Projective Geometry, Renaissance Art,* and *Roman Art.*

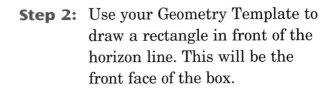

Drawing a Rectangular Solid

Follow these steps to draw a rectangular solid or box.

Step 1: Draw a horizontal line segment as the horizon line. Mark a point on the horizon line to be the **vanishing point.**

Step 2: Use your Geometry Template to draw a rectangle in front of the horizon line. This will be the front face of the box.

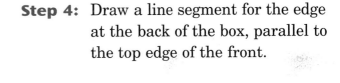

Step 3: Draw line segments from the four vertices of the rectangle to the vanishing point. These are called **vanishing lines.**

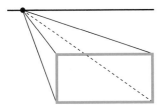

Step 4: Draw a line segment for the edge at the back of the box, parallel to the top edge of the front.

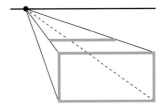

Step 5: Complete the top face of the box.

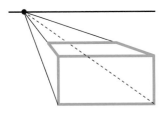

Step 6: Draw a side edge for the back of the box, parallel to the side edge of the front.

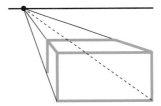

Step 7: Complete the side face of the box.

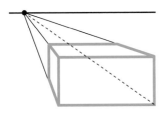

Drawing a Person

Materials   ☐ piece of cardboard measuring at least $8\frac{1}{2}$ inches by 11 inches

☐ transparency with a 7-inch square grid (from your teacher)

☐ masking or transparent tape

☐ scissors

☐ Geometry Template

☐ dark-colored transparency marker

Directions

Step 1:   Near the top of the piece of cardboard, draw a square with sides that are 5 inches long. Draw it so that its sides are at least 1 inch from the sides of the cardboard.

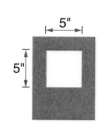

Step 2:   Cut out the square, leaving a 5-inch "window" in the cardboard.

Step 3:   Cut out the grid from the transparency. Be sure to cut along the dashed line segments. You will get a square piece about 7 inches on a side.

Step 4:   Tape the grid to the cardboard to cover the cardboard window.

Step 5:   Tape the bottom of the cardboard to a chair back so that the window is above the chair.

Step 6:   Make a perspective drawing of your partner.

Your partner sits in a chair with his or her legs on a desk or tabletop.

Place the chair with the attached cardboard 3 to 4 feet in front of your partner's feet.

Sit or kneel and look at your partner through the grid in the window. Adjust your distance away from the window. Your partner's feet and head should be completely visible and should fill up most of the window.

Artist adjusts distance from window so that view is similar to this.

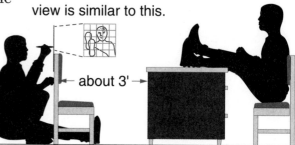

← about 3' →

Try to keep your head in the same position while drawing your partner. Use the transparency marker to trace an outline of your partner on the plastic window.

Drawing an Enlargement of a Picture

Use the following procedure to draw an enlargement.

Materials   ☐ a picture with simple shapes and lines, such as a comic-strip character
☐ 1-inch grid paper
☐ sharp pencil with eraser
☐ colored pencils, markers, or crayons

Directions

Step 1: Use a ruler and pencil to draw a grid pattern of $\frac{1}{2}$-inch squares onto the picture you chose.

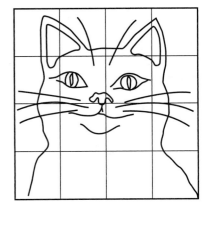

Step 2: Copy the part of the original picture in each grid square onto the corresponding grid square of the 1-inch grid. Notice where each line and shape begins and ends on the original grid, and locate its relative position on the 1-inch grid.

Try to reproduce all of the lines and shapes as accurately as you can. This method of drawing a copy is called **sighting.**

(Depending on the size of the picture you choose, you may need to tape several pieces of 1-inch grid paper together to create your enlargement.)

Step 3: After you complete the drawing in pencil, you may add color to fill in shapes or darken lines.

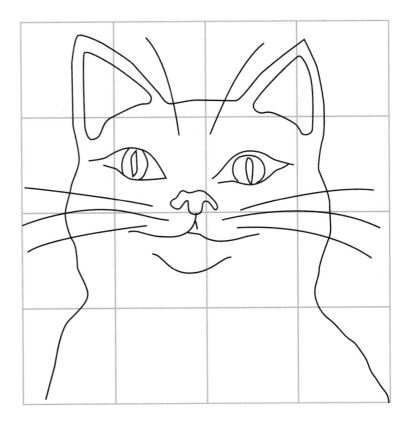

If you follow this procedure, you will draw a 2X enlargement of the original picture. Each length in the original picture will be doubled in the enlargement. The size-change factor depends on the size of the grid squares on the original picture and the size of the grid squares on the paper where you make your drawing.

The Golden Ratio

Which of the following rectangles do you like best?

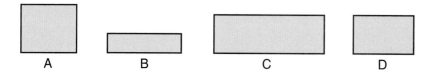

A   B   C   D

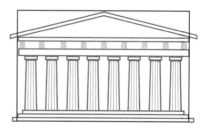

The Parthenon in Athens, Greece

It has been shown that Rectangle D, called the **Golden Rectangle,** is chosen more often than any other rectangle. In a Golden Rectangle, the ratio of the length of the longer side to the length of the shorter side is about 1.618 to 1. This ratio is known as the **Golden Ratio.**

The popularity of the Golden Ratio dates back to the ancient Greeks who used it in many of their works of art and architecture. For example, the front of the Parthenon in Athens, Greece fits almost exactly into a Golden Rectangle.

The symbol for the Golden Ratio is the Greek letter φ (phi). The letter *phi* was chosen to honor the famous Athenian sculptor Phidias (500–432 B.C.), who made frequent use of the Golden Ratio in his sculptures.

St. Jerome by Leonardo da Vinci

Throughout the ages, many artists have found that they could create a feeling of order in their works by using the Golden Ratio. For example, in the picture of St. Jerome, painted by Leonardo da Vinci in 1483, the figure of St. Jerome fits perfectly into a Golden Rectangle. It is believed that this was not just a coincidence, and that da Vinci used the Golden Ratio because of his great interest in mathematics.

The mask shown at the right was made in the Benin Kingdom in western Africa in the early sixteenth century. It was worn by the Oba, which means "king." The Oba was a sacred figure, and many ceremonies took place in his honor. If you measure the sides of the rectangles that frame some of the features in the mask, you will find that the ratio of the length of the longer side to the length of the shorter side is the Golden Ratio, about 1.618 to 1. Notice also that the mask is perfectly symmetric.

West African mask

A Classical Face

The Golden Ratio can be found in many sculptures made during the classical period of Greek art (about 480–350 B.C.). The picture at the right is a good example of a "classical Greek" face. It shows a sculpture of the head of the Greek goddess Hera. According to Greek mythology, Hera was queen of the Olympian gods and wife of the god Zeus. The sculpture was found in a temple in Argos, the second-oldest city in Greece. It was probably completed about 420 B.C. The sculpture is currently owned by the National Museum in Athens.

Nine different parts of the face are indicated on the picture shown below. By measuring various parts of the face, you will find many examples of the Golden Ratio in the sculpture.

Greek Goddess, Hera

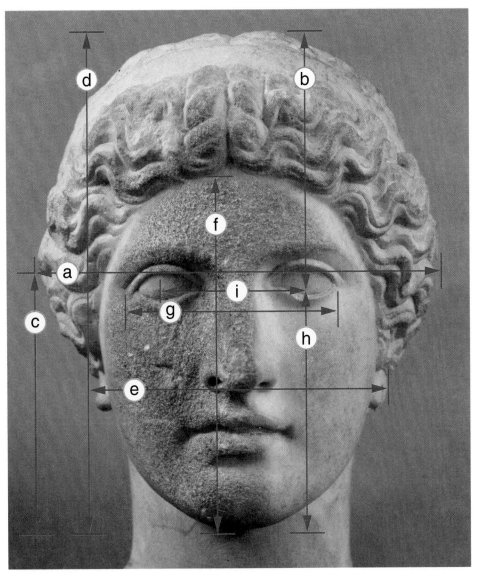

a = total width of head (including hair)

b = top of hair to pupils

c = top of eye to bottom of chin

d = top of hair to bottom of chin

e = distance from ear to ear

f = peak of hairline to bottom of chin

g = distance between outsides of eyes

h = pupil to chin

i = distance between centers of eyes

Constructing a Golden Rectangle

Step 1: Draw a square $ABCD$ on grid paper.

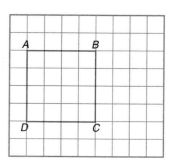

Step 2: Draw $\overline{EF}$ to divide the square in half.

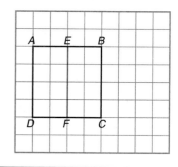

Step 3: Draw the diagonal $\overline{FB}$. Extend $\overline{DC}$.

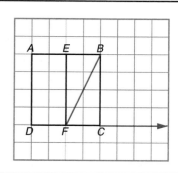

Step 4: Place the compass anchor at F and the pencil point at B. Draw an arc through point B that intersects $\overrightarrow{DC}$ at point G.

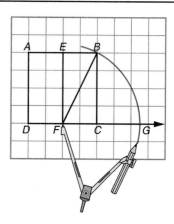

Step 5: Extend $\overline{AB}$. Draw a line segment that is perpendicular to $\overrightarrow{DC}$ at point G and that intersects the extension of $\overline{AB}$ at point H.

Rectangle $AHGD$ is a Golden Rectangle.

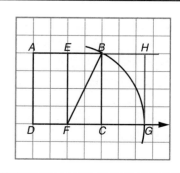

Rotation Symmetry

A Point-Symmetry Magic Trick

Setting the Stage Tell a friend that you can guess which playing card, from a set of four, she or he turns around while you are not looking.

Materials ☐ four cards from a regular deck of playing cards:

♦ 1 card with point symmetry

♦ 3 cards without point symmetry

Performing the Trick

Step 1: Place the four cards in a row, face up. Of the cards shown at the right, only the 10 of hearts has point symmetry. The other three cards do not.

Step 2: Turn your back to the cards, but first study the position of the figures on the cards. Tell your friend to rotate *one* of the cards 180 degrees.

Step 3: Turn around, study the cards, and tell which card your friend turned. If your friend has turned one of the three cards without point symmetry, it will be easy to determine that it was rotated. For example:

The middle spade is now pointing toward the bottom of the card.	The middle diamond is now at the bottom of the card.	The stem of the middle club is now pointing toward the bottom of the card.
original position 180° rotation	original position 180° rotation	original position 180° rotation

If none of the cards without point symmetry has been rotated, then the one card with point symmetry must have been rotated. *(See the margin.)*

This trick may seem simple, but that's because you know how it works. Many cards look almost the same before and after a 180° rotation.

original position   180° rotation

Drawing Shapes with Rotation Symmetry of a Given Order

Materials
☐ sharp pencil ☐ blank piece of paper
☐ scissors ☐ index card or other card stock
☐ protractor ☐ compass

Directions

Step 1: Draw a dot in the center of a blank piece of paper. Then use a pencil to *lightly* draw 3 segments from the center dot to the edge of the paper. The segments should form three 120° angles.

Step 2: Cut out any shape from an index card. Cut out a simple shape the first time you try this activity. Once you have learned the procedure, you can try a more complicated shape.

Step 3: Draw a line anywhere through the shape.

Step 4: Push the point of your compass through the shape at any point along the line you drew. Place the compass point at the point where the three segments intersect. Match the line on the shape with one of the segments.

Step 5: When the shape is lined up properly, remove the compass. Hold the shape in place, and trace around it. Replace the compass point, and rotate the shape so that the line on the shape lines up with the next segment. Remove the compass, and trace around the shape. Repeat the procedure for the third segment.

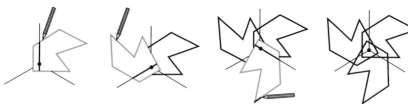

Step 6: Erase any interior lines. The outer edges of the three tracings form a shape with rotation symmetry of order 3. You can use tracing paper to copy your shape and check that it is symmetric.

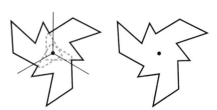

Use this procedure to create a shape with a rotation symmetry of order 2. Then create a shape with a rotation symmetry of order 5.

Tessellations

A **tessellation** is a pattern formed by repeated use of
polygons or other shapes that completely cover a surface
without any overlap. A tessellation is sometimes called a **tiling.**
A tessellation may use more than one shape.

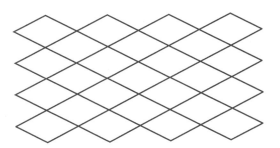 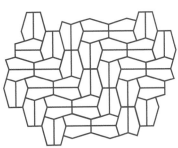 

♦ The shapes in a tessellation do not overlap.

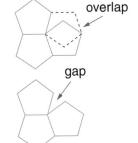

♦ There are no gaps between the shapes.

In a tessellation formed by polygons, a **vertex point**
is a point where vertices of the polygons meet.

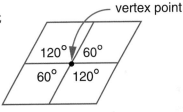

♦ The sum of the measures of the polygon angles
   around a vertex point must be exactly 360°.

120° + 60° + 120° + 60° = 360°

♦ If the sum is less than 360°, there
   will be gaps between the shapes.
   The pattern is *not* a tessellation.

♦ If the sum is greater than 360°,
   the shapes will overlap. The pattern
   is *not* a tessellation.

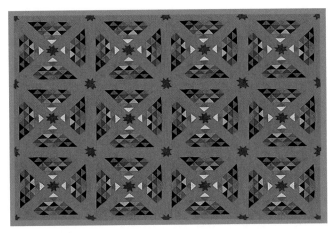

Many quilt designs use tessellations. Each of the
large triangles in this quilt is a tessellation made up
of smaller triangles.

Regular Tessellations

A tessellation made by repeating congruent copies of *one* regular polygon is called a **regular tessellation.** There are exactly three possible regular tessellations, and they are all shown below.

using congruent squares

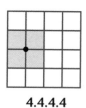

4.4.4.4

using congruent regular hexagons

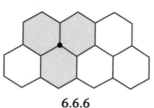

6.6.6

using congruent equilateral triangles

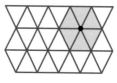

3.3.3.3.3.3

Regular tessellations are named by listing the number of sides of the polygons around any vertex point. The numbers are separated by periods. For example, the name of the hexagon tessellation above is 6.6.6 because there are three 6-sided polygons around each vertex point.

Semiregular Tessellations

Tessellations may involve more than one type of shape. (See at the right.) A tessellation made by repeating congruent copies of *two or more* different regular polygons is called a **semiregular tessellation** if the following conditions hold.

♦ If the vertex of one polygon meets another polygon, the meeting point must be a vertex of both polygons.

♦ The same polygons, in the same order, surround each vertex point.

Semiregular tessellations are named by listing the number of sides of the polygons around any vertex point. You start with the polygon with the least number of sides and then list the number of sides of each polygon as you move clockwise around the vertex point. The name of the octagon-square tessellation in the margin above is 4.8.8.

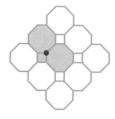

4.8.8

As you move clockwise around any vertex point, there is 1 square and 2 octagons.

Note

There are exactly eight possible semiregular tessellations. One of these is shown above, and three are shown below.

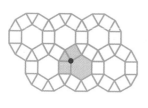

3.4.6.4

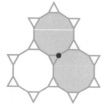

3.12.12

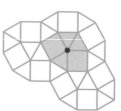

3.3.4.3.4

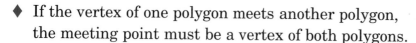
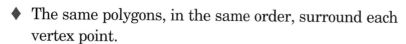

Escher Tessellations

The graphic artist M. C. Escher (Maurits Cornelius Escher; (1898–1972)) was born in a small town in the Netherlands. He became famous for his drawings that combined realistic details with optical illusions and distorted perspectives. The drawing at the right, *Hand with Reflecting Sphere,* is a self-portrait of Escher.

In 1936, Escher visited the Alhambra, a Moorish palace in Spain that was built in the thirteenth and fourteenth centuries. The Moors were Muslims from North Africa who invaded Spain in the eighth century and occupied it until 1492. They often used geometric tessellations in their art and architecture, and Escher was fascinated by the beautiful tiling patterns that covered the floors and walls of the Alhambra.

These designs inspired Escher to create tessellations like the one shown below. Unlike the Islamic artists who decorated the Alhambra, Escher did not limit himself to purely geometric designs. He built tessellations from representations of objects such as birds, fish, reptiles, and humans. Escher used translations (slides), reflections (flips), and rotations (turns) to create unusual and fantastic designs.

**Hand with Reflecting Sphere
by Escher**

**A seahorse tessellation
by Escher**

**Tessellation on a wall of
the Alhambra**

Creating an Escher-Type Translation Tessellation

Materials
- ☐ 3-inch by 3-inch square cut from card stock (such as an index card)
- ☐ large piece of white construction paper
- ☐ markers, crayons, or colored pencils
- ☐ sharp pencil
- ☐ scissors
- ☐ tape

Directions

Step 1: To create a template for your **translation tessellation,** begin with a 3-inch by 3-inch square of card stock.

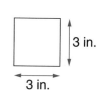

Step 2: Draw a curve from point A to point B.

Step 3: Use scissors to cut along curve AB. Tape the cutout edge of the square so point A lines up with point D and point B lines up with point C.

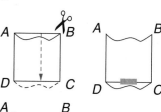

Step 4: Draw a curve from point A to point D.

Step 5: Use scissors to cut along curve AD. Tape the cutout edge of the square so point A lines up with point B and point D lines up with point C.

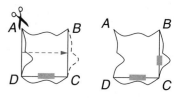

Step 6: You now have a template for your tessellation. Begin by tracing your template onto the *center* of the construction paper. Continue tracing, after interlocking your template with a previous tracing, until you have filled the entire sheet.

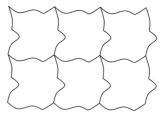

Step 7: Use markers, crayons, or colored pencils to decorate your design.

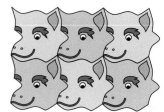

M.C. Escher worked hard to create shapes that not only tessellated but also looked like birds, reptiles, insects, and other familiar objects. You may want to repeat Steps 1 through 5 until you create a recognizable shape.

Constructing and Experimenting with Möbius Strips

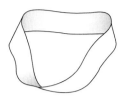

*A mathematician confided
That a Möbius band is one-sided.
And you'll get quite a laugh,
If you cut one in half,
For it stays in one piece when divided.*

August Möbius

The above limerick was inspired by the work of August Ferdinand Möbius (1790–1868), a German mathematician and astronomer. Möbius examined the properties of one-sided surfaces. One such surface, easily made from a strip of paper, became known as a **Möbius strip,** or **Möbius band.** Möbius strips are studied in the branch of mathematics known as **topology.**

You may think that Möbius strips would only interest mathematicians and magicians. However, they also have practical uses. For example, Möbius strips have been used in the design of drive belts, such as fan belts and conveyor belts.

2-sided drive belt (above) and 1-sided drive belt (below)

Friction would wear out an ordinary two-sided belt more quickly on the inside than on the outside. If a belt with a half-twist (a Möbius strip) is used, it wears more evenly and slowly because it has only one side.

Möbius strips are also recognized for their artistic properties. The artist M.C. Escher was intrigued not only by tessellations but also by Möbius strips. In his work *Möbius Strip II,* he depicts nine red ants endlessly crawling along a Möbius strip.

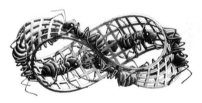

Escher's Möbius Strip II

Construct a Möbius strip. Cut a strip of a newspaper about $1\frac{1}{2}$ inches wide and as long as possible. Turn over one end of the strip (give one end a half-twist), and tape the two ends together to form a loop.

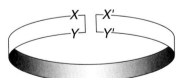

Now, poke your scissors through the paper, and cut the strip lengthwise all the way around. What happens? Is the limerick at the top of the page correct?

Designing a Paper Airplane

The First International Paper Airplane Competition

The First International Paper Airplane Competition was held in 1967 and was sponsored by *Scientific American* magazine. The 11,851 entries from 28 different countries were original designs for paper airplanes. (About 5,000 children submitted entries.) These paper airplanes were entered into one or more of the following categories:

◆ duration aloft (The winning designs spent 9.9 and 10.2 seconds in the air.)

◆ distance flown (The winning designs flew 58 feet 2 inches and 91 feet 6 inches.)

◆ aerobatics (stunts performed in flight)

◆ origami (the traditional Japanese art or technique of folding paper into a variety of decorative or representational forms)

Contestants were permitted to use paper of any weight and size. The smallest entry received, which was entered in the distance category, measured 0.08 inch by 0.00003 inch. However, this entry was found to be made of foil, not paper. The largest entry received, also entered in the distance category, was 11 feet long. It flew two times its length when tested.

Scientific American awarded winners' trophies to two designers, a professional and a nonprofessional, in each category. Professionals were "people employed in the air travel business and people who build nonpaper airplanes." Each winner received a trophy called *The Leonardo,* named after Leonardo da Vinci (1452–1519), whom *Scientific American* refers to as the "Patron Saint of Paper Airplanes."

Da Vinci, known for many accomplishments in the fields of painting and sculpture, was also an architect, engineer, and inventive builder. Studying the flight of birds, da Vinci believed that it would be possible to build a flying machine that would enable humans to soar through the air. He designed several wing-flapping machines, suggested the use of rotating wings similar to those of the modern helicopter, and invented the "air screw," similar to the modern propeller, to pull a machine through the air.

Note

More information about the First International Paper Airplane Competition, as well as templates and directions for making each of the winning designs, can be found in *The Great International Paper Airplane Book* by Jerry Mander, George Dippel, and Howard Gossage; Galahad Books, 1998.

The Leonardo

Leonardo da Vinci

A Winning Paper Airplane Design

The design plan shown below was submitted by Louis W. Schultz, an engineer. Schultz's paper airplane flew 58 feet 2 inches and was a winner in the distance category for nonprofessionals. The professional winner in the distance category was Robert Meuser. His paper airplane flew 91 feet 6 inches before it hit the rear wall of the testing site.

1. Follow the directions below to make an accurate copy of Schultz's design plan on an $8\frac{1}{2}$-inch by 11-inch sheet of paper.

   a. Use a ruler to find the midpoints at the top and bottom of the paper. Mark these points. Draw a line connecting the midpoints.

   b. Mark two points that are $\frac{1}{4}$ inch away from the midpoint at the top of the paper.

   c. Use a protractor to make two 45° angles as shown.

   d. Use a protractor to make two 82° angles as shown.

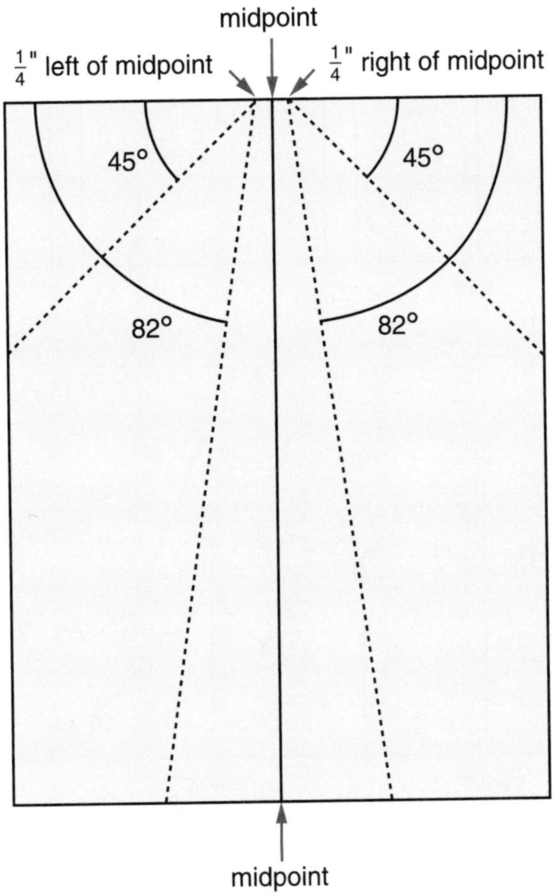

midpoint

$\frac{1}{4}$" left of midpoint $\frac{1}{4}$" right of midpoint

45°   45°

82°   82°

Louis W. Schultz's paper airplane design plan

midpoint

three hundred sixty-three

2. Assemble the paper airplane as shown below. Be very careful to make precise folds. Make the folds on a table. When making a fold, first press down on the paper with your finger. Then, go over this fold with a pen or a ruler on its side. **Do not** use your fingernails to make folds.

a. Fold the paper back and forth along the center line to make a sharp crease. Then unfold.

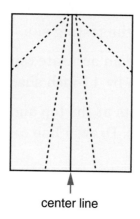

center line

b. Fold the corners along the dashed lines as shown. Use a small piece of tape to secure each corner as shown in the sketch.

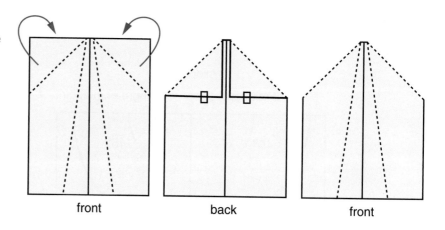

front   back   front

c. With the back of the paper facing you, fold the top-right side of the paper toward the center so that the edges highlighted in the sketch meet. Use a small piece of tape to secure the flap in the position shown in the sketch. Do the same to the other side.

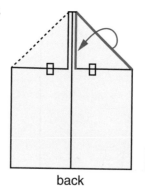
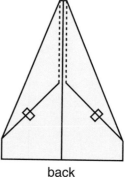
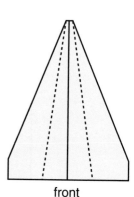

back   back   front

d. Flip the paper to the front side. Fold it in half along the center line so that the front side is now on the inside. Your paper airplane should now look like this:

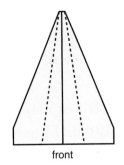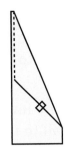

front

e. Take one of the top flaps of the paper and fold it outward along the dashed line. (Look for the dashed line on the inside of the plane.) Do the same to the other flap. Your paper airplane should now look like this:

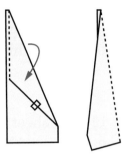

f. First, tape the wings together on top of the airplane. Then, tape the bottom as shown, making sure that all loose flaps are secured.

top view

bottom view

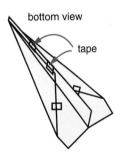

tape

Air is a real substance, just as water, earth, and maple syrup are real substances. Because air is a substance, it offers **resistance,** or opposition, to the movement of objects through it.

Imagine dropping a penny into a bottle of maple syrup. The penny will eventually fall to the bottom of the bottle, but the maple syrup will slow its progress; the maple syrup offers resistance to the movement of the penny. Air works in much the same way; objects can move through it, but the air offers resistance to the movement of those objects.

Did you know that this resistance can serve a helpful purpose? Try the following experiments to see how resistance can be used to help an object, such as an airplane, move through the air efficiently.

The "Kite Effect"

♦ Hold one end of an $8\frac{1}{2}$-inch by 11-inch sheet of paper as illustrated—forefinger on top, supported by the thumb and second finger on the bottom. Notice that the paper in the illustration is tilted slightly so that the opposite end of the paper is a bit higher than the end you are holding.

♦ *Push* the paper directly forward as illustrated.

You will notice that the end of the paper that is opposite the end you are holding tilts up. When the tilted surface of the paper pushes against the air, the air pushes back. This partially slows the paper down and partially lifts it up.

The sheet of paper has some of the characteristics of an airplane wing. The wing of an airplane is set at an angle so that its front edge is higher than its back edge. In this way, the lower surface of the airplane wing uses the air resistance to achieve a small amount of lift.

The "Vacuum Effect"

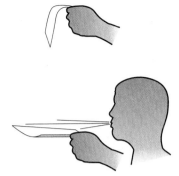

◆ Hold the small end of a 2-inch by 6-inch strip of paper between your thumb and forefinger as illustrated—thumb on top. The paper should fall forward in a curve.

◆ Blow over the top of the paper as illustrated.

As you blow over the top of the paper, you will notice that the end of the paper opposite the end you are holding tilts up. Air rushing over the upper surface of the paper causes the air pressure on that surface to decrease. When the air pressure on the upper surface is less than the air pressure on the lower surface, the higher pressure underneath lifts the paper.

The strip of paper has some characteristics of an airplane wing. Only the lower surface of an airplane wing is flat. The upper surface is curved or arched, like the paper when you are not blowing over it. In this way, the upper surface of the airplane wing creates a pressure difference that lifts the plane.

Both the "kite effect" and the "vacuum effect" contribute to the total lift of an airplane. The "vacuum effect" is responsible for about 80% of the total lift.

Additional Sources of Information about Paper Airplanes

If you are interested in finding paper airplane designs, visit this Web site:

http://bestpaperairplanes.com This site allows you to print out directions for folding a variety of planes.

If you are interested in the Second Great International Paper Airplane Contest held in 1985, visit the following site:

http://www.yale.edu/ynhti/curriculum/units/1988/6/ 88.06.02.x.html

Keep in mind that Web sites come and go. The ones cited here may no longer exist, but there will probably be many new ones to take their place.

You can find many additional Web sites of interest by simply searching the Internet under the topic "paper airplanes."

Note

For more information about paper airplanes, read *The World Record Paper Airplane Book* by Ken Blackburn and Jeff Lammers, Workman Publishing Company, 1994. This book contains designs and instructions for folding and fine-tuning paper airplanes, details on throwing techniques, and ideas on the best places to fly paper airplanes.

How to Balance a Mobile

A **mobile** is a sculpture constructed of rods and other objects that are suspended in midair by wire, twine, or thread. The rods and objects are connected in such a way that the sculpture is balanced when it is suspended. The rods and objects move independently when they are stirred by air currents.

The simplest mobile consists of a single rod with two objects hanging from it. The point at which the rod is suspended is called the **fulcrum.** The fulcrum may be at any point on the rod. Objects may be hung at the ends of the rod or at points between the ends of the rod and the fulcrum.

Suppose the fulcrum is the center point of the rod and you hang one object on each side of the fulcrum.

Let W = the weight of one object

D = the distance of this object from the fulcrum

w = the weight of the second object

d = the distance of this object from the fulcrum

The mobile will balance if $W * D = w * d$.

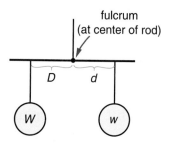

Example   The mobile below is balanced. What is the missing distance, x?

Replace the variables in the formula $W * D = w * d$ with the values shown in the diagram. Then, solve the equation.

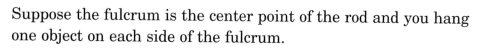

$W = 6$ $D = 7$ $w = 10$ $d = x$

Solution: $W * D = w * d$

$6 * 7 = 10 * x$

$42 = 10 * x$

$4.2 = x$

Therefore, the distance to the fulcrum is 4.2 units. $x = 4.2$ units.

Check Your Understanding

Decide whether the mobiles are in balance.

1.
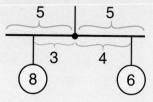

2.
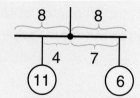

Suppose the fulcrum is *not* the center point of the rod and you hang one object on each side of the fulcrum.

Let R = the weight of the rod

$\quad L$ = the distance from the center of the rod to the fulcrum

$\quad W$ = the weight of the object that is on the same side of the fulcrum as the center of the rod

$\quad D$ = the distance of this object from the fulcrum

$\quad w$ = the weight of the other object

$\quad d$ = the distance of this object from the fulcrum

The mobile will balance if $(W * D) + (R * L) = w * d$.

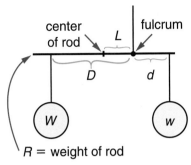

R = weight of rod

Example   The mobile below is balanced. What is the missing weight, $5x$?

Replace the variables in the formula $(W * D) + (R * L) = w * d$ with the values shown in the diagram. Then, solve the equation.

$$R = 25 \quad L = 2 \quad W = 7 \quad D = 10 \quad w = 5x \quad d = 6$$

Solution :

$$(W * D) + (R * L) = w * d$$
$$(7 * 10) + (25 * 2) = 5x * 6$$
$$70 + 50 = 30x$$
$$120 = 30x$$
$$4 = x$$

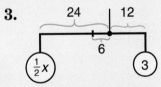

weight of rod = 25

Since $x = 4$, $5x = 5 * 4 = 20$.

Therefore, the weight of the object suspended to the right of the fulcrum is 20 units.

Check Your Understanding

In Problems 1 and 2, decide whether the mobiles are balanced. In Problem 3, find the weight of the object on the left of the fulcrum if the mobile is balanced.

1.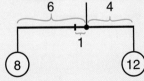

weight of rod = 5

2.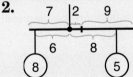

weight of rod = 4

3.

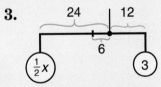

weight of rod = 2

Check your answers on page 424.

Place-Value Chart

	billions	100 millions	10 millions	millions	100 thousands	10 thousands	thousands	hundreds	tens	ones	.	tenths	hundredths	thousandths
	1,000 millions	100,000,000s	10,000,000s	1,000,000s	100,000s	10,000s	1,000s	100s	10s	1s	.	0.1s	0.01s	0.001s
	10^9	10^8	10^7	10^6	10^5	10^4	10^3	10^2	10^1	10^0	.	10^{-1}	10^{-2}	10^{-3}

Prefixes

uni-. . . . . one	tera- . . .trillion (10^{12})
bi-. . . . . two	giga- . . .billion (10^9)
tri-. . . . . three	mega- . .million (10^6)
quad- . . . . four	kilo- . . . .thousand (10^3)
penta-. . . . . . . five	hecto- . .hundred (10^2)
hexa- . . . . six	deca- . . .ten (10^1)
hepta-. . . . . . . seven	uni- . . . .one (10^0)
octa-. . . . . eight	deci- . . .tenth (10^{-1})
nona- . . . . nine	centi- . . .hundredth (10^{-2})
deca-. . . . . ten	milli- . . .thousandth (10^{-3})
dodeca- . . . . . twelve	micro- . .millionth (10^{-6})
icosa- . . . . twenty	nano- . .billionth (10^{-9})

Multiplication and Division Table

*,/	1	2	3	4	5	6	7	8	9	10	11	12
1	1	2	3	4	5	6	7	8	9	10	11	12
2	2	4	6	8	10	12	14	16	18	20	22	24
3	3	6	9	12	15	18	21	24	27	30	33	36
4	4	8	12	16	20	24	28	32	36	40	44	48
5	5	10	15	20	25	30	35	40	45	50	55	60
6	6	12	18	24	30	36	42	48	54	60	66	72
7	7	14	21	28	35	42	49	56	63	70	77	84
8	8	16	24	32	40	48	56	64	72	80	88	96
9	9	18	27	36	45	54	63	72	81	90	99	108
10	10	20	30	40	50	60	70	80	90	100	110	120
11	11	22	33	44	55	66	77	88	99	110	121	132
12	12	24	36	48	60	72	84	96	108	120	132	144

The numbers on the diagonal are square numbers.

Rules for Order of Operations

1. Do operations within parentheses or other grouping symbols before doing anything else.
2. Calculate all powers.
3. Do multiplications and divisions in order, from left to right.
4. Then do additions and subtractions in order, from left to right.

Metric System

Units of Length

1 kilometer (km) = 1,000 meters (m)

1 meter = 10 decimeters (dm)

= 100 centimeters (cm)

= 1,000 millimeters (mm)

1 decimeter = 10 centimeters

1 centimeter = 10 millimeters

Units of Area

1 square meter (m^2) = 100 square decimeters (dm^2)

= 10,000 square centimeters (cm^2)

1 square decimeter = 100 square centimeters

1 square kilometer = 1,000,000 square meters

Units of Volume

1 cubic meter (m^3) = 1,000 cubic decimeters (dm^3)

= 1,000,000 cubic centimeters (cm^3)

1 cubic decimeter = 1,000 cubic centimeters

Units of Capacity

1 kiloliter (kL) = 1,000 liters (L)

1 liter = 1,000 milliliters (mL)

1 cubic centimeter = 1 milliliter

Units of Weight

1 metric ton (t) = 1,000 kilograms (kg)

1 kilogram = 1,000 grams (g)

1 gram = 1,000 milligrams (mg)

U.S. Customary System

Units of Length

1 mile (mi) = 1,760 yards (yd)

= 5,280 feet (ft)

1 yard = 3 feet

= 36 inches (in.)

1 foot = 12 inches

Units of Area

1 square yard (yd^2) = 9 square feet (ft^2)

= 1,296 square inches ($in.^2$)

1 square foot = 144 square inches

1 acre = 43,560 square feet

1 square mile (mi^2) = 640 acres

Units of Volume

1 cubic yard (yd^3) = 27 cubic feet (ft^3)

1 cubic foot = 1,728 cubic inches ($in.^3$)

Units of Capacity

1 gallon (gal) = 4 quarts (qt)

1 quart = 2 pints (pt)

1 pint = 2 cups (c)

1 cup = 8 fluid ounces (fl oz)

1 fluid ounce = 2 tablespoons (tbs)

1 tablespoon = 3 teaspoons (tsp)

Units of Weight

1 ton (T) = 2,000 pounds (lb)

1 pound = 16 ounces (oz)

System Equivalents

1 inch is about 2.5 cm (2.54)

1 kilometer is about 0.6 mile (0.621)

1 mile is about 1.6 kilometers (1.609)

1 meter is about 39 inches (39.37)

1 liter is about 1.1 quarts (1.057)

1 ounce is about 28 grams (28.350)

1 kilogram is about 2.2 pounds (2.205)

Units of Time

1 century = 100 years

1 decade = 10 years

1 year (yr) = 12 months

= 52 weeks (plus one or two days)

= 365 days (366 days in a leap year)

1 month (mo) = 28, 29, 30, or 31 days

1 week (wk) = 7 days

1 day (d) = 24 hours

1 hour (hr) = 60 minutes

1 minute (min) = 60 seconds (sec)

Decimal and Percent Equivalents for "Easy" Fractions

"Easy" Fractions	Decimals	Percents
$\frac{1}{2}$	0.50	50%
$\frac{1}{3}$	$0.\overline{3}$	$33\frac{1}{3}\%$
$\frac{2}{3}$	$0.\overline{6}$	$66\frac{2}{3}\%$
$\frac{1}{4}$	0.25	25%
$\frac{3}{4}$	0.75	75%
$\frac{1}{5}$	0.20	20%
$\frac{2}{5}$	0.40	40%
$\frac{3}{5}$	0.60	60%
$\frac{4}{5}$	0.80	80%
$\frac{1}{6}$	$0.1\overline{6}$	$16\frac{2}{3}\%$
$\frac{5}{6}$	$0.8\overline{3}$	$83\frac{1}{3}\%$
$\frac{1}{8}$	0.125	$12\frac{1}{2}\%$
$\frac{3}{8}$	0.375	$37\frac{1}{2}\%$
$\frac{5}{8}$	0.625	$62\frac{1}{2}\%$
$\frac{7}{8}$	0.875	$87\frac{1}{2}\%$
$\frac{1}{10}$	0.10	10%
$\frac{3}{10}$	0.30	30%
$\frac{7}{10}$	0.70	70%
$\frac{9}{10}$	0.90	90%

The Global Grid

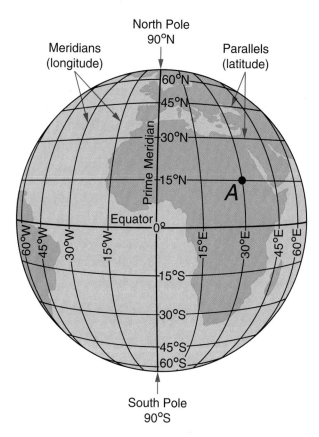

Point *A* is located at 15°N, 30°E.

Fraction-Decimal Number Line

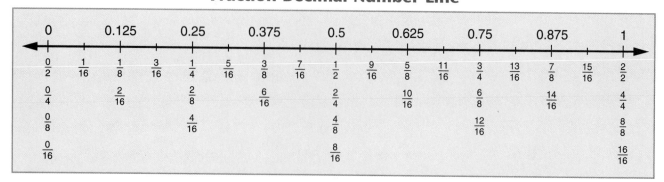

Fraction-Stick and Decimal Number-Line Chart

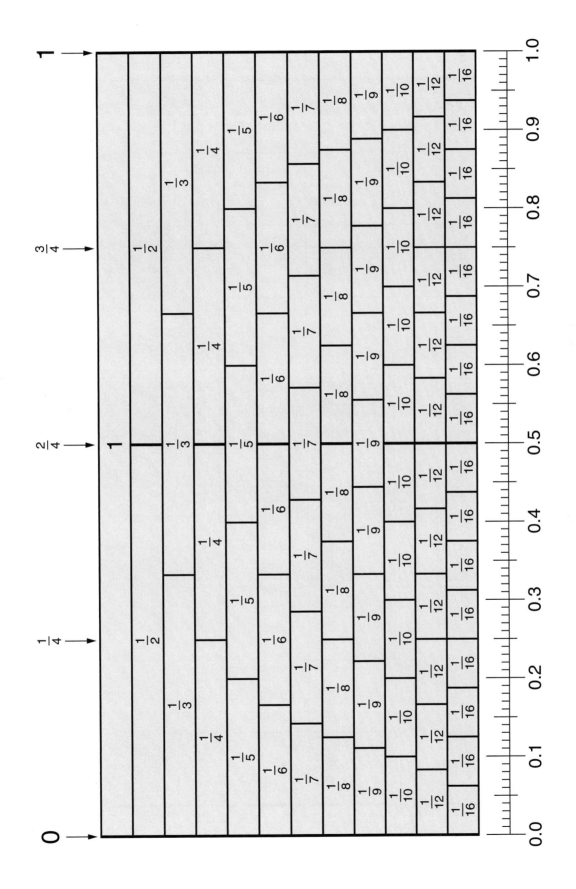

Table of Decimal Equivalents for Fractions

	Numerator									
Denominator	**1**	**2**	**3**	**4**	**5**	**6**	**7**	**8**	**9**	**10**
1	1.0	2.0	3.0	4.0	5.0	6.0	7.0	8.0	9.0	10.0
2	0.5	1.0	1.5	2.0	2.5	3.0	3.5	4.0	4.5	5.0
3	$0.\overline{3}$	$0.\overline{6}$	1.0	$1.\overline{3}$	$1.\overline{6}$	2.0	$2.\overline{3}$	$2.\overline{6}$	3.0	$3.\overline{3}$
4	0.25	0.5	0.75	1.0	1.25	1.5	1.75	2.0	2.25	2.5
5	0.2	0.4	0.6	0.8	1.0	1.2	1.4	1.6	1.8	2.0
6	$0.1\overline{6}$	$0.\overline{3}$	0.5	$0.\overline{6}$	$0.8\overline{3}$	1.0	$1.1\overline{6}$	$1.\overline{3}$	1.5	$1.\overline{6}$
7	$0.\overline{142857}$	$0.\overline{285714}$	$0.\overline{428571}$	$0.\overline{571428}$	$0.\overline{714285}$	$0.\overline{857142}$	1.0	$1.\overline{142857}$	$1.\overline{285714}$	$1.\overline{428571}$
8	0.125	0.25	.375	0.5	0.625	0.75	0.875	1.0	1.125	1.25
9	$0.\overline{1}$	$0.\overline{2}$	$0.\overline{3}$	$0.\overline{4}$	$0.\overline{5}$	$0.\overline{6}$	$0.\overline{7}$	$0.\overline{8}$	1.0	$1.\overline{1}$
10	0.1	0.2	0.3	0.4	0.5	0.6	0.7	0.8	0.9	1.0

Equivalent Fractions, Decimals, and Percents

															Decimal	Percent
$\frac{1}{2}$	$\frac{2}{4}$	$\frac{3}{6}$	$\frac{4}{8}$	$\frac{5}{10}$	$\frac{6}{12}$	$\frac{7}{14}$	$\frac{8}{16}$	$\frac{9}{18}$	$\frac{10}{20}$	$\frac{11}{22}$	$\frac{12}{24}$	$\frac{13}{26}$	$\frac{14}{28}$	$\frac{15}{30}$	0.5	50%
$\frac{1}{3}$	$\frac{2}{6}$	$\frac{3}{9}$	$\frac{4}{12}$	$\frac{5}{15}$	$\frac{6}{18}$	$\frac{7}{21}$	$\frac{8}{24}$	$\frac{9}{27}$	$\frac{10}{30}$	$\frac{11}{33}$	$\frac{12}{36}$	$\frac{13}{39}$	$\frac{14}{42}$	$\frac{15}{45}$	$0.\overline{3}$	$33\frac{1}{3}\%$
$\frac{2}{3}$	$\frac{4}{6}$	$\frac{6}{9}$	$\frac{8}{12}$	$\frac{10}{15}$	$\frac{12}{18}$	$\frac{14}{21}$	$\frac{16}{24}$	$\frac{18}{27}$	$\frac{20}{30}$	$\frac{22}{33}$	$\frac{24}{36}$	$\frac{26}{39}$	$\frac{28}{42}$	$\frac{30}{45}$	$0.\overline{6}$	$66\frac{2}{3}\%$
$\frac{1}{4}$	$\frac{2}{8}$	$\frac{3}{12}$	$\frac{4}{16}$	$\frac{5}{20}$	$\frac{6}{24}$	$\frac{7}{28}$	$\frac{8}{32}$	$\frac{9}{36}$	$\frac{10}{40}$	$\frac{11}{44}$	$\frac{12}{48}$	$\frac{13}{52}$	$\frac{14}{56}$	$\frac{15}{60}$	0.25	25%
$\frac{3}{4}$	$\frac{6}{8}$	$\frac{9}{12}$	$\frac{12}{16}$	$\frac{15}{20}$	$\frac{18}{24}$	$\frac{21}{28}$	$\frac{24}{32}$	$\frac{27}{36}$	$\frac{30}{40}$	$\frac{33}{44}$	$\frac{36}{48}$	$\frac{39}{52}$	$\frac{42}{56}$	$\frac{45}{60}$	0.75	75%
$\frac{1}{5}$	$\frac{2}{10}$	$\frac{3}{15}$	$\frac{4}{20}$	$\frac{5}{25}$	$\frac{6}{30}$	$\frac{7}{35}$	$\frac{8}{40}$	$\frac{9}{45}$	$\frac{10}{50}$	$\frac{11}{55}$	$\frac{12}{60}$	$\frac{13}{65}$	$\frac{14}{70}$	$\frac{15}{75}$	0.2	20%
$\frac{2}{5}$	$\frac{4}{10}$	$\frac{6}{15}$	$\frac{8}{20}$	$\frac{10}{25}$	$\frac{12}{30}$	$\frac{14}{35}$	$\frac{16}{40}$	$\frac{18}{45}$	$\frac{20}{50}$	$\frac{22}{55}$	$\frac{24}{60}$	$\frac{26}{65}$	$\frac{28}{70}$	$\frac{30}{75}$	0.4	40%
$\frac{3}{5}$	$\frac{6}{10}$	$\frac{9}{15}$	$\frac{12}{20}$	$\frac{15}{25}$	$\frac{18}{30}$	$\frac{21}{35}$	$\frac{24}{40}$	$\frac{27}{45}$	$\frac{30}{50}$	$\frac{33}{55}$	$\frac{36}{60}$	$\frac{39}{65}$	$\frac{42}{70}$	$\frac{45}{75}$	0.6	60%
$\frac{4}{5}$	$\frac{8}{10}$	$\frac{12}{15}$	$\frac{16}{20}$	$\frac{20}{25}$	$\frac{24}{30}$	$\frac{28}{35}$	$\frac{32}{40}$	$\frac{36}{45}$	$\frac{40}{50}$	$\frac{44}{55}$	$\frac{48}{60}$	$\frac{52}{65}$	$\frac{56}{70}$	$\frac{60}{75}$	0.8	80%
$\frac{1}{6}$	$\frac{2}{12}$	$\frac{3}{18}$	$\frac{4}{24}$	$\frac{5}{30}$	$\frac{6}{36}$	$\frac{7}{42}$	$\frac{8}{48}$	$\frac{9}{54}$	$\frac{10}{60}$	$\frac{11}{66}$	$\frac{12}{72}$	$\frac{13}{78}$	$\frac{14}{84}$	$\frac{15}{90}$	$0.1\overline{6}$	$16\frac{2}{3}\%$
$\frac{5}{6}$	$\frac{10}{12}$	$\frac{15}{18}$	$\frac{20}{24}$	$\frac{25}{30}$	$\frac{30}{36}$	$\frac{35}{42}$	$\frac{40}{48}$	$\frac{45}{54}$	$\frac{50}{60}$	$\frac{55}{66}$	$\frac{60}{72}$	$\frac{65}{78}$	$\frac{70}{84}$	$\frac{75}{90}$	$0.8\overline{3}$	$83\frac{1}{3}\%$
$\frac{1}{7}$	$\frac{2}{14}$	$\frac{3}{21}$	$\frac{4}{28}$	$\frac{5}{35}$	$\frac{6}{42}$	$\frac{7}{49}$	$\frac{8}{56}$	$\frac{9}{63}$	$\frac{10}{70}$	$\frac{11}{77}$	$\frac{12}{84}$	$\frac{13}{91}$	$\frac{14}{98}$	$\frac{15}{105}$	0.143	14.3%
$\frac{2}{7}$	$\frac{4}{14}$	$\frac{6}{21}$	$\frac{8}{28}$	$\frac{10}{35}$	$\frac{12}{42}$	$\frac{14}{49}$	$\frac{16}{56}$	$\frac{18}{63}$	$\frac{20}{70}$	$\frac{22}{77}$	$\frac{24}{84}$	$\frac{26}{91}$	$\frac{28}{98}$	$\frac{30}{105}$	0.286	28.6%
$\frac{3}{7}$	$\frac{6}{14}$	$\frac{9}{21}$	$\frac{12}{28}$	$\frac{15}{35}$	$\frac{18}{42}$	$\frac{21}{49}$	$\frac{24}{56}$	$\frac{27}{63}$	$\frac{30}{70}$	$\frac{33}{77}$	$\frac{36}{84}$	$\frac{39}{91}$	$\frac{42}{98}$	$\frac{45}{105}$	0.429	42.9%
$\frac{4}{7}$	$\frac{8}{14}$	$\frac{12}{21}$	$\frac{16}{28}$	$\frac{20}{35}$	$\frac{24}{42}$	$\frac{28}{49}$	$\frac{32}{56}$	$\frac{36}{63}$	$\frac{40}{70}$	$\frac{44}{77}$	$\frac{48}{84}$	$\frac{52}{91}$	$\frac{56}{98}$	$\frac{60}{105}$	0.571	57.1%
$\frac{5}{7}$	$\frac{10}{14}$	$\frac{15}{21}$	$\frac{20}{28}$	$\frac{25}{35}$	$\frac{30}{42}$	$\frac{35}{49}$	$\frac{40}{56}$	$\frac{45}{63}$	$\frac{50}{70}$	$\frac{55}{77}$	$\frac{60}{84}$	$\frac{65}{91}$	$\frac{70}{98}$	$\frac{75}{105}$	0.714	71.4%
$\frac{6}{7}$	$\frac{12}{14}$	$\frac{18}{21}$	$\frac{24}{28}$	$\frac{30}{35}$	$\frac{36}{42}$	$\frac{42}{49}$	$\frac{48}{56}$	$\frac{54}{63}$	$\frac{60}{70}$	$\frac{66}{77}$	$\frac{72}{84}$	$\frac{78}{91}$	$\frac{84}{98}$	$\frac{90}{105}$	0.857	85.7%
$\frac{1}{8}$	$\frac{2}{16}$	$\frac{3}{24}$	$\frac{4}{32}$	$\frac{5}{40}$	$\frac{6}{48}$	$\frac{7}{56}$	$\frac{8}{64}$	$\frac{9}{72}$	$\frac{10}{80}$	$\frac{11}{88}$	$\frac{12}{96}$	$\frac{13}{104}$	$\frac{14}{112}$	$\frac{15}{120}$	0.125	$12\frac{1}{2}\%$
$\frac{3}{8}$	$\frac{6}{16}$	$\frac{9}{24}$	$\frac{12}{32}$	$\frac{15}{40}$	$\frac{18}{48}$	$\frac{21}{56}$	$\frac{24}{64}$	$\frac{27}{72}$	$\frac{30}{80}$	$\frac{33}{88}$	$\frac{36}{96}$	$\frac{39}{104}$	$\frac{42}{112}$	$\frac{45}{120}$	0.375	$37\frac{1}{2}\%$
$\frac{5}{8}$	$\frac{10}{16}$	$\frac{15}{24}$	$\frac{20}{32}$	$\frac{25}{40}$	$\frac{30}{48}$	$\frac{35}{56}$	$\frac{40}{64}$	$\frac{45}{72}$	$\frac{50}{80}$	$\frac{55}{88}$	$\frac{60}{96}$	$\frac{65}{104}$	$\frac{70}{112}$	$\frac{75}{120}$	0.625	$62\frac{1}{2}\%$
$\frac{7}{8}$	$\frac{14}{16}$	$\frac{21}{24}$	$\frac{28}{32}$	$\frac{35}{40}$	$\frac{42}{48}$	$\frac{49}{56}$	$\frac{56}{64}$	$\frac{63}{72}$	$\frac{70}{80}$	$\frac{77}{88}$	$\frac{84}{96}$	$\frac{91}{104}$	$\frac{98}{112}$	$\frac{105}{120}$	0.875	$87\frac{1}{2}\%$
$\frac{1}{9}$	$\frac{2}{18}$	$\frac{3}{27}$	$\frac{4}{36}$	$\frac{5}{45}$	$\frac{6}{54}$	$\frac{7}{63}$	$\frac{8}{72}$	$\frac{9}{81}$	$\frac{10}{90}$	$\frac{11}{99}$	$\frac{12}{108}$	$\frac{13}{117}$	$\frac{14}{126}$	$\frac{15}{135}$	$0.\overline{1}$	$11\frac{1}{9}\%$
$\frac{2}{9}$	$\frac{4}{18}$	$\frac{6}{27}$	$\frac{8}{36}$	$\frac{10}{45}$	$\frac{12}{54}$	$\frac{14}{63}$	$\frac{16}{72}$	$\frac{18}{81}$	$\frac{20}{90}$	$\frac{22}{99}$	$\frac{24}{108}$	$\frac{26}{117}$	$\frac{28}{126}$	$\frac{30}{135}$	$0.\overline{2}$	$22\frac{2}{9}\%$
$\frac{4}{9}$	$\frac{8}{18}$	$\frac{12}{27}$	$\frac{16}{36}$	$\frac{20}{45}$	$\frac{24}{54}$	$\frac{28}{63}$	$\frac{32}{72}$	$\frac{36}{81}$	$\frac{40}{90}$	$\frac{44}{99}$	$\frac{48}{108}$	$\frac{52}{117}$	$\frac{56}{126}$	$\frac{60}{135}$	$0.\overline{4}$	$44\frac{4}{9}\%$
$\frac{5}{9}$	$\frac{10}{18}$	$\frac{15}{27}$	$\frac{20}{36}$	$\frac{25}{45}$	$\frac{30}{54}$	$\frac{35}{63}$	$\frac{40}{72}$	$\frac{45}{81}$	$\frac{50}{90}$	$\frac{55}{99}$	$\frac{60}{108}$	$\frac{65}{117}$	$\frac{70}{126}$	$\frac{75}{135}$	$0.\overline{5}$	$55\frac{5}{9}\%$
$\frac{7}{9}$	$\frac{14}{18}$	$\frac{21}{27}$	$\frac{28}{36}$	$\frac{35}{45}$	$\frac{42}{54}$	$\frac{49}{63}$	$\frac{56}{72}$	$\frac{63}{81}$	$\frac{70}{90}$	$\frac{77}{99}$	$\frac{84}{108}$	$\frac{91}{117}$	$\frac{98}{126}$	$\frac{105}{135}$	$0.\overline{7}$	$77\frac{7}{9}\%$
$\frac{8}{9}$	$\frac{16}{18}$	$\frac{24}{27}$	$\frac{32}{36}$	$\frac{40}{45}$	$\frac{48}{54}$	$\frac{56}{63}$	$\frac{64}{72}$	$\frac{72}{81}$	$\frac{80}{90}$	$\frac{88}{99}$	$\frac{96}{108}$	$\frac{104}{117}$	$\frac{112}{126}$	$\frac{120}{135}$	$0.\overline{8}$	$88\frac{8}{9}\%$

Note: The decimals for sevenths have been rounded to the nearest thousandth.

Symbols

$+$	plus or positive
$-$	minus or negative
$*, \times$	multiplied by
$\div, /$	divided by
$=$	is equal to
$\neq$	is not equal to
$<$	is less than
$>$	is greater than
$\leq$	is less than or equal to
$\geq$	is greater than or equal to
$\approx$	is about equal to
$x^n, x \wedge n$	nth power of x
$\sqrt{x}$	square root of x
$\%$	percent
$a{:}b, a / b, \frac{a}{b}$	ratio of a to b or a divided by b or the fraction $\frac{a}{b}$
$a \, [bs]$	a groups, b in each group
$n / d \rightarrow a \, \text{R} b$	n divided by d is a with remainder b
$\{ \}, (\ ), [\ ]$	grouping symbols
∞	infinity
$n!$	n factorial
$°$	degree
(a,b)	ordered pair
$\overleftrightarrow{AS}$	line AS
$\overline{AS}$	line segment AS
$\overrightarrow{AS}$	ray AS
$\llcorner$	right angle
$\perp$	is perpendicular to
$\parallel$	is parallel to
$\triangle ABC$	triangle ABC
$\angle ABC$	angle ABC
$\angle B$	angle B
$\cong$	is congruent to
$\sim$	is similar to
$\equiv$	is equivalent to

Probability Meter

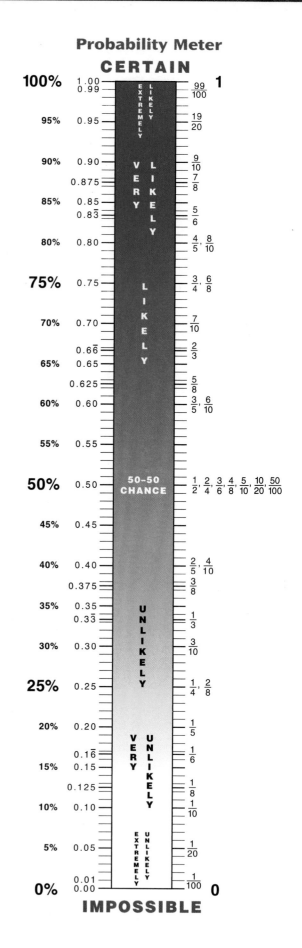

CERTAIN

100%	1.00 / 0.99	1 / $\frac{99}{100}$
95%	0.95	$\frac{19}{20}$
90%	0.90	$\frac{9}{10}$
	0.875	$\frac{7}{8}$
85%	0.85	
	0.8\overline{3}	$\frac{5}{6}$
80%	0.80	$\frac{4}{5}, \frac{8}{10}$
75%	0.75	$\frac{3}{4}, \frac{6}{8}$
70%	0.70	$\frac{7}{10}$
	0.6\overline{6}	$\frac{2}{3}$
65%	0.65	
	0.625	$\frac{5}{8}$
60%	0.60	$\frac{3}{5}, \frac{6}{10}$
55%	0.55	
50%	0.50	50–50 CHANCE $\frac{1}{2}, \frac{2}{4}, \frac{3}{6}, \frac{4}{8}, \frac{5}{10}, \frac{10}{20}, \frac{50}{100}$
45%	0.45	
40%	0.40	$\frac{2}{5}, \frac{4}{10}$
	0.375	$\frac{3}{8}$
35%	0.35 / 0.33\overline{3}	$\frac{1}{3}$
30%	0.30	$\frac{3}{10}$
25%	0.25	$\frac{1}{4}, \frac{2}{8}$
20%	0.20	$\frac{1}{5}$
	0.16\overline{6}	$\frac{1}{6}$
15%	0.15	
	0.125	$\frac{1}{8}$
10%	0.10	$\frac{1}{10}$
5%	0.05	$\frac{1}{20}$
	0.01 / 0.00	$\frac{1}{100}$ / 0
0%		

EXTREMELY LIKELY
VERY LIKELY
LIKELY
UNLIKELY
VERY UNLIKELY
EXTREMELY UNLIKELY

IMPOSSIBLE

Formulas	Meaning of Variables
Rectangles • Perimeter: $p = (2 * l) + (2 * w)$ • Area: $A = b * h$	p = perimeter; l = length; w = width A = area; b = length of base; h = height
Squares • Perimeter: $p = 4 * s$ • Area: $A = s^2$	p = perimeter; s = length of side A = area
Parallelograms • Area: $A = b * h$	A = area; b = length of base; h = height
Triangles • Area: $A = \frac{1}{2} * b * h$	A = area; b = length of base; h = height
Regular Polygons • Perimeter: $p = n * s$	p = perimeter; n = number of sides; s = length of side
Circles • Circumference: $c = \pi * d$, or $c = 2 * \pi * r$ • Area: $A = \pi * r^2$	c = circumference; d = diameter; r = radius A = area
Polygons • Pick's formula: $A = (\frac{1}{2} * P) + I - 1$	A = area; P = number of grid points on polygon; I = number of grid points in the interior
Polyhedrons • Euler's Formula: $e = (f + v) - 2$	e = number of edges; f = number of faces; v = number of vertices
Mobiles (1 rod, with 2 objects on opposite sides of the fulcrum) • When the fulcrum *is* at the center of the rod, the mobile will balance if $W * D = w * d$. • When the fulcrum is *not* at the center of the rod, the mobile will balance if $(W * D) + (R * L) = w * d$	W = weight of one object D = distance of this object from the fulcrum w = weight of the second object d = distance of second object from the fulcrum R = weight of rod L = distance from center of rod to fulcrum W = weight of object on same side of fulcrum as center D = distance of this object from the fulcrum w = weight of the second object d = distance of second object from the fulcrum

Formulas	Meaning of Variables
Rectangular Prisms • Volume: $V = B * h$, or $V = l * w * h$ • Surface area: $S = 2 * ((l * w) + (l * h) + (w * h))$ The surface area formula is true only when all of the faces of the prism are rectangles.	V = volume; B = area of base; l = length; w = width; h = height S = surface area
Cubes • Volume: $V = e^3$ • Surface area: $S = 6 * e^2$	V = volume; e = length of edge S = surface area
Cylinders • Volume: $V = B * h$, or $V = \pi * r^2 * h$ • Surface area: $S = (2 * \pi * r^2) + ((2 * \pi * r) * h)$ The surface area formula is true only when the line through the centers of the circular bases is perpendicular to the bases.	V = volume; B = area of base; h = height; r = radius of base S = surface area
Pyramids • Volume: $V = \frac{1}{3} * B * h$	V = volume; B = area of base; h = height
Cones • Volume: $V = \frac{1}{3} * B * h$, or $V = \frac{1}{3} * \pi * r^2 * h$	V = volume; B = area of base; h = height; r = radius of base
Spheres • Volume: $V = \frac{4}{3} * \pi * r^3$ • Surface area: $S = 4 * \pi * r^2$	V = volume; r = radius S = surface area

Temperatures
 	
• Fahrenheit to Celsius conversion:
  $C = \frac{5}{9} * (F - 32°)$
 	
• Celsius to Fahrenheit conversion:
  $F = (\frac{9}{5} * C) + 32°$
	C = degrees Celsius; F = degrees Fahrenheit

Distances • $d = r * t$	d = distance traveled; r = rate of speed; t = time of travel

Roman Numerals

A **numeral** is a symbol used to represent a number. **Roman numerals,** developed about 500 B.C., use letters to represent numbers.

Seven different letters are used in Roman numerals. Each letter stands for a different number.

A string of letters means that their values should be added together. For example, CCC = 100 + 100 + 100 = 300, and CLXII = 100 + 50 + 10 + 1 + 1 = 162.

If a smaller value is placed *before* a larger value, the smaller value is subtracted instead of added. For example, IV = 5 − 1 = 4, and CDX = 500 − 100 + 10 = 410.

There are several **rules for subtracting letters.**

Roman Numeral	Number
I	1
V	5
X	10
L	50
C	100
D	500
M	1,000

♦ The letters I (1), X (10), C (100), and M (1,000) represent powers of ten. These are the only letters that may be subtracted. For example, 95 in Roman numerals is XCV (VC for 95 is incorrect because V is not a power of ten).

♦ One letter may not be subtracted from a second letter if the value of the second letter is more than 10 times the value of the first. The letter I may be subtracted only from V or X. The letter X may be subtracted only from L or C. For example, 49 in Roman numerals is XLIX (IL for 49 is incorrect). And 1990 in Roman numerals is MCMXC (MXM for 1990 is incorrect).

♦ Only a *single* letter may be subtracted from another letter that follows. For example, 7 in Roman numerals is VII (IIIX for 7 is incorrect). And 300 in Roman numerals is CCC (CCD for 300 is incorrect).

The largest Roman numeral, M, stands for 1,000. One way to write large numbers is to write a string of Ms. For example, MMMM stands for 4,000. Another way to write large numbers is to write a bar above a numeral. The bar means that the numeral beneath should be multiplied by 1,000. So, $\overline{\text{IV}}$ also stands for 4,000. And $\overline{\text{M}}$ stands for 1,000 ∗ 1,000 = 1 million.

A

Absolute value

The distance between a number and 0 on the number line. The absolute value of a positive number is the number itself. The absolute value of a negative number is the *opposite* of the number. For example, the absolute value of 3 is 3, and the absolute value of −6 is 6. The absolute value of 0 is 0. The notation for the absolute value of a number n is $|n|$.

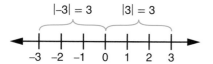

$|-3| = 3$ $|3| = 3$

Abundant number

A counting number whose *proper factors* add up to more than the number itself. For example, 12 is an abundant number because the sum of its proper factors is $1 + 2 + 3 + 4 + 6 = 16$, and 16 is greater than 12. See also *deficient number* and *perfect number*.

Acre

A unit of *area* equal to 43,560 square feet in the U.S. customary system of measurement. An acre is roughly the size of a football field. A square mile equals 640 acres.

Addend

Any one of a set of numbers that are added. For example, in $5 + 3 + 1 = 9$, the addends are 5, 3, and 1.

Adjacent angles

Angles that are next to each other; adjacent angles have a common vertex and common side but no other overlap. In the diagram, angles 1 and 2 are adjacent angles. So are angles 2 and 3, angles 3 and 4, and angles 4 and 1.

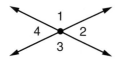

Algebraic expression

An *expression* that contains a variable. For example, if Maria is 2 inches taller than Joe and if the variable M represents Maria's height, then the algebraic expression $M - 2$ represents Joe's height.

Algorithm

A set of step-by-step instructions for doing something, such as carrying out a computation or solving a problem.

Angle

A figure that is formed by two rays or two line segments with a common endpoint. The rays or segments are called the *sides* of the angle. The common endpoint is called the *vertex* of the angle. Angles are measured in *degrees* (°).

An *acute angle* has a measure greater than 0° and less than 90°. An *obtuse angle* has a measure greater than 90° and less than 180°. A *reflex angle* has a measure greater than 180° and less than 360°. A *right angle* measures 90°. A *straight angle* measures 180°.

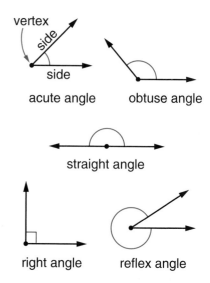

Apex

In a pyramid or a cone, the vertex opposite the base. In a pyramid, all the faces except the base meet at the *apex*. See also *base of a pyramid or a cone*.

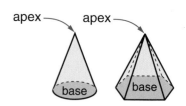

Arc Part of a circle, from one point on the circle to another. For example, a *semicircle* is an arc whose endpoints are the endpoints of a diameter of the circle.

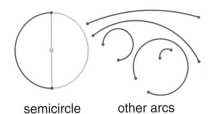

semicircle   other arcs

Area The amount of surface inside a closed boundary. Area is measured in square units, such as square inches or square centimeters.

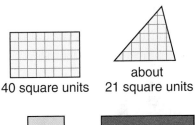

40 square units   about 21 square units

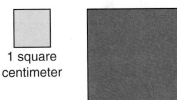

1 square centimeter

1 square inch

Area model (1) A model for multiplication problems in which the length and width of a rectangle represent the factors, and the area of the rectangle represents the product. (2) A model for showing fractions as parts of circles, rectangles, or other geometric figures.

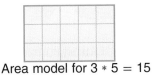

Area model for 3 * 5 = 15

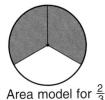

Area model for $\frac{2}{3}$

Array (1) An arrangement of objects in a regular pattern, usually in rows and columns. (2) A *rectangular array.* In *Everyday Mathematics,* an array is a rectangular array unless specified otherwise.

Associative Property A property of addition and multiplication (but not of subtraction or division) that says that when you add or multiply three numbers, it does not matter which two you add or multiply first. For example:

$(4 + 3) + 7 = 4 + (3 + 7)$
and $(5 * 8) * 9 = 5 * (8 * 9)$.

Average A typical value for a set of numbers. The word *average* usually refers to the *mean* of a set of numbers.

Axis (plural: **axes**)
(1) Either of the two number lines that intersect to form a *coordinate grid.*

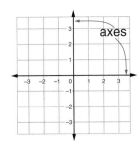

axes

(2) A line about which a solid figure rotates.

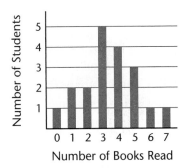

North Pole

axis   South Pole

B

Bar graph A graph that uses horizontal or vertical bars to represent data.

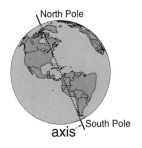

Number of Students
Number of Books Read

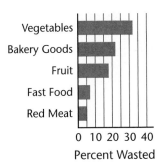

Vegetables
Bakery Goods
Fruit
Fast Food
Red Meat
0 10 20 30 40
Percent Wasted

Base (in exponential notation) The number that is raised to a power. For example, in 5^3, the base is 5. See also *exponential notation* and *power of a number*.

Base of a polygon A side on which a polygon "sits." The height of a polygon may depend on which side is called the base. See also *height of a parallelogram* and *height of a triangle*.

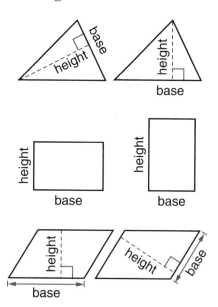

Base of a prism or a cylinder Either of the two parallel and congruent faces that define the shape of a prism or a cylinder.

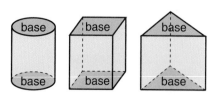

Base of a pyramid or a cone The face of a pyramid or a cone that is opposite its apex. The base of a pyramid is the only face that does not include the apex.

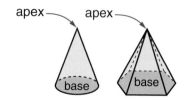

Base-ten Our system for writing numbers that uses only 10 symbols, called *digits*. The digits are 0, 1, 2, 3, 4, 5, 6, 7, 8, and 9. You can write any number using only these 10 digits. Each digit has a value that depends on its place in the number. In this system, moving a digit one place to the left makes that digit worth 10 times as much. And moving a digit one place to the right makes that digit worth one-tenth as much. See also *place value*.

Benchmark A well-known count or measure that can be used to check whether other counts, measures, or estimates make sense. For example, a benchmark for land area is that a football field is about one acre. A benchmark for length is that the width of a man's thumb is about one inch. Benchmarks are sometimes called *personal references*.

Bisect To divide a segment, an angle, or another figure into two equal parts.

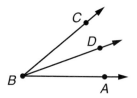

Ray *BD* bisects angle *ABC*.

Bisector A line, segment, or ray that divides a segment, an angle, or a figure into two equal parts. See also *bisect*.

Broken-line graph A graph in which data points are connected by line segments. Broken-line graphs are often used to show how something has changed over a period of time. Same as *line graph*.

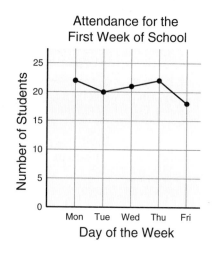

Box-and-whiskers plot A plot displaying the spread, or distribution, of a data set using 5 *landmarks*: the *minimum, lower quartile, median, upper quartile,* and *maximum*. Also called a box plot.

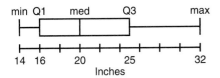

Capacity (1) The amount a container can hold. The *volume* of a container. Capacity is usually measured in units such as gallons, pints, cups, fluid ounces, liters, and milliliters. (2) The heaviest weight a scale can measure.

Change diagram A diagram used in *Everyday Mathematics* to represent situations in which quantities are increased or decreased.

Circle The set of all points in a plane that are the same distance from a fixed point in the plane. The fixed point is the *center* of the circle, and the distance is the *radius*. The center and *interior* of a circle are not part of the circle. A circle together with its interior is called a *disk* or a *circular region*. See also *diameter*.

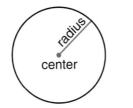

disk

Circle graph A graph in which a circle and its interior are divided by radii into parts (*sectors*) to show the parts of a set of data. The whole circle represents the whole set of data. Same as *pie graph*.

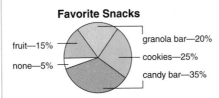

Favorite Snacks

fruit—15%
none—5%
granola bar—20%
cookies—25%
candy bar—35%

Circumference The distance around a circle; the perimeter of a circle.

Column-addition method A method for adding numbers in which the addends' digits are first added in each place-value column separately, and then 10-for-1 trades are made until each column has only one digit. Lines are drawn to separate the place-value columns.

100s	10s	1s
2	4	8
+ 1	8	7
3	12	15
3	13	5
4	3	5

$248 + 187 = 435$

Column-division method A division procedure in which vertical lines are drawn between the digits of the dividend. As needed, trades are made from one column into the next column at the right. The lines make the procedure easier to carry out.

	1	7	2
5)	8	6	3
	−5	36	13
	3	−35	−10
		1	3

$863 / 5 \rightarrow 172 \text{ R}3$

Common denominator
(1) If two fractions have the same denominator, that denominator is called a common denominator.
(2) For two or more fractions, any number that is a *common multiple* of their denominators. For example, the fractions $\frac{1}{2}$ and $\frac{2}{3}$ have the common denominators 6, 12, 18, and so on. See also *quick common denominator*.

Common factor A counting number is a common factor of two or more counting numbers if it is a *factor* of each of those numbers. For example, 4 is a common factor of 8 and 12. See also *factor of a counting number* n.

Common multiple
A number is a common multiple of two or more numbers if it is a *multiple* of each of those numbers. For example, the multiples of 2 are 2, 4, 6, 8, 10, 12, and so on; the multiples of 3 are 3, 6, 9, 12, and so on; and the common multiples of 2 and 3 are 6, 12, 18, and so on.

Commutative Property
A property of addition and multiplication (but not of subtraction or division) that says that changing the order of the numbers being added or multiplied does not change the answer. These properties are often called *turn-around facts* in *Everyday Mathematics*. For example:
$5 + 10 = 10 + 5$ and
$3 * 8 = 8 * 3$.

Comparison diagram
A diagram used in *Everyday Mathematics* to represent situations in which two quantities are compared.

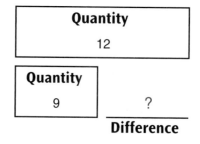

Complementary angles
Two angles whose measures total 90°.

∠1 and ∠2 are complementary angles

Composite number
A counting number that has more than 2 different factors. For example, 4 is a composite number because it has three factors: 1, 2, and 4.

Concave polygon A polygon in which at least one vertex is "pushed in." At least one inside angle of a concave polygon is a *reflex angle* (has a measure greater than 180°). Same as *nonconvex polygon*.

Concentric circles Circles that have the same center but radii of different lengths.

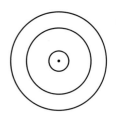

Cone A 3-dimensional shape that has a circular *base*, a curved surface, and one vertex, which is called the *apex*. The points on the curved surface of a cone are on straight lines connecting the apex and the boundary of the base.

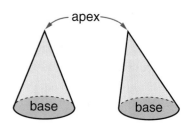

Congruent Having the same shape and size. Two 2-dimensional figures are congruent if they match exactly when one is placed on top of the other. (It may be necessary to flip one of the figures over.)

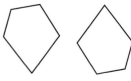

congruent pentagons

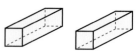

congruent prisms

Consecutive angles Two angles in a polygon that share a common side.

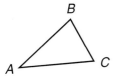

Three pairs of consecutive angles: angles *A* and *B*, *B* and *C*, and *C* and *A*

Constant A quantity that does not change.

Contour line A curve on a map through places where a certain measurement (such as temperature or elevation) is the same. Often contour lines separate regions that have been colored differently to show a range of conditions.

Contour map A map that uses *contour lines* to show a particular feature (such as elevation or climate).

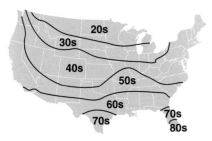

Convex polygon A polygon in which all vertices are "pushed outward." Each inside angle of a convex polygon has a measure less than 180°.

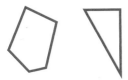

Coordinate (1) A number used to locate a point on a number line. (2) One of the two numbers in an ordered number pair. The number pair is used to locate a point on a *coordinate grid.*

Coordinate grid See *rectangular coordinate grid.*

Corresponding Having the same relative position in *similar* or *congruent figures.* In the diagram, pairs of *corresponding sides* are marked with the same number of slash marks and *corresponding angles* are marked with the same number of arcs.

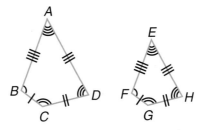

corresponding sides and angles

Counting numbers The numbers used to count things. The set of counting numbers is {1, 2, 3, 4, …}. Compare to *whole numbers.*

Cover-up method A method for solving equations by covering up key expressions.

Cross multiplication The process of finding the cross products of a *proportion.* Cross multiplication can be used in solving proportions. In the example below, the cross products are 60 and 4z.

$$3 * 20 = 60 \qquad 4 * z = 4z$$

Cross products The cross products of a *proportion* are found by multiplying the numerator of each fraction by the denominator of the other fraction. For example, in the proportion $\frac{2}{3} = \frac{6}{9}$, the cross products $2 * 9$ and $3 * 6$ are both 18. The cross products of a proportion are always equal.

$$2 * 9 = 18 \qquad 3 * 6 = 18$$

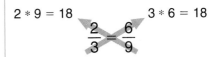

Cross section A shape formed by the intersection of a plane and a geometric solid.

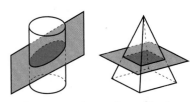

Cross sections of a cylinder and a pyramid

Cube A polyhedron with 6 square faces. A cube has 8 vertices and 12 edges.

Cubic centimeter A metric unit of volume equal to the volume of a cube with 1 cm edges. 1 cm^3 = 1 mL.

Cubic unit A unit used in measuring volume, such as a cubic centimeter or a cubic foot.

Cubit An ancient unit of length, measured from the point of the elbow to the end of the middle finger. A cubit is about 18 inches.

Curved surface A surface that is rounded rather than flat. Spheres, cylinders, and cones each have one curved surface.

Cylinder A 3-dimensional shape that has two circular bases that are parallel and congruent and are connected by a curved surface. A soup can is shaped like a cylinder.

Data Information that is gathered by counting, measuring, questioning, or observing.

Decimal A number written in *standard, base-10 notation* that contains a decimal point, such as 2.54. A whole number is a decimal, but it is usually written without a decimal point.

Decimal point A dot used to separate the ones and tenths places in decimal numbers.

Deficient number A counting number whose *proper factors* add up to less than the number itself. For example, 10 is a deficient number because the sum of its proper factors is $1 + 2 + 5 = 8$, and 8 is less than 10. See also *abundant number* and *perfect number*.

Degree (°) (1) A unit of measure for angles based on dividing a circle into 360 equal parts. Latitude and longitude are measured in degrees, and these degrees are based on angle measures. (2) A unit of measure for temperature. In all cases, a small raised circle (°) is used to show degrees.

Denominator The number below the line in a fraction. A fraction may be used to name part of a whole. If the *whole* (the *ONE*, or the *unit*) is divided into equal parts, the denominator represents the number of equal parts into which the whole is divided. In the fraction $\frac{a}{b}$, b is the denominator.

Density A *rate* that compares the *weight* of an object with its *volume*. For example, suppose a ball has a weight of 20 grams and a volume of 10 cubic centimeters. To find its density, divide its weight by its volume: $20 \text{ g} / 10 \text{ cm}^3 = 2 \text{ g} / \text{cm}^3$, or 2 grams per cubic centimeter.

Dependent variable (1) A *variable* whose value is dependent on the value of at least one other variable. (2) The output of a function machine. See also *independent variable*.

Diameter (1) A line segment that passes through the center of a circle or sphere and has endpoints on the circle or sphere. (2) The length of this line segment. The diameter of a circle or sphere is twice the length of its *radius*.

Difference The result of subtracting one number from another. See also *minuend* and *subtrahend*.

Digit One of the number symbols 0, 1, 2, 3, 4, 5, 6, 7, 8, and 9 in the standard, *base-ten* system.

Discount The amount by which the regular price of an item is reduced.

Distributive Property
A property that relates multiplication and addition or subtraction. This property gets its name because it "distributes" a factor over terms inside parentheses.

Distributive property of multiplication over addition:
$a * (b + c) = (a * b) + (a * c)$,
so $2 * (5 + 3) = (2 * 5) + (2 * 3)$
$= 10 + 6 = 16$.

Distributive property of multiplication over subtraction:
$a * (b - c) = (a * b) - (a * c)$,
so $2 * (5 - 3) = (2 * 5) - (2 * 3)$
$= 10 - 6 = 4$.

Dividend The number in division that is being divided. For example, in $35 \div 5 = 7$, the dividend is 35.

Divisible by If one counting number can be divided by a second counting number with a remainder of 0, then the first number is divisible by the second number. For example, 28 is divisible by 7 because 28 divided by 7 is 4, with a remainder of 0.

Divisibility test A test to find out whether one counting number is *divisible by* another counting number without actually doing the division. A divisibility test for 5, for example, is to check the digit in the 1s place: if that digit is 0 or 5, then the number is divisible by 5.

Division of Fractions Property A fact that makes division with fractions easier: division by a fraction is the same as multiplication by that fraction's *reciprocal*. For example, because the reciprocal of $\frac{1}{2}$ is 2, the division problem $4 \div \frac{1}{2}$ is equivalent to the multiplication problem $4 * 2$. See also *multiplicative inverses*.

Divisor In division, the number that divides another number. For example, in $35 \div 5 = 7$, the divisor is 5.

Dodecahedron
A polyhedron with 12 faces.

Edge A line segment or curve where two surfaces meet.

edges

edge

Ellipse An oval curve in the plane where the sum of the distances from any point on the curve to 2 fixed points is constant. Each of the fixed points is called a *focus* of the ellipse.

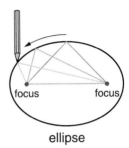

focus   focus

ellipse

Endpoint A point at the end of a *line segment* or *ray*. A line segment is named using the letter labels of its endpoints. A ray is named using the letter labels of its endpoint and another point on the ray.

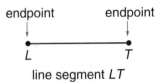

endpoint   endpoint

L T

line segment LT

Enlarge To increase the size of an object or a figure without changing its shape. See also *size-change factor*.

Equally likely outcomes If all of the possible *outcomes* for an experiment or situation have the same *probability*, they are called equally likely outcomes. In the case of equally likely outcomes, the probability of an *event* is equal to this fraction:

$$\frac{\text{number of favorable outcomes}}{\text{number of possible outcomes}}$$

See also *favorable outcome*.

Equation A number sentence that contains an equal sign. For example, $15 = 10 + 5$ is an equation.

Equilateral triangle A triangle with all three sides equal in length. In an equilateral triangle, all three angles have the same measure.

Equivalent Equal in value but possibly in a different form. For example, $\frac{1}{2}$, 0.5, and 50% are all equivalent.

Equivalent equations Equations that have the same *solution set*. For example, $2 + x = 4$ and $6 + x = 8$ are equivalent equations because the solution set for each is $x = 2$.

Equivalent fractions Fractions with different denominators that name the same number. For example, $\frac{1}{2}$ and $\frac{4}{8}$ are equivalent fractions.

Equivalent rates *Rates* that make the same comparison. For example, the rate $\frac{60 \text{ miles}}{1 \text{ hour}}$ and $\frac{1 \text{ mile}}{1 \text{ minute}}$ are equivalent. Two rates named as fractions using the *same units* are equivalent if the fractions (ignoring the units) are equivalent. For example, $\frac{12 \text{ pages}}{4 \text{ minutes}}$ and $\frac{6 \text{ pages}}{2 \text{ minutes}}$ are equivalent rates because $\frac{12}{4}$ and $\frac{6}{2}$ are equivalent.

Equivalent ratios *Ratios* that make the same comparison. Two or more ratios are equivalent if they can be named as equivalent fractions. For example, the ratios 12 to 20, 6 to 10, and 3 to 5 are equivalent ratios because $\frac{12}{20} = \frac{6}{10} = \frac{3}{5}$.

Estimate An answer that should be close to an exact answer. *To estimate* means to give an answer that should be close to an exact answer.

Evaluate To find a value for. To evaluate a mathematical *expression*, carry out the operations. If there are variables, first replace them with numbers. To evaluate a *formula*, find the value of one variable in the formula when the values of the other variables are given.

Even number A counting number that can be divided by 2 with no remainder. The even numbers are 2, 4, 6, 8, and so on. $0, -2, -4, -6,$ and so on are also usually considered even.

Event Something that happens. The *probability* of an event is the chance that the event will happen. For example, rolling a number smaller than 4 with a die is an event. The possible *outcomes* of rolling a die are 1, 2, 3, 4, 5, and 6. The event "roll a number smaller than 4" will happen if the outcome is 1 or 2 or 3. And the chance that this will happen is $\frac{3}{6}$ If the probability of an event is 0, the event is *impossible*. If the probability is 1, the event is *certain*.

Expanded notation A way of writing a number as the sum of the values of each digit. For example, in expanded notation, 356 is written $300 + 50 + 6$. See also *standard notation, scientific notation,* and *number-and-word notation.*

Exponent A small raised number used in *exponential notation* to tell how many times the *base* is used as a *factor.* For example, in 5^3, the base is 5, the exponent is 3, and $5^3 = 5 * 5 * 5 = 125$. See also *power of a number.*

Exponential notation
A way to show repeated multiplication by the same factor. For example, 2^3 is exponential notation for $2 * 2 * 2$. The small raised 3 is the *exponent*. It tells how many times the number 2, called the *base*, is used as a factor.

Expression A group of mathematical symbols that represents a number—or can represent a number if values are assigned to any variables in the expression. An expression may include numbers, variables, operation symbols, and grouping symbols – but *not* relation symbols ($=$, $>$, $<$, and so on). Any expression that contains one or more variables is called an *algebraic expression*.

2π $3 + 4$ $5 * (7 - 3)$

expressions

x $\pi * r$ (or πr) $a^2 + (a / 5)$

algebraic expressions

Extended multiplication fact A multiplication fact involving multiples of 10, 100, and so on. For example, $6 * 70$, $60 * 7$, and $60 * 70$ are extended multiplication facts.

Face A flat surface on a 3-dimensional shape.

Fact family A set of related addition and subtraction facts, or related multiplication and division facts. For example, $5 + 6 = 11$, $6 + 5 = 11$, $11 - 5 = 6$, and $11 - 6 = 5$ are a fact family. $5 * 7 = 35$, $7 * 5 = 35$, $35 \div 5 = 7$, and $35 \div 7 = 5$ are another fact family.

Factor (in a product) Whenever two or more numbers are multiplied to give a product, each of the numbers that is multiplied is called a factor. For example, in $4 * 1.5 = 6$, 6 is the product and 4 and 1.5 are called factors. See also *factor of a counting number* n.

$$\underset{\text{factors}}{4 * 1.5} = \underset{\text{product}}{6}$$

Note: This definition of *factor* is much less important than the next definition.

Factor of a counting number *n* A counting number whose product with some other counting number equals *n*. For example, 2 and 3 are factors of 6 because $2 * 3 = 6$. But 4 is not a factor of 6 because $4 * 1.5 = 6$, and 1.5 is not a counting number.

$$\underset{\text{factors}}{2 * 3} = \underset{\text{product}}{6}$$

Note: This definition of *factor* is much more important than the previous definition.

Factor pair Two factors of a counting number whose product is the number. A number may have more than one factor pair. For example, the factor pairs for 18 are 1 and 18, 2 and 9, and 3 and 6.

Factor rainbow A way to show factor pairs in a list of all the factors of a counting number. A factor rainbow can be used to check whether a list of factors is correct.

Factor rainbow for 24

Factor string A counting number written as a product of two or more of its factors. The number 1 is never part of a factor string. For example, a factor string for 24 is $2 * 3 * 4$. This factor string has three factors, so its length is 3. Another factor string for 24 is $2 * 3 * 2 * 2$ (length 4).

Factor tree A way to get the *prime factorization* of a counting number. Write the original number as a product of counting-number factors. Then write each of these factors as a product of factors, and so on, until the factors are all prime numbers. A factor tree looks like an upside-down tree, with the root (the original number) at the top and the leaves (the factors) beneath it.

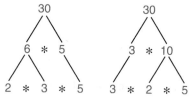

Two factor trees for 30

Factorial A product of a whole number and all the smaller whole numbers except 0. An exclamation point (!) is used to write factorials. For example, "three factorial" is written as 3! and is equal to $3 * 2 * 1 = 6$. $10! = 10 * 9 * 8 * 7 * 6 * 5 * 4 * 3 * 2 * 1 = 3,628,800$. 0! is defined to be equal to 1.

Fair Free from bias. Each side of a fair die or coin will come up about equally often.

Fair game A game in which every player has the same chance of winning.

False number sentence A number sentence that is not true. For example, $8 = 5 + 5$ is a false number sentence.

Fathom A unit used by people who work with boats and ships to measure depths underwater and lengths of cables. A fathom is now defined as 6 feet.

Favorable outcome An *outcome* that satisfies the conditions of an *event* of interest. For example, suppose a 6-sided die is rolled and the event of interest is rolling an even number. There are 6 possible outcomes: 1, 2, 3, 4, 5, or 6. There are 3 favorable outcomes: 2, 4, or 6. See also *equally likely outcomes*.

Figurate numbers Numbers that can be shown by specific geometric patterns. Square numbers and triangular numbers are examples of figurate numbers.

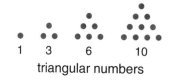

triangular numbers

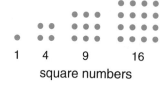

square numbers

Formula A general rule for finding the value of something. A formula is often written using letters, called *variables*, which stand for the quantities involved. For example, the formula for the area of a rectangle may be written as $A = l * w$, where A represents the area of the rectangle, l represents its length, and w represents its width.

Fraction (primary definition) A number in the form $\frac{a}{b}$ where a and b are whole numbers and b is not 0. A fraction may be used to name part of a whole, or to compare two quantities. A fraction may also be used to represent division. For example, $\frac{2}{3}$ can be thought of as 2 divided by 3. See also *numerator* and *denominator*.

Fraction (other definitions) (1) A fraction that satisfies the definition above, but includes a unit in both the numerator and denominator. This definition of fraction includes any rate that is written as a fraction.

For example, $\frac{50 \text{ miles}}{1 \text{ gallon}}$ and $\frac{40 \text{ pages}}{10 \text{ minutes}}$. (2) Any number written using a fraction bar, where the fraction bar is used to indicate division.

For example, $\frac{2.3}{6.5}$, $\frac{1\frac{4}{5}}{12}$, and $\frac{\frac{3}{4}}{\frac{5}{8}}$.

Fraction-Stick Chart A diagram used in *Everyday Mathematics* to represent simple fractions.

Fulcrum (1) The point or place around which a lever pivots. (2) A point on a mobile at which a rod is suspended.

Genus In *topology,* the number of holes in a geometric shape. Shapes with the same genus are *topologically equivalent.* For example, a doughnut and a teacup are both genus 1.

Geometric solid A 3-dimensional shape, such as a prism, pyramid, cylinder, cone, or sphere. Despite its name, a geometric solid is hollow; it does not contain the points in its interior.

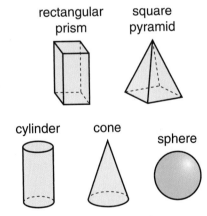

rectangular prism   square pyramid

cylinder   cone   sphere

Geometry Template An *Everyday Mathematics* tool that includes a millimeter ruler, a ruler with sixteenth-inch intervals, half-circle and full-circle protractors, a percent circle, pattern-block shapes, and other geometric figures. The template can also be used as a compass.

Golden Ratio A ratio of approximately 1.618 to 1. The golden ratio is sometimes denoted by the Greek letter *phi:* ϕ. The golden ratio is an *irrational number.* Since ancient times, rectangles whose sides have the golden ratio have been thought to be pleasing and used in art and design.

Golden Rectangle A rectangle in which the ratio of the length of the longer side to the length of the shorter side is the *Golden Ratio,* or about 1.618 to 1. A 5-inch by 3-inch index card has sides that nearly form a golden rectangle.

Great span The distance from the tip of the thumb to the tip of the little finger (pinkie) when the hand is stretched as far as possible.

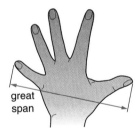

great span

Greatest common factor (GCF) The largest factor that two or more counting numbers have in common. For example, the common factors of 24 and 36 are 1, 2, 3, 4, 6, and 12. The greatest common factor of 24 and 36 is 12.

Grouping symbols Symbols such as parentheses ( ), brackets [ ], and braces { } that tell the order in which operations in an expression are to be done. For example, in the expression $(3 + 4) * 5$, the operation in the parentheses should be done first. The expression then becomes $7 * 5 = 35$.

Height of a parallelogram The length of the shortest line segment between the *base* of a parallelogram and the line containing the opposite side. That shortest segment is perpendicular to the base and is also called the *height.* See also *base of a polygon.*

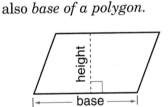

height   base

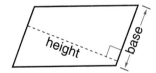

height   base

Height of a prism or a cylinder The length of the shortest line segment between the base of a prism or a cylinder and the plane containing the opposite base. That shortest segment is perpendicular to the base and is also called the *height*. See also *base of a prism or a cylinder*.

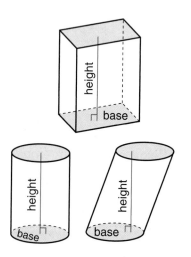

Height of a pyramid or a cone The length of the shortest line segment between the vertex of a pyramid or a cone and the plane containing its base. That shortest segment is also perpendicular to the plane containing the base and is called the *height*. See also *base of a pyramid or a cone*.

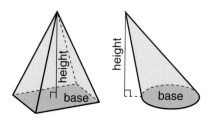

Height of a triangle The length of the shortest line segment between the line containing a base of a triangle and the vertex opposite that base. That shortest segment is perpendicular to the line containing the base and is also called the *height*. See also *base of a polygon*.

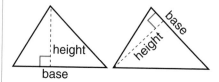

Hemisphere Half of Earth's surface. Also, half of a sphere.

Hexagon A polygon with six sides.

Hexagram A 6-pointed star formed by extending the sides of a regular hexagon.

Horizon Where the earth and sky appear to meet if nothing is in the way. When looking out to sea, the horizon looks like a line.

Horizontal In a left-right orientation; parallel to the horizon.

Hypotenuse In a right triangle, the side opposite the right angle.

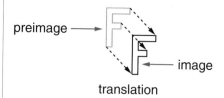

Icosahedron A polyhedron with 20 faces.

Image The reflection of an object that you see when you look in a mirror. Also, a figure that is produced by a *transformation* (a *reflection*, *translation*, or *rotation*, for example) of another figure. See also *preimage*.

preimage

image

translation

Improper fraction A fraction whose numerator is greater than or equal to its denominator. For example, $\frac{4}{3}$, $\frac{5}{2}$, $\frac{4}{4}$, and $\frac{24}{12}$ are improper fractions. In *Everyday Mathematics*, improper fractions are sometimes called "top-heavy" fractions.

Independent variable (1) A *variable* whose value does not rely on the values of other variables. (2) The input of a function machine. See also *dependent variable*.

Indirect measurement
Determining heights, distances, and other quantities that cannot be measured directly.

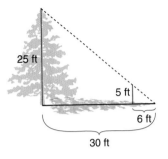

Inequality A number sentence with >, <, ≥, ≤, or ≠. For example, the sentence 8 < 15 is an inequality.

Inscribed polygon
A polygon whose vertices are all on the same circle.

inscribed square

Integer A number in the set {..., −4, −3, −2, −1, 0, 1, 2, 3, 4, ...}; a *whole number* or the *opposite* of a whole number, where 0 is its own opposite.

Interior The inside of a closed 2-dimensional or 3-dimensional figure. The interior is usually not considered to be part of the figure.

Interquartile range (IQR)
(1) The length of the interval between the *lower* and *upper quartiles* in a data set. Illustrated by the box in a *box-and-whiskers plot*. (2) The interval itself. The middle half of the data is *in* the interquartile range.

Intersect To meet or cross.

Intersecting Meeting or crossing one another. For example, lines, segments, rays, and planes can intersect.

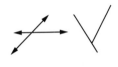

intersecting lines and segments   intersecting planes

Interval (1) The set of all numbers between two numbers, *a* and *b,* which may include *a* or *b* or both. (2) A part of a line, including all points between two specific points.

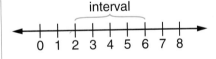

interval

Interval estimate
An estimate that places an unknown quantity in a range. For example, an interval estimate of a person's weight might be "between 100 and 110 pounds."

Irrational number
A number that cannot be written as a fraction, where both the numerator and the denominator are *integers* and the denominator is not zero. For example, π (pi) is an irrational number.

Isometry transformation
A transformation such as a *translation* (slide), *reflection* (flip), or *rotation* (turn) that changes the position or orientation of a figure but does not change its size or shape.

slide   flip   turn

Isosceles triangle
A triangle with at least two sides equal in length. In an isosceles triangle, at least two angles have the same measure. A triangle with all three sides the same length is an isosceles triangle, but is usually called an *equilateral triangle.*

Glossary

Kite A quadrilateral with two pairs of adjacent equal sides. The four sides cannot all have the same length, so a rhombus is not a kite.

Landmark A notable feature of a data set. Landmarks include the *median, mode, maximum, minimum,* and *range.* The *mean* can also be thought of as a landmark.

Latitude A measure, in degrees, that tells how far north or south of the equator a place is.

Lattice method A very old way to multiply multidigit numbers.

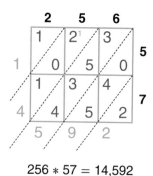

$256 * 57 = 14,592$

Leading-digit estimation A way to estimate in which the left-most, nonzero digit in a number is not changed, but all other digits are replaced by zeros. For example, to estimate $432 + 76$, use the leading-digit estimates 400 and 70: $400 + 70 = 470$.

Least common denominator (LCD) The *least common multiple* of the denominators of every fraction in a given collection. For example, the least common denominator of $\frac{1}{2}, \frac{4}{5}$, and $\frac{3}{8}$ is 40.

Least common multiple (LCM) The smallest number that is a multiple of two or more numbers. For example, while some common multiples of 6 and 8 are 24, 48, and 72, the least common multiple of 6 and 8 is 24.

Left-to-right subtraction A subtraction method in which you start at the left and subtract column by column. For example, to subtract $932 - 356$:

```
              9 3 2
Subtract the 100s.  − 3 0 0
              ─────
              6 3 2
Subtract the 10s.   −   5 0
              ─────
              5 8 2
Subtract the 1s.    −     6
              ─────
              5 7 6
```

$932 - 356 = 576$

Leg of a right triangle A side of a right triangle that is not the *hypotenuse.*

Like In some situations, like means *the same.* The fractions $\frac{2}{5}$ and $\frac{3}{5}$ have like denominators. The measurements 23 cm and 52 cm have like units.

Like terms In an *algebraic expression,* either the constant terms or any terms that contain the same variable(s) raised to the same power(s). For example, $4y$ and $7y$ are like terms in the expression $4y + 7y - z$.

Line A straight path that extends infinitely in opposite directions.

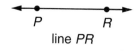

line PR

Line graph See *broken-line graph.*

Line of reflection (mirror line) A line halfway between a figure (preimage) and its reflected image. In a *reflection,* a figure is "flipped over" the line of reflection.

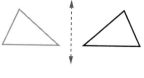

line of reflection

Line of symmetry A line drawn through a figure so that it is divided into two parts that are mirror images of each other. The two parts look alike but face in opposite directions. See also *line symmetry*.

line of symmetry

Line plot A sketch of data in which check marks, Xs, or other marks above a labeled line show the frequency of each value.

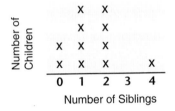

Line segment A straight path joining two points. The two points are called *endpoints* of the segment.

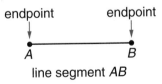

line segment *AB*

Line symmetry A figure has line symmetry if a line can be drawn through it so that it is divided into two parts that are mirror images of each other. The two parts look alike but face in opposite directions. See also *line of symmetry*.

Lines of latitude Lines that run east-west on a map or globe and locate a place with reference to the equator, which is also a line of latitude. On a globe, lines of latitude are circles, and are called *parallels* because the planes containing these circles are parallel to the plane containing the equator.

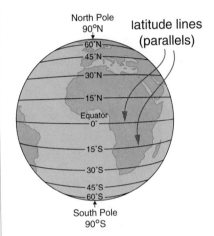

Lines of longitude Lines that run north-south on a map or globe and locate a place with reference to the *prime meridian,* which is also a line of longitude. On a globe, lines of longitude are *semicircles* that meet at the North and South Poles. They are also called *meridians*.

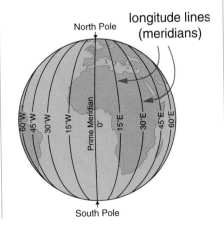

Longitude A measure, in degrees, that tells how far east or west of the *prime meridian* a place is.

Lower quartile In an ordered data set, the middle value of the *data* below the *median*. Data values at the median are not included when finding the lower quartile. See also *upper quartile*.

Lowest terms See *simplest form*.

M

Magnitude estimate A rough *estimate*. A magnitude estimate tells whether an answer should be in the tens, hundreds, thousands, ten-thousands, and so on.

Map legend (map key) A diagram that explains the symbols, markings, and colors on a map.

Map scale A tool that helps you estimate real distances between places shown on a map. It relates distances on the map to distances in the real world. For example, a map scale may show that 1 inch on a map represents 100 miles in the real world. See also *scale*.

Maximum The largest amount; the greatest number in a set of data.

Mean The sum of a set of numbers divided by the number of numbers in the set. The mean is often referred to simply as the *average*.

Mean absolute deviation (m.a.d.) In a data set, the *mean* distance between individual data values and the mean of those values. The distance between a data value and the mean of the set is the *absolute value* of their difference.

Median The middle value in a set of data when the data are listed in order from smallest to largest, or from largest to smallest. If there are an even number of data points, the median is the *mean* of the two middle values.

Metric system of measurement A measurement system based on the *base-ten* numeration system. It is used in most countries around the world.

Midpoint A point halfway between two other points. The midpoint of a line segment is the point halfway between the endpoints.

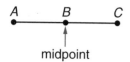
midpoint

Minimum The smallest amount; the smallest number in a set of data.

Minuend In subtraction, the number from which another number is subtracted. For example, in $19 - 5 = 14$, the minuend is 19. See also *subtrahend*.

Mixed number A number that is written using both a whole number and a fraction. For example, $2\frac{1}{4}$ is a mixed number equal to $2 + \frac{1}{4}$.

Möbius strip (Möbius band) A shape with only one side and one edge. The Möbius strip is named for the mathematician August Ferdinand Möbius.

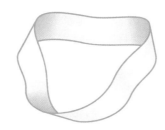

Mode The value or values that occur most often in a set of data.

Multiple of a number *n* (1) A product of *n* and a counting number. For example, the multiples of 7 are 7, 14, 21, 28, ... (2) a product of *n* and an integer. The multiples of 7 are ..., $-21, -14, -7, 0, 7, 14, 21, ...$.

Multiplication counting principle A way of determining the total number of possible outcomes for two or more separate choices. For example, suppose you roll a die and then flip a coin. There are 6 choices for which face of the die shows and 2 choices for which side of the coin shows. So there are 6 * 2, or 12 possible outcomes in all: (1,H), (1,T), (2,H), (2,T), (3,H), (3,T), (4,H), (4,T), (5,H), (5,T), (6,H), (6,T).

Multiplication diagram A diagram used for problems in which there are several equal groups. The diagram has three parts: a number of groups, a number in each group, and a total number. Also called *multiplication/division diagram*. See also *rate diagram*.

rows	chairs per row	total chairs
15	25	?

Multiplication property of −1 A property of multiplication that says that for any number *a*, $(-1) * a = \text{OPP}(a)$, or $-a$. For $a = 5$: $(-1) * 5 = \text{OPP}(5) = -5$. For $a = -3$: $(-1) * (-3) = \text{OPP}(-3) = -(-3) = 3$. See *opposite of a number*.

Multiplicative inverses
Two numbers whose product is 1. The multiplicative inverse of 5 is $\frac{1}{5}$, and the multiplicative inverse of $\frac{3}{5}$ is $\frac{5}{3}$. Multiplicative inverses are also called *reciprocals* of each other.

Mystery plot An unlabeled graph or plot. The viewer is asked to give a situation that the plot or graph might represent.

Name-collection box A diagram that is used for writing equivalent names for a number.

25	37 − 12	20 + 5
⫴⫴⫴ ⫴⫴⫴ ⫴⫴⫴ ⫴⫴⫴ ⫴⫴⫴		5^2
twenty-five		*veinticinco*

Negative number
A number that is less than zero; a number to the left of zero on a horizontal number line or below zero on a vertical number line. The symbol − may be used to write a negative number. For example, "negative 5" is usually written as −5.

***n*-gon** A polygon with *n* sides. For example, a 5-gon is a pentagon, and an 8-gon is an octagon.

Nonagon A polygon with nine sides.

Nonconvex polygon
See *concave polygon.*

***n*-to-1 ratio** A ratio with 1 in the denominator.

Number-and-word notation A way of writing a number using a combination of numbers and words. For example, 27 billion is number-and-word notation for 27,000,000,000.

Number model A *number sentence* or *expression* that models or fits a number story or situation. For example, the story *Sally had $5, and then she earned $8,* can be modeled as the number sentence 5 + 8 = 13, or as the expression 5 + 8.

Number sentence At least two *numbers* or *expressions* separated by a relation symbol (=, >, <, ≥, ≤, ≠). Most number sentences contain at least one *operation symbol* (+, −, ×, *, ÷, or /). Number sentences may also have *grouping symbols,* such as parentheses and brackets.

Number story A story with a problem that can be solved using arithmetic.

Numeral A word, symbol, or figure that represents a number. For example, six, VI, and 6 are numerals that represent the same number.

Numerator The number above the line in a fraction. A fraction may be used to name part of a whole. If the *whole* (the *ONE,* or the *unit*) is divided into equal parts, the numerator represents the number of equal parts being considered. In the fraction $\frac{a}{b}$, a is the numerator.

Octagon A polygon with eight sides.

Octahedron A polyhedron with 8 faces.

Odd number A counting number that cannot be evenly divided by 2. When an odd number is divided by 2, there is a remainder of 1. The odd numbers are 1, 3, 5, and so on.

ONE See *whole* and *unit.*

Open sentence A *number sentence* which has *variables* in place of one or more missing numbers. An open sentence is usually neither true nor false. For example, 5 + x = 13 is an open sentence. The sentence is true if 8 is substituted for x. The sentence is false if 4 is substituted for x.

Operation symbol A symbol used to stand for a mathematical operation. Common operation symbols are +, −, ×, *, ÷, and /.

Opposite angles

(1) of a *quadrilateral:* Angles that do not share a common side.

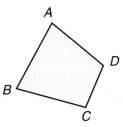

Angles *A* and *C* and angles *B* and *D* are pairs of opposite angles.

(2) of a *triangle:* An angle is opposite the side of a triangle that is not one of the sides of the angle.

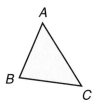

Angle *C* is opposite side *AB*.

(3) of two *lines that intersect:* The angles that do not share a common side are opposite angles. Opposite angles have equal measures. Same as *vertical angles.*

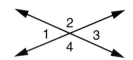

Angles 2 and 4 and angles 1 and 3 are pairs of opposite, or vertical, angles.

Opposite of a number

A number that is the same distance from 0 on the number line as a given number, but on the opposite side of 0. The opposite of a number *n* may be written OPP(*n*) or $-n$. For example, the opposite of $+3$ is OPP(3) or -3, and the opposite of -5 is OPP(-5) or $-(-5)$ or $+5$.

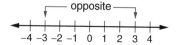

Order of operations

Rules that tell in what order to perform operations in arithmetic and algebra. The order of operations is as follows:

1. Do the operations in parentheses first. (Use rules 2-4 inside the parentheses.)
2. Calculate all the expressions with exponents.
3. Multiply and divide in order from left to right.
4. Add and subtract in order from left to right.

Ordered number pair

(ordered pair) Two numbers that are used to locate a point on a *rectangular coordinate grid.* The first number gives the position along the horizontal axis, and the second number gives the position along the vertical axis. The numbers in an ordered pair are called *coordinates.* Ordered pairs are usually written inside parentheses: (5,3). See *rectangular coordinate grid* for an illustration.

Origin

(1) The 0 point on a number line. (2) The point (0,0) where the two axes of a coordinate grid meet.

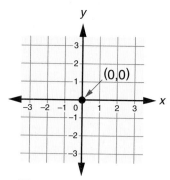

The ordered pair (0,0) names the origin.

Outcome

A possible result of an experiment or situation. For example, heads and tails are the two possible outcomes of tossing a coin. See also *event* and *equally likely outcomes.*

Pan balance

A tool used to weigh objects or compare weights. The pan balance is also used as a model in balancing and solving equations.

Parallel Lines, line segments, or rays in the same plane are parallel if they never cross or meet, no matter how far they are extended. Two planes are parallel if they never cross or meet. A line and a plane are parallel if they never cross or meet. The symbol ‖ means *is parallel to*.

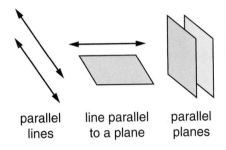

| parallel lines | line parallel to a plane | parallel planes |

Parallelogram
A quadrilateral with two pairs of parallel sides. Opposite sides of a parallelogram are congruent. Opposite angles in a parallelogram have the same measure.

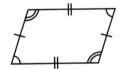

Parentheses Grouping symbols, ( ), used to tell which parts of an expression should be calculated first.

Partial-differences method A way to subtract in which differences are computed for each place (ones, tens, hundreds, and so on) separately. The partial differences are then combined to give the final answer.

```
                        9 3 2
                      − 3 5 6
900 − 300      →        6 0 0
30 − 50        →  −       2 0
2 − 6          →  −         4
600 − 20 − 4 → 	        5 7 6
```

932 − 356 = 576

Partial-products method
A way to multiply in which the value of each digit in one factor is multiplied by the value of each digit in the other factor. The final product is the sum of these partial products.

```
                    6 7
                ×   5 3
50 × 60 →   3 0 0 0
50 × 7  →     3 5 0
3 × 60  →     1 8 0
3 × 7   →       2 1
Add.         3,5 5 1
```

67 * 53 = 3,551

Partial-quotients method A way to divide in which the dividend is divided in a series of steps. The quotients for each step (called partial quotients) are added to give the final answer.

```
    6)1010
  −  600  | 100
     410
  −  300  |  50
     110
  −   60  |  10
      50
  −   48  |   8
       2    168
```
remainder   quotient

1,010 ÷ 6 → 168 R2

Partial-sums method A way to add in which sums are computed for each place (ones, tens, hundreds, and so on) separately. The partial-sums are then added to give the final answer.

```
                              2 6 8
                            + 4 8 3
Add the 100s.        →        6 0 0
Add the 10s.         →        1 4 0
Add the 1s.          →          1 1
Add the partial sums. →       7 5 1
```

268 + 483 = 751

Parts-and-total diagram
A diagram used in *Everyday Mathematics* to represent situations in which two or more quantities are combined to form a total quantity.

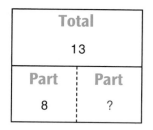

Total	
13	
Part	Part
8	?

Part-to-part ratio A *ratio* that compares a part of a whole to another part of the same whole. For example, the statement "There are 8 boys for every 12 girls" expresses a part-to-part ratio. See also *part-to-whole ratio*.

Part-to-whole ratio
A *ratio* that compares a part of a whole to the whole. For example, the statements "8 out of 20 students are boys" and "12 out of 20 students are girls," both express part-to-whole ratios. See also *part-to-part ratio*.

Pentagon A polygon with five sides.

Percent (%) Per hundred or out of a hundred. For example, "48% of the students in the school are boys" means that 48 out of every 100 students in the school are boys; $48\% = \frac{48}{100} = 0.48$.

Percent Circle A tool on the *Geometry Template* that is used to measure and draw figures that involve percents (such as circle graphs).

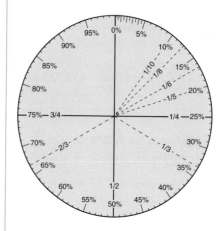

Perfect number A counting number whose *proper factors* add up to the number itself. For example, 6 is a perfect number because the sum of its proper factors is $1 + 2 + 3 = 6$. See also *abundant number*, and *deficient number*.

Perimeter The distance around a 2-dimensional shape, along the boundary of the shape. The perimeter of a circle is called its *circumference*. A formula for the perimeter P of a rectangle with length l and width w is $P = 2 * (l + w)$.

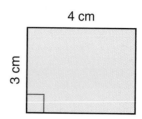

$$P = 2 * (4 \text{ cm} + 3 \text{ cm})$$
$$= 2 * 7 \text{ cm} = 14 \text{ cm}$$

Perpendicular Crossing or meeting at *right angles*. Lines, rays, line segments, or planes that cross or meet at right angles are perpendicular. The symbol ⊥ means *is perpendicular to*.

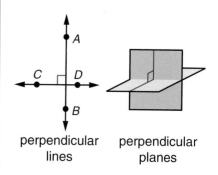

perpendicular lines	perpendicular planes

Personal reference
See *benchmark*.

Perspective drawing
A method of drawing that realistically represents a 3-dimensional object on a 2-dimensional surface.

Per-unit rate A *rate* with 1 in the denominator. Per-unit rates tell how many of one thing there are for one of another thing. For example, "2 dollars per gallon" is a per-unit rate. "12 miles per hour" and "4 words per minute" are also examples of per-unit rates.

Pi (π) The ratio of the *circumference* of a circle to its *diameter*. Pi is also the ratio of the area of a circle to the square of its radius. Pi is the same for every circle and is an irrational number that is approximately equal to 3.14. Pi is the sixteenth letter of the Greek alphabet and is written π.

Pictograph A graph constructed with pictures or symbols. The *key* for a pictograph tells what each picture or symbol is worth.

Number of Cars Washed

Friday	🚗 🚗
Saturday	🚗 🚗 🚗 🚗 🚗
Sunday	🚗 🚗 🚗 🚗

KEY: 🚗 = 6 cars

Pie graph See *circle graph.*

Place value A system that gives a digit a value according to its position in a number. In our *base-ten* system for writing numbers, moving a digit one place to the left makes that digit worth 10 times as much, and moving a digit one place to the right makes that digit worth one-tenth as much. For example, in the number 456, the 4 in the hundreds place is worth 400; but in the number 45.6, the 4 in the tens place is worth 40.

Plane A flat surface that extends forever.

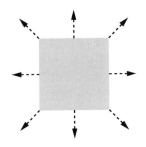

Point symmetry A figure has point symmetry if it can be rotated 180° about a point in such a way that the resulting figure (the *image*) exactly matches the original figure (the *preimage*). Point symmetry is *rotation symmetry* in which the turn is 180°.

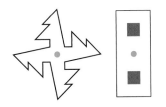

Figures with point symmetry

Polygon A 2-dimensional figure that is made up of three or more line segments joined end to end to make one closed path. The line segments of a polygon may not cross.

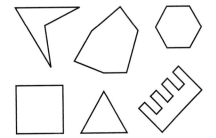

Polyhedron A geometric solid whose surfaces (*faces*) are all flat and formed by polygons. Each face consists of a polygon and the interior of that polygon. A polyhedron does not have any curved surface.

Population In data collection, the collection of people or objects that is the focus of study.

Positive number A number that is greater than zero; a number to the right of zero on a horizontal number line, or above zero on a vertical number line. A positive number may be written using the + symbol, but is usually written without it. For example, +10 = 10 and π = +π.

Power of a number The product of factors that are all the same. For example, $5 * 5 * 5$ (or 125) is called "5 to the third power" or "the third power of 5" because 5 is a factor three times. $5 * 5 * 5$ can also be written as 5^3. See also *exponent.*

Power of 10 A whole number that can be written as a *product of 10s.* For example, 100 is equal to 10 * 10, or 10^2. 100 is called "the second power of 10" or "10 to the second power." A number that can be written as a *product of* $\frac{1}{10}s$ is also a power of 10. For example, $10^{-2} = \frac{1}{10^2} = \frac{1}{10 * 10} = \frac{1}{10} * \frac{1}{10}$ is a power of 10.

Precise Exact or accurate. The smaller the unit or fraction of a unit used in measuring, the more precise the measurement is. For example, a measurement to the nearest inch is more precise than a measurement to the nearest foot. A ruler with $\frac{1}{16}$-inch markings is more precise than a ruler with $\frac{1}{4}$-inch markings.

Preimage A geometric figure that is changed (by a *reflection, rotation,* or *translation,* for example) to produce another figure. See also *image.*

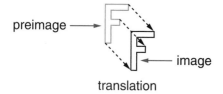

preimage ⟶
image ⟵
translation

Prime factorization A counting number expressed as a product of prime factors. Every counting number greater than 1 can be written as a product of prime factors in only one way. For example, the prime factorization of 24 is 2 * 2 * 2 * 3. (The order of the factors does not matter; 2 * 3 * 2 * 2 is also the prime factorization of 24.) The prime factorization of a prime number is that number. For example, the prime factorization of 13 is 13.

Prime meridian An imaginary semicircle on Earth that connects the North and South Poles and passes through Greenwich, England.

Prime number A counting number that has exactly two different *factors:* itself and 1. For example, 5 is a prime number because its only factors are 5 and 1. The number 1 is not a prime number because that number has only a single factor, the number 1 itself.

Prism A polyhedron with two parallel *faces,* called *bases* that are the same size and shape. All other *faces* connect the bases and are shaped like parallelograms. The *edges* that connect the bases are parallel to each other. Prisms get their names from the shape of their bases.

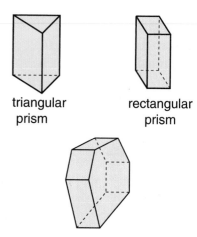

triangular prism   rectangular prism

hexagonal prism

Probability A number from 0 through 1 that tells the chance that an event will happen. The closer a probability is to 1, the more likely the event is to happen. See also *equally likely outcomes.*

Probability tree diagram
A drawing used to analyze a probability situation that consists of two or more choices or stages. For example, the branches of the probability tree diagram below represent the four equally likely outcomes when one coin is flipped two times.

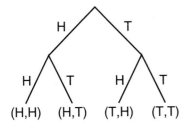

Product The result of multiplying two numbers, called *factors*. For example, in $4 * 3 = 12$, the product is 12.

Proper factor Any *factor of a counting number* except the number itself. For example, the *factors* of 10 are 1, 2, 5, and 10, and the *proper factors* of 10 are 1, 2, and 5.

Proper fraction A fraction in which the numerator is less than the denominator; a proper fraction names a number that is less than 1. For example, $\frac{3}{4}$, $\frac{2}{5}$ and $\frac{12}{24}$ are proper fractions.

Proportion A number model that states that two fractions are equal. Often the fractions in a proportion represent rates or ratios. For example, the problem *Alan's speed is 12 miles per hour. At the same speed, how far can he travel in 3 hours?* can be modeled by the proportion

$$\frac{12 \text{ miles}}{1 \text{ hour}} = \frac{n \text{ miles}}{3 \text{ hours}}$$

Protractor A tool on the *Geometry Template* that is used to measure and draw angles. The half-circle protractor can be used to measure and draw angles up to 180°; the full-circle protractor, to measure angles up to 360°.

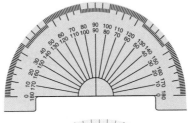

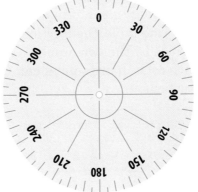

Pyramid A polyhedron in which one face, the *base*, may have any polygon shape. All of the other faces have triangle shapes and come together at a vertex called the *apex*. A pyramid takes its name from the shape of its base.

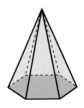

hexagonal pyramid   square pyramid

Pythagorean Theorem
The following famous *theorem:* If the *legs of a right triangle* have lengths *a* and *b* and the *hypotenuse* has length *c,* then $a^2 + b^2 = c^2$.

Q

Quadrangle A polygon that has four angles. Same as *quadrilateral*.

Quadrilateral A polygon that has four sides. Same as *quadrangle*.

Quick common denominator The product of the denominators of two or more fractions. For example, the quick common denominator of $\frac{1}{4}$ and $\frac{3}{6}$ is 4 * 6, or 24. As the name suggests, this is a quick way to get a *common denominator* for a collection of fractions, but it does not necessarily give the *least common denominator*.

Quotient The result of dividing one number by another number. For example, in $35 \div 5 = 7$, the quotient is 7.

Radius (plural: **radii**) (1) A line segment from the center of a circle (or sphere) to any point on the circle (or sphere). (2) The length of this line segment.

Random numbers Numbers produced by an experiment, such as rolling a die or spinning a spinner, in which all *outcomes* are *equally likely*. For example, rolling a *fair* die produces random numbers because each of the six possible numbers 1, 2, 3, 4, 5, and 6 has the same chance of coming up.

Random sample A *sample* that gives all members of the *population* the same chance of being selected.

Range The difference between the *maximum* and the *minimum* in a set of data.

Rate A comparison by division of two quantities with *unlike units*. For example, a speed such as 55 miles per hour is a rate that compares distance with time. See also *ratio*.

Rate diagram A diagram used to model *rate* situations. See also *multiplication diagram*.

number of pounds	cost per pound	total cost
3	79¢	$2.37

Rate table A way of displaying *rate* information. In a rate table, the fractions formed by the two numbers in each column are equivalent fractions.

miles	35	70	105
gallons	1	2	3

Ratio A comparison by division of two quantities with *like units*. Ratios can be expressed with fractions, decimals, percents, or words. Sometimes they are written with a colon between the two numbers that are being compared. For example, if a team wins 3 games out of 5 games played, the ratio of wins to total games can be written as $\frac{3}{5}$, 0.6, 60%, 3 to 5, or 3:5. See also *rate*.

Rational number Any number that can be written or renamed as a *fraction* or the *opposite* of a fraction. Most of the numbers you have used are rational numbers. For example, $\frac{2}{3}$, $-\frac{2}{3}$, $60\% = \frac{60}{100}$, and $-1.25 = -\frac{5}{4}$ are all rational numbers.

Ray A straight path that starts at one point (called the *endpoint*) and continues forever in one direction.

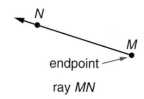

ray *MN*

Real number Any *rational* or *irrational* number.

Reciprocal Same as *multiplicative inverse*.

Rectangle A parallelogram with four right angles. See *square*.

Rectangle method A method for finding area in which rectangles are drawn around a figure or parts of a figure. The rectangles form regions with boundaries that are rectangles or triangular halves of rectangles. The area of the original figure can be found by adding or subtracting the areas of these regions.

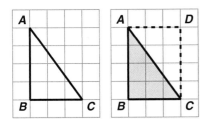

Rectangular array An arrangement of objects into rows and columns that form a rectangle. All rows and columns must be filled. Each row has the same number of objects. And each column has the same number of objects.

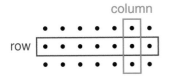

Rectangular coordinate grid A device for locating points in a plane using *ordered number pairs,* or *coordinates*. A rectangular coordinate grid is formed by two number lines that intersect at their zero points and form right angles. Also called a *coordinate grid*.

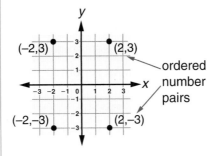

Rectangular prism A *prism* with rectangular *bases*. The four faces that are not bases are either rectangles or other parallelograms.

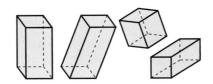

Reduce (1) To decrease the size of an object or figure, without changing its shape. See also *size-change factor*. (2) To put a fraction in *simpler form*.

Reflection The "flipping" of a figure over a line (the *line of reflection*) so that its *image* is the mirror image of the original figure *(preimage)*. A reflection of a solid figure is a mirror-image "flip" over a plane.

Regular polygon A polygon whose sides are all the same length and whose interior angles are all equal.

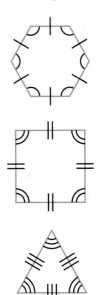

Regular polyhedron

A polyhedron whose faces are congruent and formed by *regular polygons,* and whose *vertices* all look the same. There are five regular polyhedrons:

regular tetrahedron	4 faces, each formed by an equilateral triangle
cube	6 faces, each formed by a square
regular octahedron	8 faces, each formed by an equilateral triangle
regular dodecahedron	12 faces, each formed by a regular pentagon
regular icosahedron	20 faces, each formed by an equilateral triangle

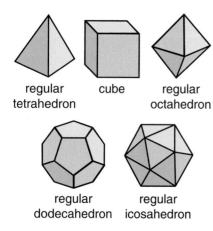

regular tetrahedron   cube   regular octahedron

regular dodecahedron   regular icosahedron

Regular tessellation

A *tessellation* made by repeating *congruent* copies of one *regular polygon.* (Each *vertex point* must be a vertex of *every* polygon around it.) There are only three regular tessellations.

The three regular tessellations

Relation symbol A symbol used to express a relationship between two quantities.

Symbol	Meaning
=	is equal to
≠	is not equal to
>	is greater than
<	is less than
≥	is greater than or equal to
≤	is less than or equal to

Remainder An amount left over when one number is divided by another number. For example, in $38 \div 7 \rightarrow 5 \text{ R}3$, where R3 stands for the remainder.

Repeating decimal A *decimal* in which one digit or a group of digits is repeated without end. For example, 0.3333… and $23.\overline{147} = 23.147147…$ are repeating decimals. See also *terminating decimal.*

Rhombus A quadrilateral whose sides are all the same length. All rhombuses are parallelograms. Every square is a rhombus, but not all rhombuses are squares.

Right angle A 90° angle.

Right cone A cone whose base is perpendicular to the line joining the apex and the center of the base.

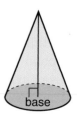
base

Right cylinder A cylinder whose bases are perpendicular to the line joining the centers of the bases.

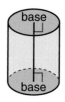
base
base

Right prism A prism whose bases are perpendicular to all of the edges that connect the two bases.

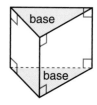
base
base

Right triangle A triangle that has a right angle (90°).

Rotation A movement of a figure around a fixed point, or axis; a *turn*.

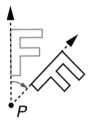

Rotation symmetry A figure has rotation symmetry if it can be rotated less than a full turn around a point or an axis so that the resulting figure (the *image*) exactly matches the original figure (the *preimage*). See also *point symmetry*.

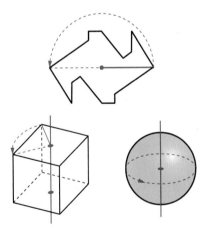

Shapes with rotation symmetry

Round To adjust a number to make it easier to work with or to make it better reflect the level of precision of the data. Often numbers are rounded to the nearest multiple of 10, 100, 1,000, and so on. For example, 12,964 rounded to the nearest thousand is 13,000.

Rubber-sheet geometry See *topology*.

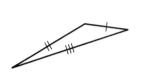

Sample A part of a group chosen to represent the whole group. See also *population* and *random sample*.

Scale (1) The *ratio* of a distance on a map, globe, or drawing to an actual distance. (2) A system of ordered marks at fixed intervals used in measurement; or any instrument that has such marks. For example, a ruler with scales in inches and centimeters, and a thermometer with scales in °F and °C. See also *map scale* and *scale drawing*.

Scale drawing A drawing of an object or a region in which all parts are drawn to the same *scale*. Architects and builders use scale drawings.

Scale factor The *ratio* of the size of a drawing or model of an object to the actual size of the object. See also *scale model* and *scale drawing*.

Scale model A model of an object in which all parts are in the same proportions as in the actual object. For example, many model trains and airplanes are scale models of actual vehicles.

Scalene triangle A triangle with sides of three different lengths. In a scalene triangle, all three angles have different measures.

Scientific notation A system for writing numbers in which a number is written as the product of a *power of 10* and a number that is at least 1 and less than 10. Scientific notation allows you to write big and small numbers with only a few symbols. For example, $4 * 10^{12}$ is scientific notation for 4,000,000,000,000.

Sector A region bounded by an *arc* and two *radii* of a circle. The arc and 2 radii are part of the sector. A sector resembles a slice of pizza. The word *wedge* is sometimes used instead of sector.

sector

Semicircle Half of a circle. Sometimes the diameter joining the endpoints of the circle's arc is included. And sometimes the interior of this closed figure is also included.

Semiregular tessellation A *tessellation* made with *congruent* copies of two or more different *regular polygons*. The same combination of polygons must meet in the same order at each *vertex point*. (Each vertex point must be a vertex of *every* polygon around it.) There are 8 semiregular tessellations. See also *regular tessellation*.

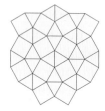

Side (1) One of the rays or segments that form an angle. (2) One of the line segments of a polygon. (3) One of the faces of a polyhedron.

Significant digits The *digits* in a number that convey useful and reliable information. A number with more significant digits is more *precise* than a number with fewer significant digits.

Similar Figures that have the same shape, but not necessarily the same size.

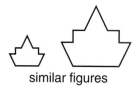

similar figures

Simpler form An equivalent fraction with a smaller numerator and smaller denominator. A fraction can be put in simpler form by dividing its numerator and denominator by a common factor greater than one. For example, dividing the numerator and denominator of $\frac{18}{24}$ by 2 gives the simpler form $\frac{9}{12}$.

Simplest form A fraction that cannot be renamed in simpler form. Also known as *lowest terms*. A *mixed number* is in simplest form if its fractional part is in simplest form.

Simplify (1) For a fraction: To express a fraction in *simpler form*. (2) For an equation or expression: To rewrite by removing parentheses and combining like terms and constants. For example, $7y + 4 + 5 + 3y$ simplifies to $10y + 9$, and $2(a + 4) = 4a + 1 + 3$ simplifies to $2a + 8 = 4a + 4$.

Simulation A model of a real situation. For example, tossing a fair coin can be used to simulate a series of games between two evenly matched teams.

Size-change factor A number that tells the amount of enlargement or reduction. See also *enlarge, reduce, scale,* and *scale factor*.

Skew lines Lines in space that do not lie in the same plane. Skew lines *do not intersect* and are *not parallel*. For example, an east-west line on the floor and a north-south line on the ceiling are skew.

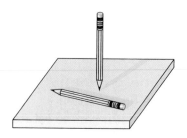

Slanted prism or cone or cylinder A prism (or cone, or cylinder) that is *not* a right prism (or cone, or cylinder).

Slide See *translation*.

Slide rule An *Everyday Mathematics* tool used for adding and subtracting integers and fractions.

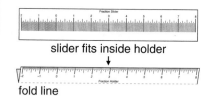

slider fits inside holder

fold line

Solution of an open sentence A value that makes an open sentence *true* when it is substituted for the variable. For example, 7 is a solution of $5 + n = 12$.

Solution set The set of all solutions of an equation or inequality. For example, the solution set of $x^2 = 25$ is $\{5, -5\}$ because substitution of either 5 or -5 for x makes the sentence true.

Span The distance from the tip of the thumb to the tip of the first (index) finger of an outstretched hand. Also called *normal span*.

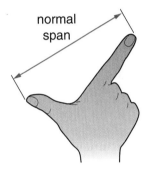

normal span

Speed A *rate* that compares a distance traveled with the time taken to travel that distance. For example, if a car travels 100 miles in 2 hours, then its speed is $\frac{100 \text{ mi}}{2 \text{ hr}}$, or 50 miles per hour.

Sphere The set of all points in space that are the same distance from a fixed point. The fixed point is the *center* of the sphere, and the distance is the *radius*.

radius

center

Spreadsheet program A computer application in which numerical information is arranged in cells in a grid. The computer can use the information in the grid to perform mathematical operations and evaluate formulas. When a value in a cell changes, the values in all other cells that depend on it are automatically changed.

	A	B	C	D
		Class Picnic ($$)		
1		budget for class picnic		
2				
3	quantity	food items	unit price	cost
4	6	packages of hamburgers	2.79	16.74
5	5	packages of hamburger buns	1.29	6.45
6	3	bags of potato chips	3.12	9.36
7	3	quarts of macaroni salad	4.50	13.50
8	4	bottles of soft drinks	1.69	6.76
9			subtotal	52.81
10			8% tax	4.22
11			total	57.03

Square A rectangle whose sides are all the same length.

Square number A number that is the product of a counting number with itself. For example, 25 is a square number because $25 = 5 * 5$. The square numbers are 1, 4, 9, 16, 25, and so on.

Square of a number The product of a number with itself. For example, 81 is the square of 9 because $81 = 9 * 9$. And 0.64 is the square of 0.8 because $0.64 = 0.8 * 0.8$.

Square root of a number The square root of a number n is a number that, when multiplied by itself, gives n. For example, 4 is the square root of 16 because $4 * 4 = 16$.

Square unit A unit used in measuring area, such as a square centimeter or a square foot.

Standard notation The most familiar way of representing whole numbers, integers, and decimals. In standard notation, numbers are written using the *base-ten place-value* system. For example, standard notation for three hundred fifty-six is 356. See also *expanded notation, scientific notation,* and *number-and-word notation.*

Stem-and-leaf plot A display of data in which digits with larger *place values* are "stems" and digits with smaller place values are "leaves."

Data List: 24, 24, 25, 26, 27, 27, 31, 31, 32, 32, 36, 36, 41, 41, 43, 45, 48, 50, 52

Stems (10s)	Leaves (1s)
2	4 4 5 6 7 7
3	1 1 2 2 6 6
4	1 1 3 5 8
5	0 2

Step graph A graph that looks like steps because the y values are the same for an interval of x values, then change (or "step") for the next interval of x values.

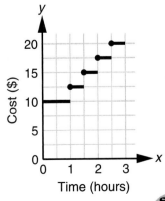

Glossary

Straightedge A tool used to draw line segments. A straightedge does not need to have ruler marks on it; if you use a ruler as a straightedge, ignore the ruler marks.

Substitute To replace one thing with another. In a formula, to replace variables with numerical values.

Subtrahend In subtraction, the number being subtracted. For example, in $19 - 5 = 14$, the subtrahend is 5. See also *minuend.*

Sum The result of adding two or more numbers. For example, in $5 + 3 = 8$, the sum is 8. See also *addend.*

Supplementary angles Two angles whose measures total 180°.

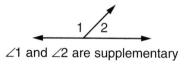

∠1 and ∠2 are supplementary

Surface (1) The boundary of a 3-dimensional object. The part of an object that is next to the air. Common surfaces include the top of a body of water, the outermost part of a ball, and the topmost layer of ground that covers Earth. (2) Any 2-dimensional layer, such as a *plane* or any one of the faces of a *polyhedron.*

Surface area The total area of all of the surfaces of a 3-dimensional object. The surface area of a *rectangular prism* is the sum of the areas of its six faces. The surface area of a *cylinder* is the sum of the area of its curved surface and the areas of its two circular bases.

Survey A study that collects data.

Symmetric (1) Having two parts that are mirror images of each other. (2) Looking the same when turned by some amount less than 360°. See also *line symmetry, point symmetry,* and *rotation symmetry.*

Tally chart A table that uses marks, called *tallies,* to show how many times each value appears in a set of data.

Number of Pull-ups	Number of Children
0	༒ ࠫ
1	ࠫ
2	ࠫ
3	//

Term In an *algebraic expression,* a number or a product of a number and one or more *variables.* For example, in the expression $5y + 3k - 8$, the terms are $5y$, $3k$, and 8. 8 is called a *constant term,* or simply a *constant,* because it has no variable part.

Terminating decimal A *decimal* that ends. For example, 0.5 and 2.125 are terminating decimals. See also *repeating decimal.*

Tessellate To make a *tessellation;* to tile.

Tessellation An arrangement of shapes that covers a surface completely without overlaps or gaps. Also called a *tiling.*

Test number A number used to replace a variable when solving an equation using the *trial-and-error method.* Test numbers are useful for "closing in" on an exact solution.

Tetrahedron A polyhedron with 4 faces. A tetrahedron is a triangular pyramid.

Theorem A mathematical statement that can be proved to be true.

3-dimensional (3-D) Having length, width, and thickness. Solid objects take up volume and are 3-dimensional. A figure whose points are not all in a single plane is 3-dimensional.

Time graph A graph that is constructed from a story that takes place over time. A time graph shows what has happened during a period of time.

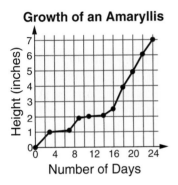

Growth of an Amaryllis

Topological properties Properties of a figure that are not changed by *topological transformations*. See also *topology*.

Topological transformation A shrinking, stretching, twisting, bending, or turning inside out of one shape that changes it into another. Tearing, breaking, and sticking together are *not* topological transformations. See also *topology*.

Topologically equivalent Shapes that can be changed into one another by a *topological transformation* are called topologically equivalent shapes.

Topology The study of the properties of shapes that are unchanged by shrinking, stretching, twisting, bending, and turning inside out. (Tearing, breaking, and sticking together are not allowed.)

Trade-first subtraction method A subtraction method in which all trades are done before any subtractions are carried out.

Transformation Something done to a geometric figure that produces a new figure. The most common transformations are *translations* (slides), *reflections* (flips), and *rotations* (turns). See also *isometry transformation* and *topological tranformation*.

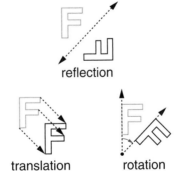
reflection
translation   rotation

Transformation geometry The study of *transformations*.

Translation A movement of a figure along a straight line; a *slide*. In a translation, each point of the figure slides the same distance in the same direction.

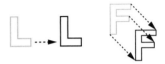

Transversal A line that crosses two or more other lines.

transversal

Trapezoid A quadrilateral that has exactly one pair of parallel sides.

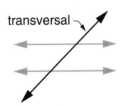

Tree diagram A diagram such as a *factor tree* or *probability tree diagram*. A tree diagram is a network of points connected by line segments. Tree diagrams can be used to factor numbers and to represent probability situations that consist of two or more choices or stages.

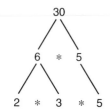

Prime factorization of 30

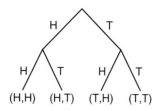

Flipping a coin twice

Trial-and-error method A method for finding the solution of an equation by trying several *test numbers*.

Triangle A polygon with three sides.

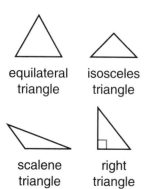

equilateral triangle   isosceles triangle

scalene triangle   right triangle

Triangular numbers Counting numbers that can be shown by triangular arrangements of dots. The triangular numbers are 1, 3, 6, 10, 15, 21, 28, 36, 45, and so on.

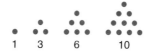

1   3   6   10

Triangular prism A prism whose bases are triangles.

Triangular pyramid A pyramid in which all of the faces are triangles; also called a *tetrahedron*. Any one of the four faces of a triangular pyramid can be called the base. If all of the faces are equilateral triangles, the pyramid is a *regular tetrahedron*.

regular tetrahedron

True number sentence A number sentence in which the relation symbol accurately connects the two sides. For example, $15 = 5 + 10$ and $25 > 20 + 3$ are both true number sentences.

Truncate (1) In a decimal, to cut off all digits after the decimal point, or after a particular place to the right of the decimal point. Also called *rounding down*. For example, 6.543 can be truncated to 6.54 or 6.5 or 6. Truncating a decimal number will always make the number smaller (unless all of the digits cut off are 0s). (2) To cut off part of a solid figure.

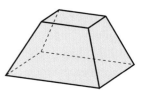

truncated pyramid

Turn See *rotation*.

Turn-around facts A pair of multiplication or addition facts in which the order of the factors (or addends) is reversed. For example, $3 * 9 = 27$ and $9 * 3 = 27$ are turn-around multiplication facts. And $4 + 5 = 9$ and $5 + 4 = 9$ are turn-around addition facts. There are no turn-around facts for division or subtraction. See also *commutative property*.

Turn-around rule A rule for solving addition and multiplication problems based on the *commutative property*. For example, if you know that $6 * 8 = 48$, then, by the turn-around rule, you also know that $8 * 6 = 48$.

Twin primes Two *prime numbers* that have a difference of 2. For example, 3 and 5 are twin primes, and 11 and 13 are twin primes.

2-dimensional (2-D) Having length and width but not thickness. A figure whose points are all in one plane is 2-dimensional. Circles and polygons are 2-dimensional. 2-dimensional shapes have area but not volume.

Unit A label used to put a number in context. The *ONE*. In measuring length, for example, the inch and the centimeter are units. In a problem about 5 apples, *apple* is the unit. See also *whole*.

Unit fraction A fraction whose numerator is 1. For example, $\frac{1}{2}$, $\frac{1}{3}$, $\frac{1}{8}$, and $\frac{1}{20}$ are unit fractions.

Unit percent One percent (1%).

Unlike denominators Denominators that are different, as in $\frac{1}{2}$ and $\frac{1}{3}$.

"Unsquaring" a number Finding the *square root* of a number.

Upper quartile In an ordered data set, the middle value of the *data* above the *median*. Data values at the median are not included when finding the upper quartile. See also *lower quartile*.

U.S. customary system of measurement The measuring system most frequently used in the United States.

Vanishing line A line connecting a point on a figure in a *perspective drawing* with the *vanishing point*.

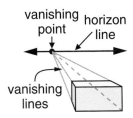

Vanishing point In a *perspective drawing*, the point at which parallel lines moving away from the viewer seem to meet. It is located on the *horizon* line. See also *vanishing line*.

Variable A letter or other symbol that represents a number. In the number sentence $5 + n = 9$, any number may be substituted for the variable n, but only 4 makes the sentence true. In the inequality $x + 2 < 10$, any number may be substituted for the variable x, but only numbers less than 8 make the sentence true. In the equation $a + 3 = 3 + a$, any number may be substituted for the variable a, and every number makes the sentence true.

Variable term A *term* that contains at least one variable.

Venn diagram A picture that uses circles or rings to show relationships between sets.

Girls on Sports Teams

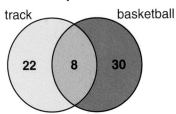

Vertex (plural: **vertices**) The point where the sides of an angle, the sides of a polygon, or the edges of a polyhedron meet.

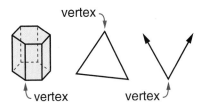

Vertex point A point where corners of shapes in a *tessellation* meet.

Vertical Upright; perpendicular to the horizon.

Vertical (opposite) angles When two lines intersect, the angles that do not share a common side. Vertical angles have equal measures.

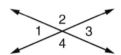

Angles 1 and 3 and angles 2 and 4 are pairs of vertical angles.

Volume A measure of how much space a solid object takes up. Volume is measured in cubic units, such as cubic centimeters or cubic inches. The volume or *capacity* of a container is a measure of how much the container will hold. Capacity is measured in units such as gallons or liters.

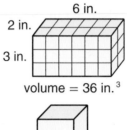

volume = 36 in.^{3}

1 cubic centimeter (actual size)

If the cubic centimeter were hollow, it would hold exactly 1 milliliter.
1 milliliter (mL) = 1 cm^3.

"What's My Rule?" Problem A type of problem that asks for a rule connecting two sets of numbers. Also, a type of problem that asks for one of the sets of numbers, given a rule and the other set of numbers.

Whole (or ONE or unit) The entire object, collection of objects, or quantity being considered. The ONE, the *unit,* 100%.

Whole numbers The *counting numbers,* together with 0. The set of whole numbers is {0, 1, 2, 3, …}

Page 4
1. 6,000
2. 600,000
3. 60
4. 60,000

Page 6
1. 36
2. 64
3. 100,000
4. 9
5. 50,625
6. 161,051

Page 8
1. $2 * 10^5$
2. $1 * 10^7$
3. $4.3 * 10^8$
4. $6 * 10^{-5}$
5. $3.5 * 10^{-2}$
6. 40,000,000
7. 28,000
8. 662,000,000
9. 0.03
10. 0.00123

Page 9
1. false
2. true
3. false
4. false

Page 10
1. 1, 2, 7, 14
2. 1, 2, 4, 8, 16
3. 1, 3, 5, 9, 15, 45
4. 1, 2, 3, 4, 6, 8, 9, 12, 18, 24, 36, 72
5. 1, 19
6. 1, 2, 4, 5, 10, 20, 25, 50, 100

Page 11
1. 105: 3, 5
2. 4,470: 2, 3, 5, 6, 10
3. 526: 2
4. 621: 3, 9
5. 13,680: 2, 3, 5, 6, 9, 10

Page 12
Sample answers:

1.
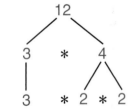

2.
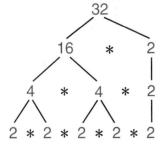

3.
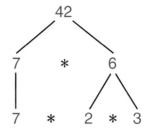

4.
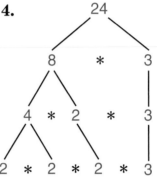

5.
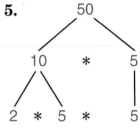

6.
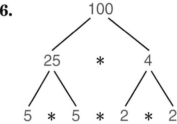

Page 14
1. 289
2. 113
3. 963
4. 1,727
5. 6,927

Page 15
1. 46
2. 291
3. 242
4. 72
5. 5,545

Page 16
1. 282
2. 82
3. 778
4. 324

Page 17
1. 593
2. 404
3. 174
4. 3,798

Page 18
1. 600
2. 89,000
3. 6,300
4. 64,000
5. 480,000
6. 36,000

Page 19

1. 716   **2.** 2,368   **3.** 3,540

4. 3,915   **5.** 19,110

Page 20

1.

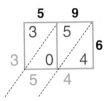

$6 * 59 = 354$

2.

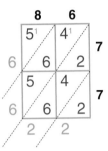

$77 * 86 = 6,622$

3.

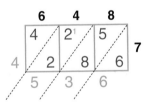

$76 * 98 = 7,448$

4.

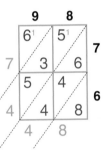

$7 * 648 = 4,536$

5.

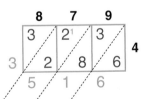

$879 * 4 = 3,516$

Page 21

1. 84   **2.** 7,000   **3.** 50   **4.** 50

Page 23

1. 15 R2   **2.** 129   **3.** 91   **4.** 208 R2

Page 27

1. 4 tenths and 0 point 4

2. 1 and 65 hundredths and 1 point six five

3. 872 thousandths and 0 point eight seven two

4. 16 and 4 hundredths and 16 point zero four

5. 3 thousandths and 0 point zero zero three

6. 59 and 61 thousandths and 59 point zero six one

Page 30

1. $0.78 > 0.079$ **2.** $1.099 < 1.1$

3. $0.99 > 0.10$ **4.** $\frac{5}{4} < 1.3$

Page 33

1. 10.04   **2.** 12.7   **3.** 10.482   **4.** 8.21

Page 34

1. 4.7 meters

2. a. 1.1 m   **b.** 0.04 sec

Page 36

1. 456   **2.** 0.0456   **3.** 2,600

4. 0.407   **5.** 0.00044   **6.** $7,500

7. 10.1   **8.** 0.453

Page 38

1. 12.88   **2.** 19.825

3. 1.5756   **4.** 0.0044

Page 39

1.

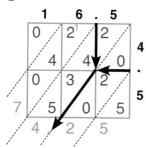

$16.5 * 4.5 = 74.25$

2.

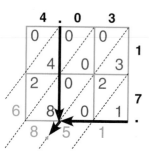

$4.03 * 17 = 68.51$

3.

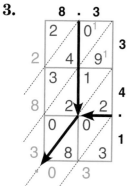

$8.3 * 34.1 = 283.03$

Page 40

1. 7.98   **2.** 798   **3.** 0.0078   **4.** $0.36
5. 6,930   **6.** 0.008   **7.** 2,650   **8.** 30

Page 41

1. 3.45   **2.** 345   **3.** 0.0013
4. $0.76   **5.** 8,250   **6.** 0.005
7. 6,250   **8.** 10   **9.** 0.1

Page 42

1. 24.8   **2.** 2.11   **3.** 6.2   **4.** 44.4

Page 43

1. 4   **2.** 3.2   **3.** 473   **4.** 0.38

Page 46

1. 4.7, 4.67, 4.674   **2.** 8.3, 8.31, 8.310
3. 0.1, 0.05, 0.053   **4.** 0.0, 0.00, 0.002

Page 49

1. $3.40   **2.** $13.75, $41.25

Page 50

$250

Page 52

1. $44   **2.** 21 free throws

Page 53

1. 0.75   **2.** 0.4   **3.** 1.5
4. 0.55   **5.** 0.12

Page 54

1. 0.7   **2.** about 0.62 or 0.63
3. about 3.33   **4.** 0.75
5. about 1.44   **6.** about 0.43

Page 56

1. 0.375   **2.** 0.833   **3.** 0.7778

Page 57

1. 0.375   **2.** 0.8333…   **3.** 0.5555…

Page 58

1. 0.875   **2.** $0.\overline{3}$ **3.** $0.41\overline{6}$
4. 0.3125   **5.** $0.\overline{2}$ **6.** $1.1\overline{6}$

Page 60

1. a. 60%   **b.** 90%   **c.** 37.5%
d. 120%   **e.** 175%

2. a. $\frac{4}{5}$ **b.** $\frac{3}{20}$ **c.** $1\frac{1}{4}$

3. a. $\frac{3}{5}$ **b.** $\frac{9}{20}$
c. $5\frac{43}{100}$ **d.** $1\frac{19}{1,000}$

Page 71

1. $\frac{10}{3}$ **2.** $\frac{3}{2}$ **3.** $\frac{13}{5}$
4. $\frac{11}{6}$ **5.** $\frac{15}{4}$ **6.** $\frac{5}{1}$

Page 72

1. $1\frac{1}{5}$ **2.** $2\frac{5}{8}$ **3.** 4
4. $5\frac{1}{2}$ **5.** $3\frac{3}{4}$ **6.** $6\frac{2}{3}$

Page 73

Sample answers:

1. $\frac{2}{8}$ 2. $\frac{10}{20}$ 3. $\frac{10}{8}$

4. $\frac{2}{3}$ 5. $\frac{3}{5}$ 6. $\frac{3}{5}$

Page 74

Sample answers for Problems 1–3:

1. $\frac{3}{6}$, or $\frac{2}{4}$, or $\frac{1}{2}$ 2. $\frac{8}{10}$, or $\frac{4}{5}$ 3. $\frac{8}{14}$, or $\frac{4}{7}$

4. $\frac{3}{5}$ 5. $\frac{3}{7}$ 6. $\frac{5}{6}$

Page 75

1. $>$ 2. $>$ 3. $<$ 4. $>$ 5. $>$

Page 77

Sample answers for Problems 1–3:

1. $\frac{1}{3}$ 2. $\frac{1}{2}$ 3. $\frac{2}{3}$ 4. $\frac{3}{8}$

5. $\frac{6}{10}$ 6. $\frac{1}{9}$ 7. $\frac{7}{9}$ 8. $\frac{3}{7}$

9. $\frac{13}{16}$ 10. $\frac{11}{12}$

Page 78

1. 12   2. 20   3. 12

4. 24   5. 18   6. 45

Page 79

Sample answers:

1. $\frac{2}{6}$ and $\frac{5}{6}$ 2. $\frac{15}{20}$ and $\frac{12}{20}$ 3. $\frac{7}{10}$ and $\frac{15}{10}$

4. $\frac{5}{20}$ and $\frac{6}{20}$ 5. $\frac{32}{48}$ and $\frac{42}{48}$

Page 80

1. 1   2. 2   3. 8

4. 7   5. 6   6. 3

Page 82

1. 32 posters   2. 20 counters

3. $4,000 per month

Page 83

1. $\frac{8}{9}$ 2. $\frac{5}{8}$ 3. $\frac{11}{12}$

4. $\frac{5}{12}$ 5. $\frac{29}{24}$

Page 84

1. 10   2. $6\frac{3}{10}$ 3. $10\frac{5}{12}$

4. $23\frac{19}{63}$ or $22\frac{82}{63}$

Page 86

1. $4\frac{1}{15}$ 2. $3\frac{1}{8}$ 3. $3\frac{7}{9}$ 4. $4\frac{5}{6}$

Page 87

1. 8   2. 24   3. 18   4. 60

5. Gina gets $42; Robert gets $14.

Page 88

1. $3\frac{3}{4}$ 2. 4   3. $3\frac{1}{5}$

4. $2\frac{2}{5}$ 5. $4\frac{1}{2}$

Page 89

1. $\frac{3}{10}$ 2. $\frac{15}{24}$, or $\frac{5}{8}$ 3. $\frac{80}{15}$, or $5\frac{1}{3}$

4. $\frac{0}{49}$, or 0   5. $\frac{60}{75}$, or $\frac{4}{5}$

Page 90

1. $\frac{3}{8}$ 2. $\frac{40}{3}$, or $13\frac{1}{3}$ 3. $\frac{85}{10}$, or $8\frac{1}{2}$

Page 92

1. $9 \div \frac{1}{2} = \frac{18}{2} \div \frac{1}{2} = 18$, 18 people

2. $10 \div \frac{1}{3} = \frac{30}{3} \div \frac{1}{3} = 30$, 30 bracelets

3. Rewrite $\frac{1}{4} * \square = 7$ as $7 \div \frac{1}{4} = \square$.

   $7 \div \frac{1}{4} = \frac{28}{4} \div \frac{1}{4} = 28$

Page 93

1. $\frac{12}{5}$, or $2\frac{2}{5}$ 2. $\frac{15}{5}$, or 3   3. $\frac{7}{21}$, or $\frac{1}{3}$

4. $\frac{8}{12}$, or $\frac{2}{3}$ 5. $\frac{28}{10}$, or $2\frac{4}{5}$

Page 95

1. -1 2. -9

3. 5   4. $-\frac{3}{2}$, or $-1\frac{1}{2}$

Page 96

1. -6 2. 3   3. 7   4. -4

5. 33   6. -0.4 7. 0   8. -2

Page 97

1. -63 2. -320 3. -63

4. -4 5. -7 6. 4

Page 98
1. 0 **2.** 0
3. 0 **4.** No solution

Page 101
1. 1, 2, 3, 4 **2.** −1, 0, 1, 2, 3, 4
3. yes (sample: $\frac{2}{10}$)
4. yes
5. yes **6.** yes
7. $1.1\overline{6}$, or $1\frac{1}{6}$; $1.\overline{3}$, or $1\frac{1}{3}$; 1.5, or $1\frac{3}{6}$; $1.\overline{6}$, or $1\frac{2}{3}$
8. yes (sample: $\frac{7}{6}$)
9. tan 30°, $\sqrt{2}$, $\sqrt{3}$, $\sqrt{5}$, e, π, $\sqrt{12}$

Page 103
Sample answers for Problems 1 and 2:
1. for $r = 1$: $2 * (1) + (1) = 3 * (1)$
  for $r = -1$: $2 * (-1) + (-1) = 3 * (-1)$
  for $r = 0$: $2 * (0) + (0) = 3 * (0)$
2. for $b = 1$: $1 + (1 + 1) + (1 + 2) = 3 * (1 + 1)$
  for $b = -1$: $-1 + (-1 + 1) + (-1 + 2) = 3 * (-1 + 1)$
  for $b = 0$: $0 + (0 + 1) + (0 + 2) = 3 * (0 + 1)$
3. $x + y = y + x$ **4.** $\frac{a}{b} * \frac{b}{a}$

Page 109
1. 5 dollars/hour, or $\frac{5 \text{ dollars}}{1 \text{ hour}}$

dollars	5	10	15	20	25	30	35
hours	1	2	3	4	5	6	7

2. 45 words/minute, or $\frac{45 \text{ words}}{1 \text{ minute}}$

words	45	90	135	180	225	270	315
minutes	1	2	3	4	5	6	7

Page 112
1. 24 quarts **2.** $7
3. $30 **4.** 25 laps

Page 115
1. $a = 18$ **2.** $t = 40$ **3.** $m = 1\frac{1}{2}$
4. $b = 8$ **5.** $x = 8$ **6.** $c = 6$

Page 116
about 52 minutes

Page 119
1. $\frac{\text{money spent}}{\text{total allowance}} = \frac{3}{5}$ **2.** $\frac{\text{money saved}}{\text{total spent}} = \frac{2}{3}$
3. 1.5 times **4.** 40%

Page 120
20 cousins

Page 123
1. a. 7.5 cm **b.** 30 cm

Page 124
1. a. about 600 miles **b.** about 200 miles
  c. about 475 miles **d.** about 625 miles

Page 134
1.

Number of Hits	Number of Players
0	ЖЖ //
1	///
2	ЖЖ
3	//
4	/

2.

```
               X
               X
               X        X
Number         X        X
  of           X  X  X
Players        X  X  X  X
               X  X  X  X  X
              ─────────────────
               0  1  2  3  4
              Number of Hits
```

Page 135

1.

Age of Space Travelers	Number of Travelers
20–29	////
30–39	//// /
40–49	/

2.

Age of Space Travelers	
Stems (10s)	Leaves (1s)
2	7 5 8 6
3	7 2 1 9 6
4	3

Page 136

min = 0; max = 4; range = 4;
mode = 2; median = 1.5

Page 137

Megan's mean (average) score is 80.91.

Page 138

1. 20 years
2. 25 percent

Page 140

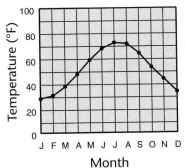

Average Temperatures for Boston, Massachusetts

Page 141

1. a. $2.00   b. $3.50   c. $3.50
   d. $6.50   e. $6.50
2. a. $2.00   b. $1.50

Page 144

1. labels
2. 4.50; a number representing the unit price of 1 qt of macaroni salad
3. B1
4. 5; a number representing the number of packages of hamburger buns
5. A4, A5, A6, A7, A8, C4, C5, C6, C7, C8
6. Column D

Page 145

3rd grade represents 62% − 45%, or 17%;
4th grade represents 85% − 62%, or 23%;
5th grade represents 100% − 85%, or 15%.

Page 146

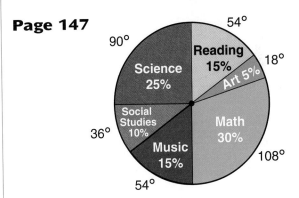

Points Scored

Page 147

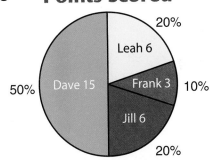

Page 153

1. $\frac{4}{13}$ 2. $\frac{9}{13}$ 3. $\frac{3}{13}$
4. $\frac{6}{13}$ 5. $\frac{7}{13}$ 6. $\frac{13}{13}$, or 1

Answer Key

Page 155

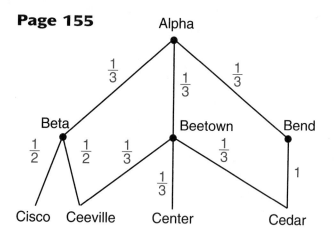

1. $\frac{1}{6}$

2. $\frac{1}{6} + \frac{1}{9} = \frac{5}{18}$

3. $\frac{1}{9}$

4. $\frac{1}{9} + \frac{1}{3} = \frac{4}{9}$

Page 156

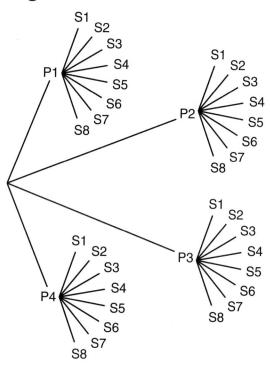

Page 161

1.

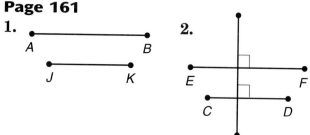

2.

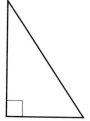

Page 162

1. X

2. 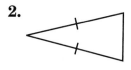 P Q

3.  E D F

4. L M

5. S T / J K

6. U T

Page 163

1. a. same measure   **b.** supplementary

c. supplementary   **d.** supplementary

e. same measure   **f.** same measure

2. a. 120°   **b.** 60°   **c.** 120°

Page 165

1. a. octagon

b. quadrangle or quadrilateral

c. hexagon

2. a.   **b.**

3. The sides of a journal page are not all the same length.

Page 166

1. J K L

2.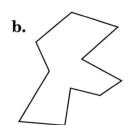

KLJ, LJK, JLK, LKJ, and *KJL*

3.

Page 168

1.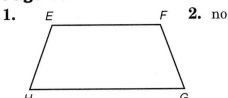

2. no

3. *FGHE, GHEF, HEFG, EHGF, HGFE, GFEH, FEHG*

Page 169

Sample answers:

1. The sides of a rhombus all have the same length. The sides of a rectangle may not all have the same length.

2. A trapezoid has exactly 1 pair of parallel sides. All four sides can have different lengths. A square has 2 pairs of parallel sides, and all sides have the same length.

3. A kite has no parallel sides. A parallelogram has two pairs of parallel sides.

Page 171

Sample answers:

1. **a.** They do not have any vertices. They each have one curved surface.

   b. A cylinder has two flat surfaces (faces); a sphere has no flat surfaces. A cylinder has two edges; a sphere has no edges.

2. **a.** They each have at least one vertex. They each have a flat base.

   b. A cone has a curved surface; the surfaces of a pyramid are all flat surfaces (faces). A cone has one circular face; the faces of a pyramid are all shaped like polygons. A cone has only one vertex; a pyramid has at least four vertices.

Page 172

1. **a.** 5   **b.** 1
2. **a.** 5   **b.** 2

Page 173

1. **a.** 7   **b.** 15   **c.** 10
2. octagonal prism

Page 174

1. **a.** 4   **b.** 6   **c.** 4
2. pentagonal pyramid
3. Sample answers:

   a. The surfaces of each are all formed by polygons. The shape of their base is used to name them.

   b. A prism has at least one pair of parallel faces; no two faces of a pyramid are parallel. The faces of a prism that are not bases are all parallelograms; the faces of a pyramid that are not the base are all triangles.

Page 175

1. **a.** 6   **b.** 4
2. Sample answers:

   a. Their faces are equilateral triangles that all have the same size.

   b. The tetrahedrons have 4 faces, 6 edges, and 4 vertices. The icosahedrons have 20 faces, 30 edges, and 12 vertices.

Page 178

C

Page 179

1. 10 inches   2. 15 inches

Page 181

1. WHAT   2. *A*'(6,15); *T*'(11,11)
3. B

Page 182

1.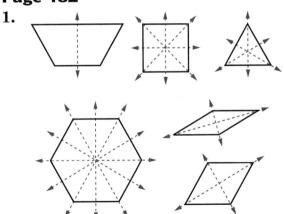

2. infinite; any line drawn directly through its center is a line of symmetry

Page 183

Sample answers:

1.   **2.**

Page 185

1. a. false   **b.** false   **c.** true
    d. false   **e.** true

2. Sample answer:

Page 209

1. millimeter, gram, kilometer, and decimeter
2. a. one thousand (1,000)   **b.** 0.002 kg

Page 212

1. 24 ft   **2.** 90 yd

Page 213

1. about 27 mm   **2.** about 84.8 mm
3. about 44 in.

Page 215

1. 6 square units   **2.** 40 in.^{2}
3. 121 m^2

Page 216

1. 352 ft^2 **2.** 108 in.^{2} **3.** 10.4 cm^2

Page 217

1. 6 in.^{2} **2.** 27 cm^2 **3.** 45.08 yd^2

Page 218

1. about 21 mm   **2.** about 10.5 mm
3. about 346 mm^2

Page 221

1. 336 yd^3 **2.** 1,728 cm^3 **3.** 2,304 ft^3

Page 223

1. 72 in.^{3} **2.** 5,120 cm^3 **3.** 600 ft^3

Page 224

1. 33.5 in.^{3} **2.** 268.1 cm^3

Page 225

1,360 cm^2

Page 226

377.0 in.^{2}

Page 228

1. 180 grams; 170.1 grams
2. 588 ounces

Page 232

1. 60°   **2.** 290°   **3.** 75°

4.   **5.**

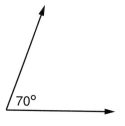
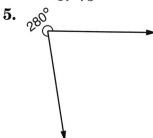

6.

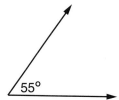

Page 233

1. a. 2   **b.** 3   **c.** 6   **d.** 10
2. 540°   **3.** 135°
4. number of sides of a polygon − 2 = number
of triangles into which it can be divided

Page 234

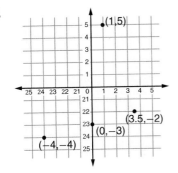

Page 239

1. $2n = 40$

2. $n - 9 = 11$

3. 137.5 miles

4. 3 min 40 sec

Page 240

1. $m - 3$ inches

2. $2H$ minutes

3. $\$3 + \$0.50 * R$

Page 241

1. true

2. true

3. false

4. true

5. false

6. false

Page 243

1. $c = 14$

2. $z = 7$

3. $f = 10.5$

4. $\$20 - \$12.49 = c$, or $\$20 = \$12.49 + c$; $c = \$7.51$

5. $\$10 * w = \$90$, or $\frac{\$90}{\$10} = w$; $w = 9$ (weeks)

Page 244

1. any number less than 8

2. any number greater than 0

3.

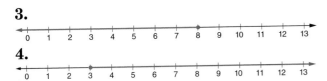

4.

Page 246

1. 576 in.^{2}, or 4 ft^2

2. 1,200 feet

3. 72 m^2

4. $V = \frac{1}{3} * B * h$

5. $i = \$1,000 * r * t$

Page 247

1. 31

2. 13.5

3. 0

4. 6

Page 249

1. $(6 * 100) + (6 * 40) = 840$

2. $(35 * 6) - (15 * 6) = 120$

3. $(4 * 80) - (4 * 7) = 292$

4. $1.23 * (456 + 789) = 1.23 * (1,245) = 1,531.35$;
$(1.23 * 456) + (1.23 * 789) = 560.88 + 970.47 = 1,531.35$

Page 250

1. $4B + 2M = 1M + 8B$

2. $6C + 2P = 8P + 3C$

Page 252

1. Both sides of the equation are equal to 44.
$5(12 + 3) - 3(12) + 5 = 75 - 36 + 5 = 44$
$4(12 - 1) = 4 * 11 = 44$

2. $x = 4$

3. $s = -2$

4. $b = 3$

Page 253

1.

in	out
n	n / 3
9	3
36	12

2.

in	out
k	k − 4
11	7
28	24

3. Rule: multiply "in" by 30

in	out
x	x * 30
4	120
10	300

Page 254

1. 500 words

2. $31.50

Page 257

1. 5 students

2. $135

Page 259

Sample answer:
Walk for 1 minute and count your steps.
Multiply this count by 60.

Page 262
1. 35,500
2. 40,000
3. 35,000
4. 35,481.75

Page 264
1.
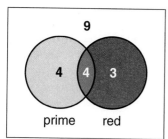
prime   red

2. $\frac{3}{20}$
3. $\frac{11}{20}$

Page 270
1. 67
2. 8
3. 11.8
4. 23

Page 272
1. 17 R3
2. 21 R0
3. 411 R28

Page 281
1. 0.1875
2. $\frac{37}{200}$, or $\frac{185}{1,000}$
3. 0.3%
4. 7.23
5. 28.125%
6. $\frac{17}{25}$, or $\frac{68}{100}$

Page 283
1. 0.7
2. 428
3. 4,384.488
4. 0.80

Page 288
1. 0.00072
2. 83,000,000
3. 0.000003726, or 0.0000037
4. −0.000034

Page 289
1. $6.0846 * 10^{15}$, or $6.085 * 10^{15}$
2. $5.3798 * 10^{15}$, or $5.38 * 10^{11}$
3. $1.5365 * 10^{16}$, or $1.537 * 10^{16}$
4. $2.7775 * 10^{17}$, or $2.778 * 10^{17}$
5. exactly $3.118752 * 10^8 = 311,875,200$ hands, or about $3.1187 * 10^8 = 311,870,000$ hands

Page 291
1. 15,394 ft^2
2. 115.3 cm

Page 294
about 800 ft^2

Page 295
$12.75; $97.75

Page 298
1. $90.50
2. $29.50

Page 300
1. 11, 18, 25, 32, 39, 46
2. 120, 107, 94, 81, 68, 55

Page 368
1. balanced
2. unbalanced

Page 369
1. unbalanced
2. balanced
3. $x = 2$; weight on left = $\frac{1}{2}x = 1$

Page 24A
1. 7,288
2. 9,441
3. 28,326
4. 16,867,725

Page 24B
1. 56
2. 185
3. 2,697
4. 810,992

Page 24C
1. 1,617
2. 2,052
3. 15,536
4. 717,160

Page 24D
1. 21,024
2. 41,495
3. 161,298
4. 133,175

Page 24E
1. 120
2. 164
3. 134 R3
4. 880

Page 24F
1. 896
2. 1,431
3. 1,730 R3
4. 2,676 R1

Page 24H
1. 26
2. 308 R20
3. 447 R8
4. 110 R6

Page 60A

1. 13.84
2. 44.26
3. 926.626
4. 14.215

Page 60B

1. 73.80
2. 0.97
3. 53.22
4. 1.25

Page 60D

1. 223.5
2. 1.7496
3. 0.15111
4. 0.315

Page 60E

1. $1.21
2. $1.23
3. $0.80 R2¢
4. $2.68

Page 60F

1. 2.07
2. 2.41
3. 1.445
4. 7.73

Page 60H

1. 1,120
2. 24.6
3. 156
4. 610

Page 60I

1. $0.\overline{6}$
2. $0.\overline{27}$
3. $0.\overline{8}$
4. $0.8\overline{3}$

Page 254B

1. Sample answer: The input or "in" number x is independent. The output or "out" variable y is dependent. $y = \frac{1}{2} * x$.
2. V is dependent. r is independent. $V = \frac{4}{3}\pi r^3$.

Index

Index

Notation. *See also* Symbols.
　absolute value, 94
　addition, 13, 240–241
　base-10 blocks, 30
　decimals, 26–27, 57, 101
　degrees, 160, 227, 230
　division, 22, 69, 240–241
　exponential, 5–6, 29
　fractions, 68–69, 100
　grouping symbols, 241
　metric measures, 371
　multiplication, 19, 240–241,
　　245
　negative numbers, 7, 94
　opposites, 94, 96, 105
　percents, 47
　powers of 10, 5, 7, 29
　rates, 109
　rational numbers, 100
　remainders, 22, 260
　repeating decimals, 57–58
　scientific, 7–8, 287–289
　standard, 5–6, 8, 29
　subtraction, 13, 240–241
　temperatures, 209, 227
　U.S. customary measures, 371
Number lines
　addition, 95–96
　decimals on, 30, 94, 100
　fractions on, 30, 69, 75, 100
　fraction-decimal, 372
　fraction/decimal/percent, 376
　fraction-stick and decimal,
　　373
　fraction-whole number
　　multiplication, 88
　inequalities, 241, 244
　multiplication models, 88
　negative numbers, 94–96, 99
　positive and negative
　　numbers, 94–96, 99
　rational numbers, 100
　real numbers, 102
　subtraction, 95–96
Number models, 10, 108, 243
Number pairs, 234, 254
Number sentences, 241–243
Number stories, 257
Numbers
　absolute value, 94
　base-ten place value system,
　　4, 28
　big/small, 4, 7–8, 287–289,
　　370
　comparing, 9
　composite, 12
　counting, 2–3, 99–100
　decimals, 27–28, 386
　exponents, 5–6, 28, 36, 247,
　　285–286, 311, 332

finding fractions of, 87
finding percents of, 49–50
fractions, 68, 100, 273, 390
integers, 100, 271–272
irrational, 100–101
kinds, 3
mixed, 71–72, 260, 273
naming, 4, 27, 68, 329
negative, 3, 94, 99–100
opposite, 94
pairs, 234
place value, 4, 28
pi, 102, 213, 218, 290–291
positive, 94, 100
powers of 10, 5, 7–8, 29
prefixes, 370
prime, 12
properties, 104–106
random, 302
rational, 100–101
reading, 4, 26, 68, 94
real, 102
rounding, 46, 261–262
in spreadsheets, 143
square, 6, 10
types, 99–100
uses, 2
whole, 3, 100
Numerators, 68, 74–75

Obtuse angles, 160
Obtuse triangles, 166
Octagons, 165
Octahedrons, 175
ONE
　decimal operations, 31
　fractions, 68
　multiplication property of, 104
　percents, 47
　representing the whole, 31,
　　47–48, 68
Open sentences, 242–243
Operation symbols, 94, 241
Opposite angles, 163
Opposite-change addition rule, 14
Opposite numbers, 94
Opposites properties, 105
Order of operations, 247
　calculators, 269
Order of rotation symmetry, 183,
　355–356
Ordered number pairs, 234, 254
Origin, 234
Outcomes, 150–153

Pan-balance problems, 250
Paper airplanes, 362–367

Parallel bases, 170, 173
Parallel lines/line segments, 161–
　163, 200
Parallelograms, 168–169, 191, 216
Parallels, 235
Parentheses
　calculators, 269
　order of operations, 242, 247,
　　252
Part-to-part/part-to-whole ratios,
　118
Partial-differences subtraction
　method, 17
Partial-products multiplication
　method, 19, 37–38, 248
Partial-quotients method of
　division, 22–23, 42–44,
　55–56
Partial-sums addition method,
　13, 32
Parts per million (PPM), 81
Pattern-block shapes, 186
Patterns, 28, 97, 103
Pentagonal prisms, 173
Pentagonal pyramids, 172, 174
Pentagons, 165
Per, 108–109
Per-unit rates, 111–112
Percent Circles, 145–146, 186, 187
Percents
　calculators, 277–281
　circle graphs, 145–147
　conversions, 59–60
　definition, 47
　describing ratios with, 117
　equivalent
　　fractions/decimals, 59–60,
　　372, 375
　finding part of a whole,
　　48–49, 52
　finding the whole, 50, 52
　finding the percent, 51
　games, 314–316, 333
　notation, 47
　of a number, 49–50
　on number lines, 376
　origin of term, 47
　probability, 47, 148, 376
　proportions, 51–52
　ratios, 117–118
　renaming fractions and
　　decimals, 59–60, 372, 375
　unit, 50
　uses, 47–48
Perimeter, 212
Perpendicular bisector, 182
　constructing, 194
Perpendicular lines/line segments,
　161–162, 216–217
　constructing, 195–196

Index

Photo Credits